Multi-Platform Wireless Web Applications

Multi-Platform Wireless Web Applications

Cracking the Code

Dreamtech Software Team

Hungry Minds™

Best-Selling Books • Digital Downloads • e-Books • Answer Networks •
e-Newsletters • Branded Web Sites • e-Learning

New York, NY ◆ Cleveland, OH ◆ Indianapolis, IN

Multi-Platform Wireless Web Applications: Cracking the Code

Published by

Hungry Minds, Inc.
909 Third Avenue
New York, NY 10022
www.hungryminds.com

Library of Congress Control Number: 2001095940

ISBN: 0-7645-4903-0

Printed in the United States of America

10 9 8 7 6 5 4 3 2 1

1B/QZ/RR/QR/IN

Distributed in the United States by Hungry Minds, Inc.

Distributed by CDG Books Canada Inc. for Canada; by Transworld Publishers Limited in the United Kingdom; by IDG Norge Books for Norway; by IDG Sweden Books for Sweden; by IDG Books Australia Publishing Corporation Pty. Ltd. for Australia and New Zealand; by TransQuest Publishers Pte Ltd. for Singapore, Malaysia, Thailand, Indonesia, and Hong Kong; by Gotop Information Inc. for Taiwan; by ICG Muse, Inc. for Japan; by Intersoft for South Africa; by Eyrolles for France; by International Thomson Publishing for Germany, Austria, and Switzerland; by Distribuidora Cuspide for Argentina; by LR International for Brazil; by Galileo Libros for Chile; by Ediciones ZETA S.C.R. Ltda. for Peru; by WS Computer Publishing Corporation, Inc., for the Philippines; by Contemporanea de Ediciones for Venezuela; by Express Computer Distributors for the Caribbean and West Indies; by Micronesia Media Distributor, Inc. for Micronesia; by Chips Computadoras S.A. de C.V. for Mexico; by Editorial Norma de Panama S.A. for Panama; by American Bookshops for Finland.

For general information on Hungry Minds' products and services please contact our Customer Care department within the U.S. at 800-762-2974, outside the U.S. at 317-572-3993 or fax 317-572-4002.

For sales inquiries and reseller information, including discounts, premium and bulk quantity sales, and foreign-language translations, please contact our Customer Care department at 800-434-3422, fax 317-572-4002 or write to Hungry Minds, Inc., Attn: Customer Care Department, 10475 Crosspoint Boulevard, Indianapolis, IN 46256.

For information on licensing foreign or domestic rights, please contact our Sub-Rights Customer Care department at 212-884-5000.

For information on using Hungry Minds' products and services in the classroom or for ordering examination copies, please contact our Educational Sales department at 800-434-2086 or fax 317-572-4005.

For press review copies, author interviews, or other publicity information, please contact our Public Relations department at 317-572-3168 or fax 317-572-4168.

For authorization to photocopy items for corporate, personal, or educational use, please contact Copyright Clearance Center, 222 Rosewood Drive, Danvers, MA 01923, or fax 978-750-4470.

Credits

Acquisitions Editor
Chris Webb

Project Editor
Neil Romanosky

Technical Editor
N. R. Parsa

Copy Editors
Lori Cates Hand
Greg Robertson

Media Development Specialist
Travis Silvers

Permissions Editor
Carmen Krikorian

Media Development Manager
Laura Carpenter VanWinkle

Project Coordinator
Dale White

Cover Design
Anthony Bunyan

Proofreader
Mary Lagu

Indexer
Johnna VanHoose Dinse

Cover
Vault door image used courtesy of
Brown Safe Manufacturing
www.BrownSafe.com

Dreamtech Software India, Inc., Team
dreamtech@mantraonline.com
www.dreamtechsoftware.com

Dreamtech Software India, Inc., is a leading provider of corporate software solutions. Based in New Delhi, India, the company is a successful pioneer of innovative solutions in e-learning technologies. Dreamtech's developers have over 50 years of combined software engineering experience in areas including Java, wireless applications, XML, voice-based solutions, .NET, COM/COM+ technologies, distributed computing, DirectX, Windows Media technologies, and security solutions.

About the Authors

Vikas Gupta is co-founder and president of Dreamtech Software. He is a software engineer and publisher actively engaged in developing and designing new technologies in wireless, e-learning, and other cutting-edge areas. He is also the managing director of IDG Books India (P) Ltd.

Avnish Dass, co-founder and CEO of Dreamtech Software, is a talented and seasoned programmer with 15 years of experience in systems and application/database programming. He has developed security systems, anti-virus programs, wireless and communication technologies, and ERP systems.

Charul Shukla, a senior software developer at Dreamtech, is an expert in Web development and server-side programming. He is intensively involved in designing and developing solutions using XML and related technologies with the goal of providing unified accessible solutions.

Jyotsna Gupta is a senior software developer at Dreamtech, specializing in managing and designing back-end solutions, including overall structure. She is also involved in developing XML-based wireless applications for mobile and handheld devices.

Ashish Baveja, M.K. Dhyani, and **Ravinder B. Srivastava** are a team of programmers with more than two years of experience in software design, analysis, development, testing, and implementation.

To our parents and family and our beloved country, India,
for providing an excellent environment and
for nurturing and creating world-class IT talent

Preface

The past few years have witnessed a dramatic change in the area of information transportation. Until the last decade, we had only the telephone, fax machines, televisions, and radios as the basic communication devices. But, with the rapid development of Information Technology, the scenario changed, and the Internet assumed command as a dynamic and interactive medium. Since its debut in the early '90s, the Internet has made its presence felt around the globe. Today, it enjoys an enviable position that is unparalleled and unsurpassed by any other channel of mass media. It has given a new dimension to the "broadcasting of information" over networks that helps your message and information reach the masses at a rapid pace that was impossible before.

Since the emergence of the Internet in the mid-'90s, desktop PCs have been the most convenient way to access content over the Internet. However, this method has one basic drawback: Every time the user wants to access information through the Internet, he or she needs to find a computer. Considering the size of a normal desktop PC, it is practically impossible to roam around with a PC. A laptop is handy, but the high price is daunting to the average person.

We have been using mobile phones and pagers for quite a long time as established communication devices. From time to time, improved technology has reinforced the utility of mobile phones and other handheld devices. These days, mobile phones and other handheld devices are used to transfer not only voice data but also text data over networks. SMS is another popular service these devices provide in addition to voice communication. Today, most mobile phones and handheld devices can transfer a considerable amount of data over the networks; besides, these devices can be connected to the Internet using the Wireless Application Protocol (WAP). Now you can check e-mail messages and browse the Web using your mobile phone or any other handheld device.

Today, with the emergence of new technologies such as palmtop computers or personal digital assistants (PDAs), pagers, mobile phones, and so on, the user market has increased dramatically because all these new technologies and devices enable people to access content from the Internet. Of late, several companies have developed a number of markup languages for content on their devices. Consequently, the traditional HTML markup language has been shown the door. As a service provider, you now have the choice of a multi-platform structure for your content markup; this is achievable only by virtue of the XML- and XSLT-based approach for content transformations. With this approach, the markup information is separated from the content. You store all the information and content in XML format as neutral data on the server and design different XSLT documents for the transformation of content for different devices.

What This Book Covers

This book, based on the unique concept of "Cracking the Code" and mastering the technology, is loaded with code and keeps theory to a minimum. The applications, for which the source code is given, were tested at the Dreamtech Software Research Lab. The source code provided in the book is based on commercial applications that were also developed by Dreamtech. Each program of the application is explained in a very detailed manner so that the reader gets insight into the implementation of the technology in a real-world situation. At the end of the book, some add-ons to this application are given so that the inquisitive reader can further explore the new developments that are taking place.

In this book, we consider in depth the XML- and XSLT-based approach for content serving on the Internet. This allows a wide range of users to access content over different networks and devices. In this book, we have undertaken a project that demonstrates the power of content transformation using the XML- and XSLT-based approach.

Although the book deals with a relatively new subject, every effort has been made to make the presentation lucid and simple. This book will facilitate an easy understanding of the approach and will equip you to readily deploy the application using the source code provided. The purpose of this book is to help Web site programmers and designers by giving them a practical demonstration. A thorough perusal of the project will equip readers with enough know-how to develop their own applications using the XML- and XSLT-based approach for specific purposes.

Who Should Read This Book?

This book is intended for experienced Web programmers, those who already possess experience in Web application development and deployment. It will help them reframe their legacy applications into the XML- and XSLT-based approach, so that user accessibility increases dramatically. The book enables developers to provide content for most of the available media on the market. Users of this book are expected to be conversant with the concepts of XML and XSLT.

How This Book Is Organized

Chapter 1 consists of a quick roundup of the XSLT language, providing about a dozen examples and code. Readers can quickly brush up on the basics of the XSLT language and prepare themselves to grasp the nuances in the book.

Chapter 2 is devoted to the design specifications of the project we develop throughout this book. It also contains the complete design specifications of the database we used in the project throughout the book.

Chapter 3 covers one of the hottest technologies introduced by Microsoft in the latest version of SQL Server 2000. In this version, Microsoft has made some improvements to the standard SQL queries so that it can support XML features. Now the user can access the data directly in XML format from the database. This chapter contains a number of working samples and detailed theoretical discussions.

Chapter 4 focuses on the content-transformation process for HTML clients using the XML- and XSLT-based approach. In this chapter, you collect neutral data from the SQL server in the form of XML and then transform it to HTML format using XSLT documents and Active Server Pages.

Chapter 5 explains the transformation process used to make the application compatible with WAP. Here we explain the XSLT-based approach in a detailed and systematic manner. This chapter calls for prior knowledge of Wireless Markup Language. It equips the reader with enough information to structure custom-made applications for WAP devices using the XSLT-based approach.

In Chapter 6, we take a quick look at the Handheld Device Markup Language introduced by Openwave (formerly phone.com) for its browser, which is very popular in the United States. The chapter also takes up in detail a case study involving HDML.

Chapter 7 explains the process of transforming XML data into HDML so that developers can target the HDML-compliant devices on the market. If you are new to the HDML world, please go through Chapter 6 thoroughly to learn the nuances of HDML and its features before you start this chapter.

In Chapter 8, we present a brief discussion on i-mode technology and the cHTML language, introduced by NTT DoCoMo and used as a content markup language for i-mode network-based devices.

Chapter 9 takes you to the world of i-mode technology with the help of the content-transformation process using XSLT. This chapter is of particular relevance to the user who proposes to convert legacy content into i-mode device–compliant data.

Chapter 10 unravels the enigma of VoiceXML. The material in this chapter explains the technical aspects of VoiceXML in a step-by-step process backed up with sample code. The case study presented in this chapter serves to enhance your skills in the area of VoiceXML.

In Chapter 11, we demonstrate the process of transforming the ongoing project into a working, interactive voice-response system (IVRS) based on VoiceXML with the XML and XSLT approach. This chapter opens a whole new world of opportunities to aspiring developers and introduces them to an Internet-based interactive voice-response system. Before taking up this chapter in detail, you must be thoroughly familiar with the nuances of VoiceXML, which is covered in Chapter 10.

Chapter 12 introduces a completely new aspect, which will be of immense interest to application developers. This chapter presents the newly introduced XUL language. This language is used to build cross-platform GUIs for desktop applications. When you decide to develop an application for more than one operating system, it gives you a clear edge in the area of the user interface of the application for most operating systems.

Acknowledgments

We would like to acknowledge the contributions of the following people for their support in making this book possible: John Kilcullen, for sharing the dream and providing the vision in making this project a reality; Mike Violano and Joe Wikert, for believing in us; and M. V. Shastri, Asim Chowdury, V. K. Rajan, Sanjeev Chatterjee, and Priti for their immense help in coordinating various activities throughout this project. We also thank technical writers Mridula Sharma and Sunil Gupta, who contributed in developing this book's content.

Contents

Chapter 1

A Rapid Tutorial of XSLT

This chapter presents the basics of XML and XSLT and shows how the combination of XML and XSLT brings power and flexibility to Web-based applications. This chapter includes the following topics:

♦ Introduction to XML

♦ Introduction to XSLT

♦ Advanced XSLT elements

Introduction to XML

Extensible Markup Language (XML) is a text-based language, which is quickly attaining popularity as a means for data interchange on the World Wide Web. There are two primary reasons for its popularity: First, its markup tags can intermingle with the data; and second, it is platform and language independent.

To provide unique solutions for data, XML employs a three-tier architecture. Structured data in XML is maintained as an entity separate from business logic. XML even separates the structured data from the user interface. After the data is separated from the formatting information, it can be used across the platforms for providing information. XML has a unique namespace quality, meaning that elements with the same name do not mix with one another. They can harmoniously coexist in the same document and can also be referred to by different schemas. Being an open-text–based format, HTTP can deliver XML, and agents can easily send and receive XML updates to both the database server and the client.

Authoring XML Documents

Authoring an XML document is not a very difficult task; you can start authoring your XML documents using any simple text editor like Notepad. These days there are several other editors available on the market for authoring the XML-based documents with a user-friendly environment – such as XML SPY.

Writing a Sample XML Document

Examine the XML document in Listing 1-1, `Invoice.xml`. This document represents the status of various client invoices in a computer consultancy.

Listing 1-1: Invoice.xml

```xml
<?xml version = "1.0"?>
<invoiceDetails>
<invoice invoiceNo = "1002" status = "paid" invoiceDate = "2000-03-31">
 <client>A. Baveja</client>
  <address>P-2</address>
 <city>New Delhi</city>
 <state>Delhi</state>
```

```
    <zip>110014</zip>
  <country>India</country>
    <phone>91-011-8888888</phone>
  <services>Networking Issues</services>
  <charges>$50</charges>
</invoice>
<invoice invoiceNo = "1001" status = "unpaid" invoiceDate = "2000-03-31">
    <client>M.K.Dhyani</client>
  <address>S-303</address>
    <city>New Delhi</city>
  <state>Delhi</state>
  <zip>110092</zip>
  <country>India</country>
  <phone>91-011-7777777</phone>
  <services>Software Consultancy</services>
    <charges>$100</charges>
</invoice>
<invoice invoiceNo = "1003" status = "unpaid" invoiceDate = "2000-03-31">
  <client>R.B.Srivastav</client>
    <address>X-79</address>
  <city>New Delhi</city>
    <state>Delhi</state>
  <zip>110002</zip>
  <country>India</country>
    <phone>91-011-9999999</phone>
  <services>Hardware Consultancy</services>
    <charges>$200</charges>
</invoice>
</invoiceDetails>
```

The invoiceDetails element contains the following child elements: an invoice that has the attributes invoiceNo, which states the invoice number; status, which specifies whether the service charges are paid or unpaid; and invoiceDate, which states the date when the invoice was issued. The invoice element has the following child elements that display the details of the client: client, address, city, zip, country, phone, services, and charges.

Opening the XML Document in the Browser

Load the XML document into the browser (for example, Microsoft Internet Explorer) by choosing File ⇨ Open or by using a hyperlinked URL. The browser gives a default view of XML documents. If an XML-style-sheet processing instruction is given in the XML document, Internet Explorer uses it to present the information. If no instruction is provided, it applies its default style sheet. Microsoft XML Parser (MSXML) parses the document after loading and checks it for accuracy. In the event of parser failure, Internet Explorer reports error messages. If parsing completes successfully, Internet Explorer shows the document structure as an outline, with small + and – icons for expanding and closing sections of the document. In long documents with large sections, you can also view the navigational information.

Validating the XML Document

You cannot validate or view an XML document just by opening it in the browser. You cannot view the output even with the help of XSL or XSLT style sheets. You can validate an XML document only by means of an embedded schema; this validation takes place while loading the document via the Internet Explorer MIME viewer. You can download the MIME viewer (the filename is iexmltls.exe) by going to the download section of the msdn site and looking for the INTERNET EXPLORER TOOLS FOR VALIDATING XML AND VIEWING XSLT OUTPUT under the **web development > xml** section.

Installing Relevant Files

The following files enable validation of an XML document browsed in Internet Explorer, as well as viewing of the XSLT output displayed in the browser:

- `msxmlval.htm`
- `msxmlval.inf`
- `msxmlvw.htm`
- `msxmlvw.inf`

You can install the preceding files by doing the following:

1. Navigate to the directory containing those files.
2. Right-click on each of the .inf files and then select the Install menu option.

After you install these files, right-click on the browser window to show added options on the drop-down menu. These entries provide the following options for validating the XML document and viewing the XSL output:

- Validate XML
- View XSL Output

You validate the XML document, `Invoice.xml`, by using the XML Validation Tools (see Figure 1-1).

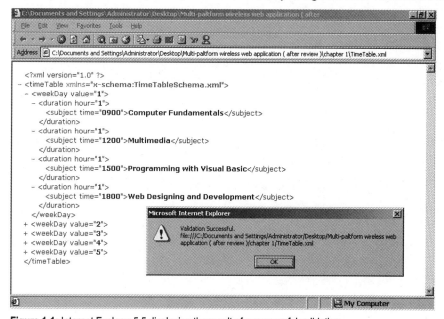

Figure 1-1: Internet Explorer 5.5 displaying the result of a successful validation

XML Namespaces

XML namespaces are a part of the XML 1.0 specification. They avoid naming redundancies by using Uniform Resource Identifiers (URIs). URIs provide a means of creating names in an XML document that uniquely identify elements or attributes. For example, consider the following XML data:

```
<address>Personal Address...</address>
<address>Official Address...</address>
```

If both of these elements are used in the same document, a naming collision occurs and it becomes difficult to determine the nature of the data each element contains. You can solve this problem by using namespaces to differentiate between these two address elements, as shown in the following examples:

```
<personal:address>Personal Address...</personal:address>
<official:address>Official Address...</official:address>
```

Both personal and official are namespace prefixes. Every namespace prefix is bound to a unique URI that identifies the namespace, as shown in the following examples:

```
<detail xmlns:personal = "Personal Address"
 xmlns:official = "Official Address">
```

Here, the detail element is the root element that uses the namespace keyword xmlns to define two namespace prefixes, personal and official, that describe the personal and official addresses of an individual person. The values assigned to attributes xmlns:personal and xmlns:official are called URIs, which can be a string of characters used to differentiate names. Usually, you use a URL to define a URI because all URLs must be unique.

The preceding format can be illustrated in an XML document in the following manner:

```
<?xml version = "1.0"?>
<detail xmlns:personal = "Personal Address"
 xmlns:official = "Official Address">
 <personal:address description = "personal">
  <personal:flatNo>23</personal:flatNo>
  <personal:street>Golf Links</personal:street>
      <personal:city>New Delhi</personal:city>
  <personal:state>Delhi</personal:state>
  <personal:zip>110015</personal:zip>
  <personal:country>India</personal:country>
 </personal:address>
 <official:address description = "official">
  <official:flatNo>39</official:flatNo>
  <official:street>South Extention</official:street>
      <official:city>New Delhi</official:city>
  <official:state>Delhi</official:state>
  <official:zip>110016</official:zip>
  <official:country>India</official:country>
 </official:address>
</detail>
```

To eliminate the need to place a namespace prefix in each element, you can specify a default namespace, using xmlns, for an element and its child elements, as shown in the following example:

```
<?xml version = "1.0"?>
<detail xmlns = "Personal Address"
 xmlns:official = "Official Address">
 <address description = "personal">
  <flatNo>23</flatNo>
  <street>Golf Links</street>
      <city>New Delhi</city>
  <state>Delhi</state>
  <zip>110015</zip>
  <country>India</country>
 </address>
 <official:address description = "official">
  <official:flatNo>39</official:flatNo>
```

```
  <official:street>South Extention</official:street>
      <official:city>New Delhi</official:city>
  <official:state>Delhi</official:state>
  <official:zip>110016</official:zip>
  <official:country>India</official:country>
 </official:address>
</detail>
```

The first namespace is the default namespace. The default namespace is used by the `detail` element and the address child element. The other child element address still uses the official namespace.

Schemas

Schemas declare the element types that can appear in an XML document, as well as what attributes an element may contain; they also define what elements and attributes are required or optional, and the number of occurrences a subelement may have. In addition, they define which element type may be child elements of a specific parent element, the data type for the contents of an element, default values for attributes, and so on.

Listing 1-2, `TimeTableSchema.xml`, models a personal schedule for five working days in a week.

Listing 1-2: TimeTableSchema.xml

© 2001 Dreamtech Software India Inc.
All Rights Reserved

```
<?xml version = "1.0"?>
<Schema xmlns = "urn:schemas-microsoft-com:xml-data"
     xmlns:dt = "urn:schemas-microsoft-com:datatypes">
<ElementType name = "timeTable"
 content = "eltOnly"
 model = "closed">
  <element type = "weekDay"
    minOccurs = "0"
    maxOccurs = "*"/>
 </ElementType>

<ElementType name = "weekDay"
 content = "eltOnly"
 model = "closed">
  <AttributeType name = "value"
 dt:type = "int"/>
  <attribute type = "value"/>
  <element type = "duration"
      minOccurs = "0"
    maxOccurs = "*"/>
 </ElementType>
<ElementType name = "duration"
 content = "eltOnly"
 model = "closed">
 <AttributeType name = "hour"
  dt:type = "int"/>
 <attribute type = "hour"/>
  <element type = "subject"
    minOccurs = "0"
    maxOccurs = "1"/>
 </ElementType>
```

```
<ElementType name = "subject"
content = "textOnly"
model = "closed"
dt:type="string">
  <AttributeType name = "time"
dt:type = "int"/>
            <attribute type = "time"/>
</ElementType>
</Schema>
```

Note the following lines of code:

```
<ElementType name = "timeTable"
content = "eltOnly"
model = "closed">
  <element type = "weekDay"
    minOccurs = "0"
    maxOccurs = "*"/>
</ElementType>
```

These lines define an element timeTable, which can contain only elements, as the attribute content is eltOnly. The fact that the Value of Attribute model is closed indicates that only elements declared in this schema are allowed in the conforming XML document. The element element indicates that the weekDay child element may be included in the timeTable element. The minOccurs and maxOccurs attributes declare the minimum and maximum number of times the element may appear in the timeTable element, respectively. The value 0 for the minOccurs attribute indicates that the timeTable element may or may not contain the child element weekDay. The value * for the maxOccurs attribute specifies that there is no limit on the maximum number of weekDay child elements that may appear in timeTable element.

The weekDay element of int data type is defined by the dt:type attribute using the following code:

```
<AttributeType name = "value"
  dt:type = "int"/>
<attribute type = "value"/>
```

Listing 1-3 shows the XML document TimeTable.xml, which conforms to the XML schema TimeTableSchema.xml.

Listing 1-3: TimeTable.xml

```
<?xml version = "1.0" ?>
<timeTable xmlns = "x-schema:TimeTableSchema.xml">
 <weekDay value = "1">
  <duration hour = "1">
   <subject time = "0900">Computer Fundamentals</subject>
  </duration>
  <duration hour = "1">
   <subject time = "1200">Multimedia</subject>
  </duration>
  <duration hour = "1">
   <subject time = "1500">Programming with Visual Basic</subject>
  </duration>
  <duration hour = "1">
   <subject time = "1800">Web Designing and Development</subject>
```

```
 </duration>
</weekDay>
<weekDay value = "2">
 <duration hour = "1">
  <subject time = "0900">Computer Fundamentals</subject>
 </duration>
 <duration hour = "1">
  <subject time = "1200"> Software Engineering</subject>
 </duration>
 <duration hour = "1">
  <subject time = "1500">Programming with Visual Basic</subject>
 </duration>
 <duration hour = "1">
  <subject time = "1800">Web Designing and Development</subject>
 </duration>
</weekDay>
<weekDay value = "3">
 <duration hour = "1">
  <subject time = "0900">Computer Fundamentals</subject>
 </duration>
 <duration hour = "1">
  <subject time = "1200"> Software Engineering</subject>
 </duration>
 <duration hour = "1">
  <subject time = "1500">Programming with Visual Basic</subject>
 </duration>
 <duration hour = "1">
  <subject time = "1800">Web Designing and Development</subject>
 </duration>
</weekDay>
<weekDay value = "4">
 <duration hour = "1">
  <subject time = "0900">Computer Fundamentals</subject>
 </duration>
 <duration hour = "1">
  <subject time = "1200"> Software Engineering</subject>
 </duration>
 <duration hour = "1">
  <subject time = "1500">Programming with Visual Basic</subject>
 </duration>
 <duration hour = "1">
  <subject time = "1800">Web Designing and Development</subject>
 </duration>
</weekDay>
<weekDay value = "5">
 <duration hour = "1">
  <subject time = "0900">Computer Fundamentals</subject>
 </duration>
 <duration hour = "1">
  <subject time = "1200">Software Engineering</subject>
 </duration>
 <duration hour = "1">
  <subject time = "1500">Programming with Visual Basic</subject>
 </duration>
 <duration hour = "1">
  <subject time = "1800">Web Designing and Development</subject>
```

```
    </duration>
  </weekDay>
</timeTable>
```

Note the following line of code in Listing 1-3:

```
<timeTable xmlns = "x-schema:TimeTableSchema.xml">
```

This code shows that the XML document `TimeTable.xml` uses an attribute `xmlns` (see Figure 1-2). This attribute begins with `x-schema`, which is followed by a colon (`:`) and then the name of the schema document `TimeTableSchema.xml`.

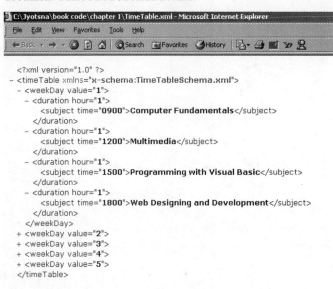

Figure 1-2: Internet Explorer 5.5 displaying TimeTable.xml in default mode

The schema is used for proper validation of the document structure on the basis of predefined structure. Proper allocation of the document structure is possible by applying the schema to an XML document. The following code clarifies the effect of applying schemas. Before the schema is constructed, you need to add the following opening and closing tags to `Invoice.xml`:

```
<myschema xmlns = "x-schema:InvoiceSchema.xml"></myschema>
```

in place of

```
<invoiceDetails></invoiceDetails>
```

Listing 1-4, `InvoiceSchema.xml`, is the schema that represents `Invoice.xml`.

Listing 1-4: InvoiceSchema.xml

© 2001 Dreamtech Software India Inc.
All Rights Reserved

```
<?xml version = "1.0" ?>
<Schema name = "InvoiceSchema"
    xmlns = "urn:schemas-microsoft-com:xml-data"
    xmlns:dt = "urn:schemas-microsoft-com:datatypes">
 <ElementType name = "myschema"
              content = "eltOnly"
              model = "closed">
```

```
    <description>Invoice Details</description>
    <element type = "invoice"
        minOccurs = "0"
        maxOccurs = "*"/>
</ElementType>
<ElementType name = "invoice"
                content = "eltOnly"
            order = "many"
            model = "closed">
    <AttributeType name = "invoiceNo"
            dt:type = "int"
                    required = "yes" />
    <attribute type = "invoiceNo"/>
    <AttributeType name = "status"
            default = "unpaid"
                    required = "no" />
    <attribute type = "status"/>
    <AttributeType name = "invoiceDate"
                    dt:type = "date"
                    required = "yes" />
    <attribute type = "invoiceDate"/>
    <element type = "client" />
    <group order = "seq">
     <element type = "address"
        minOccurs = "1"
        maxOccurs = "*" />
     <element type = "city" />
     <element type="state" />
     <element type = "zip" />
     <element type = "country" />
    <element type = "phone" />
    </group>
    <element type = "services"/>
    <element type = "charges" />
</ElementType>
<ElementType name = "client"
            content = "textOnly" />
<ElementType name = "address"
            content = "textOnly" />
<ElementType name = "city"
            content="textOnly" />
<ElementType name = "state"
            content="textOnly" />
<ElementType name = "zip"
            content = "textOnly" />
<ElementType name = "country"
            content = "textOnly" />
<ElementType name = "phone"
            content = "textOnly" />
<ElementType name = "services"
            content = "textOnly" />
<ElementType name = "charges"
        content = "textOnly" />
</Schema>
```

The <Schema> element can be classified as the root element. It has a name attribute for proper identification. The element type attribute indicates, in the inner portion of the element, whether the element contains mixed data or character data or if it is just an empty element. The element is named using the name attribute. The order attribute clearly specifies whether the element types occur one by one or in a sequence. Information (on client, address, phone, services, charge elements, and so on) is contained by the invoice element; and the subelements of the element type are synchronized in proper order by the group element.

Schemas, therefore, have many advantages in comparison to DTDs. One of the biggest advantages is that they provide a large range of data types; they also use namespaces to uniquely identify elements, and all schemas follow XML syntax.

Display XML in the Browser Using Style Sheets

The XML document in Listing 1-5, audio.xml, is a database of a person's music collection. It has an element called AUDIO, which contains complete data on each album in the collection --such as artist name, album name, record company, price, and the year of release.

Listing 1-5: audio.xml

© 2001 Dreamtech Software India Inc.
All Rights Reserved

```xml
<?xml version = "1.0"?>
<LIST>
<AUDIO>
    <ARTIST>Bryan Adams</ARTIST>
    <ALBUM>So far so good</ALBUM>
    <COMPANY>Virgin Records</COMPANY>
    <AMOUNT>$11.50</AMOUNT>
    <RELEASE>1994</RELEASE>
 </AUDIO>
 <AUDIO>
    <ARTIST>George Michael</ARTIST>
    <ALBUM>Faith</ALBUM>
    <COMPANY>CBS Records</COMPANY>
    <AMOUNT>$9.90</AMOUNT>
    <RELEASE>1988</RELEASE>
 </AUDIO>
 <AUDIO>
    <ARTIST>U2</ARTIST>
    <ALBUM>The Joshua Tree</ALBUM>
    <COMPANY>Atlantic</COMPANY>
    <AMOUNT>$8.50</AMOUNT>
    <RELEASE>1988</RELEASE>
 </AUDIO>
 <AUDIO>
    <ARTIST>Michael Jackson</ARTIST>
    <ALBUM>History</ALBUM>
    <COMPANY>SONY</COMPANY>
    <AMOUNT>$20.50</AMOUNT>
    <RELEASE>1995</RELEASE>
 </AUDIO>
</LIST>
```

Write a Cascading Style Sheet

You use a Cascading Style Sheet (CSS) to apply rules and formatting properties and to provide a simple syntax for styling the elements in an XML document. Some HTML browsers have partly utilized CSS.

The complete specification for Cascading Style Sheets 1 (CSS1) is available on the Web at `http://www.w3.org/TR/REC-CSS1`. Recommendations for CSS2 are available at `http://www.w3.org/TR/REC-CSS2`.

Listing 1-6 is a CSS, `audio_list.css`, for the audio collection example.

Listing 1-6: audio_list.css

```
{
background-color: #ccffff;
width: 100%;
}
AUDIO
{
display: block;
border-style: outset;
border-width: thin;
padding: 1em;
}
ALBUM
{
font-family: Arial, Helvetica, sans-serif;
font-size: 15pt;
color: green
}
ARTIST
{
font-family: sans-serif, Helvetica, Arial;
font-size: 20pt;
color: red;
}
AMOUNT,RELEASE,COMPANY
{
color: blue;
margin-left: 20pt;
text-align: left;
}
```

Apply a CSS to the XML Document

In accordance with the World Wide Web Consortium (W3C) recommendation, Microsoft Internet Explorer implements an XML-style–sheet processing instruction. The following XML-style-sheet processing instruction identifies a style sheet built using a CSS:

```
<?xml-stylesheet type = "text/css" href = "css_filename"?>
```

This processing instruction must appear in the prolog, before the document or root element. Each style-sheet processing instruction must have the `type` attribute, which describes the type of style sheet to apply; here, `"text/css"` indicates a CSS style sheet is being applied. The `href` attribute is a URL that links to the style sheet. Relative URLs are evaluated relative to the URL of the XML document.

Listing 1-7 shows the altered XML document `audio.xml`, renamed `audio_list_css.xml`.

Listing 1-7: audio_list_css.xml

```xml
<?xml version = "1.0"?>
<?xml-stylesheet type = "text/css" href = "audio_list.css"?>
<LIST>
<AUDIO>
    <ARTIST>Bryan Adams</ARTIST>
    <ALBUM>So far so good</ALBUM>
    <COMPANY>Virgin Records</COMPANY>
    <AMOUNT>$11.50</AMOUNT>
    <RELEASE>1994</RELEASE>
  </AUDIO>
  <AUDIO>
    <ARTIST>George Michael</ARTIST>
    <ALBUM>Faith</ALBUM>
    <COMPANY>CBS Records</COMPANY>
    <AMOUNT>$9.90</AMOUNT>
    <RELEASE>1988</RELEASE>
  </AUDIO>
  <AUDIO>
    <ARTIST>U2</ARTIST>
    <ALBUM>The Joshua Tree</ALBUM>
    <COMPANY>Atlantic</COMPANY>
    <AMOUNT>$8.50</AMOUNT>
    <RELEASE>1988</RELEASE>
  </AUDIO>
  <AUDIO>
    <ARTIST>Michael Jackson</ARTIST>
    <ALBUM>History</ALBUM>
    <COMPANY>SONY</COMPANY>
    <AMOUNT>$20.50</AMOUNT>
    <RELEASE>1995</RELEASE>
  </AUDIO>
</LIST>
```

Listing 1-7 is linked to the CSS `audio_list.css`, as indicated by the `href` and `type` attributes defined in the initial processing instructions.

Figure 1-3 shows the results of applying a CSS document on an XML document.

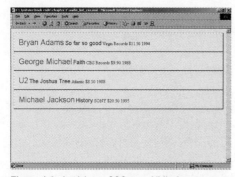

Figure 1-3: Applying a CSS to an XML document

Apply Different Styles to the XML Document

You can further modify the CSS document audio_list.css to enhance the appearance of the XML document. For example, you can change a few properties in the CSS document, as follows:

```
LIST
{
background-color: #ccffff;
width: 100%;
}
AUDIO
{
border-style: outset;
border-width: thin;
display: block;
margin-bottom: 5pt;
margin-left: 10pt;
}
ALBUM,ARTIST
{
border-style: outset;
border-width: thin;
font-family: Arial, Helvetica, sans-serif;
font-size: 15pt;
color: red
}
COUNTRY,AMOUNT,RELEASE,COMPANY
{
border-style: outset;
border-width: thin;
font-family: sans-serif, Helvetica, Arial;
font-size: 12pt;
color: blue;
}
```

Likewise, the properties defined in the preceding, modified CSS change the look and feel of the XML example document (see Figure 1-4).

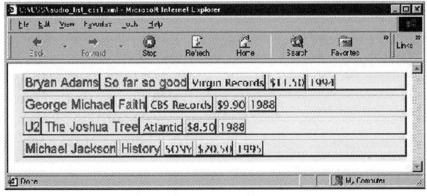

Figure 1-4: Modifications to the CSS document cause changes in the appearance of the XML data.

Both CSS and XSL transform the XML documents to give information to the client or the user. In the future, however, XSL will be used more than CSS, because CSS defines only the style of HTML elements of an XML document, whereas XSL defines the whole document.

Introduction to XSLT

Extensible Style Sheet Language (XSL) consists of three main components (as described in the W3C Recommendations):

♦ XSL Transformations (XSLT): a language for transforming XML documents

♦ XPath: an expression language for referencing specific parts of an XML document

♦ XSL Formatting Objects: a vocabulary of formatting objects and formatting properties

The XSLT style sheet contains *template rules* that specifically indicate the document parts to be selected. It also indicates the way in which they will be processed to get the result. There are two types of template rules: a pattern and a template. Using the XPath syntax, a template rule expresses a *pattern* After that, it is matched against the elements present in the *source tree* in order to be processed by the template.

XML documents use XSL for formatting by generating formatting objects. They generate HTML or XHTML pages from XML data or documents. Itcan transform one XML document into another XML document. Lastly, it is able to generate a textual representation of an XML document.

An XSL style sheet contains a set of templates. Every template has some elements that match the source tree. These elements match to form the *result tree* Most templates have the following form:

```
<xsl:template match = "emphasis">
<xsl:apply-templates/>
</xsl:template>
```

♦ The whole <xsl:template> element is a template.

♦ The match pattern indicates the element that applies to the template.

♦ Literal result element(s) come from non-XSL namespace(s).

♦ XSLT elements come from the XSL namespace.

For effective location of the specific parts of an XML document, XPath provides an extensible string-based syntax. In XPath, an XML document is viewed conceptually as a tree in which each part of the document is represented as a node. XPath has seven node types. They are the *root, element, attribute, text, comment, processing, instruction,* and *namespace* The XPath tree contains a single-root node, which possesses all other nodes in the tree.

Architecture of XSL

The XSL style sheet transforms a source tree into a result tree. Figure 1-5 illustrates its architecture. The XML source document parses into a source tree, which ultimately changes into a result tree. The result tree is copied in the result file by a special format.

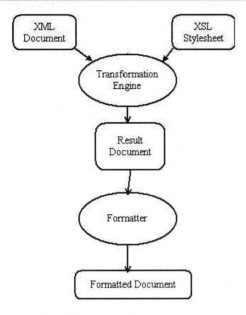

Figure 1-5: Architecture of XSL

The XML Parser first transforms the XML document into a source tree. It then reads the XSL style sheet and matches its template rules with root nodes of the source tree. Ultimately, the result tree is created when the matching is done. After that, it is copied to an output file.

The Process of XSL Formatting and Transforming XML Documents

XSL is composed mainly of two parts: XSLT, which is a language for transforming XML documents, and XSL Formatting Objects (XSL FO), which is an XML vocabulary used to format XML documents for presentation.

XSL FO itself is an XML-based markup language that helps specify in great detail the pagination, layout, and styling information that can be applied to the content of an XML document. An XML document is transformed into an XSL document that marks up the data using formatted objects.

This section gives you an introduction to the data types, variables, and expressions used in XSLT for the rendering of the document.

XSLT Variables, Data Types, and Expressions

Unlike variables in other programming languages, you can assign XSLT variables a value only once — namely, when they are declared. Using XSLT variables in a style sheet is very useful. First, it helps overall readability; second, it gets rid of long XPath expressions. Variables are prefixed with the $ symbol for differentiating them from regular XPath expressions.

```
<xsl:variable name = "sample variable" select = "expression"/>
```

A variable declaration for an XPath value is

```
<xsl:variable name = "sample variable"> template </..>
```

A variable declaration for a result tree fragment (template is instantiated to give value) is

```
<xsl:param name = "sample variable" select = "expression"/>
```

The parameter declaration for an XPath value (the value is the default value of the parameter) is

```
<xsl:param name = " sample variable "> template </..>
```

A parameter declaration for a result tree fragment is as follows:

```
$name
```

It returns the XPath value in expressions, that is, in attribute value templates.

```
 <xsl:with-param name = "sample variable" select = "sample"/>
```

and

```
<xsl:with-param name = "sample variable"> template </..>
```

The preceding code passes parameters in xsl: call-template or xsl: apply-templates. Note that the result-tree fragments held by variables or parameters can be used as the source in pattern matching and template instantiation.

XSLT style sheets are XML documents; in other words, they are namespaces (http://www.w3.org/TR/REC-xml-names) used to identify semantically significant elements. Most of the style sheets are standalone documents that are rooted at <xsl:stylesheet> or <xsl:transform>.*Single-template* style sheets/documents are also a possibility here, and <xsl:stylesheet> and <xsl:transform> are completely synonymous.

Sample XSLT Document

First, it's necessary to prepare an XML document. The simple XML document in Listing 1-8 — example1.xml — uses an XSL style sheet for transformation.

Listing 1-8: example1.xml

```
<?xml version = "1.0"?>
<?xml:stylesheet type = "text/xsl" href = "example1.xsl"?>
<message>
 The first XSL document.
  </message>
```

The stylesheet element attaches a style sheet to an XML document. The type attribute defines the type of the file being attached, and text/xsl denotes an XSL document. The href attribute holds the file — example1.xsl — that is being attached.

Now you will write some XSL for the XML document. As mentioned earlier, there are two types of node trees in an XSL transformation: the source tree and the result tree. The result tree contains the XML document, which is formed after the transformation. Listing 1-9, example1.xsl, transforms the XML document example1.xml.

Listing 1-9: example1.xsl

```
<xsl:stylesheet version = "1.0"
               xmlns:xsl = "http://www.w3.org/1999/XSL/Transform">
<xsl:template match = "/">
<html>
<head>
```

```
    <title>Introduction to XSL...</title>
  </head>
  <body>
    <h1><xsl:value-of select = "message"/></h1><hr/>
  </body>
</html>
</xsl:template>
</xsl:stylesheet>
```

The statements

```
<xsl:stylesheet version = "1.0"
                xmlns:xsl = "http://www.w3.org/1999/XSL/Transform">
<xsl:template match = "/">
```

indicate the XSLT document's root element (the `xsl:stylesheet`) and its attributes. The `version` attribute defines the XSLT specification used. The `xmlns:xsl` is the XSLT namespace definition, which is identified by its unique URI value (`http://www.w3.org/1999/XSL/Transform`). The template element matches specific XML document nodes by using an XPath expression in the attribute match. The "`/`" identifies the root node of the XML document. The text contents of the node set are placed in the result tree by using the value-of element and an XPath expression in attribute select.

Figure 1-6 shows the results of displaying the XML transformation in the browser.

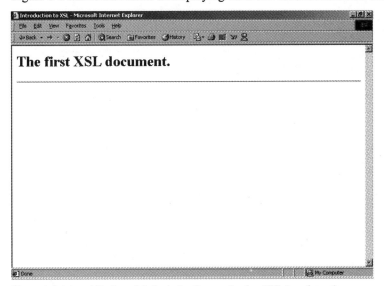

Figure 1-6: Internet Explorer 5.5 displaying the results of an XML transformation

Introduction to XSLT Elements

XSLT transforms XML data in such a way that it acts like a sieve, creating the result tree by filtering the source tree.

In an XSLT transformation process, when a source tree transforms into a result tree, a template is instantiated for a particular source element to create part of the result tree. It can contain elements that specify the final element structure. It can also contain elements from the XSLT namespace, which are instructions for creating result-tree fragments. When a template is instantiated, each instruction is executed and replaced by the result-tree fragment, which it creates. The instructions can be selected and processed as descendant source elements. Processing a descendant element, however, creates a result-tree fragment by finding the applicable template rule and instantiating its template.

The result tree is constructed by finding the template rule for the root node and instantiating its template. For simple transformations where the structure of the result tree is independent of the structure of the source tree, however, a style sheet can often consist of only a single template that functions as a template for the complete result tree. When a template is instantiated, it is always instantiated with respect to a current node and a current node list. The current node is always a member of the current node list. Many operations in XSLT are relative to the current node. XSLT makes use of the XPath expression language for selecting elements for processing, for conditional processing, and for generating text. XSLT templates are defined by using the XSLT elements listed in Table 1-1. Go to `http://www.w3c.org/tr/xsl` for more information.

Table 1-1: XSLT Elements

Element Name	Element Description
`xsl:stylesheet` or `xsl:transform`	Defines the root element of the XSLT document used for the transformation or rendering process of any given XML document.
`xsl:template`	Defines what should be output for elements matching a given on a pattern for transforming the source XML document tree.
`xsl:value-of`	Fetches the value of the selected node as text —mostly used for selecting the value of any given node.
`xsl:apply-templates`	Applies a template to the current element.
`xsl:import`	Imports one style sheet into another style sheet. Use of this increases the scalability and reusability of XSLT documents.
`xsl:apply-imports`	Applies a defined template rule from an imported style sheet into another style sheet. This element contains zero or more `xsl:with-param` child elements. `xsl:with-param` passes a parameter to a template.
`xsl:output`	Defines the format of the style sheet output, such as HTML, XML, and so on.
`xsl:for-each`	Provides a way to create a looping structure in the output stream for fetching the values of all nodes under the given root node.
`xsl:if`	Provides a way to to build a conditional structure for executing a task.
`xsl:attribute`	Adds an attribute to the nearest containing element. This element can be used within a template body or within an `xsl:attribute-set` element. The `xsl:attribute-set` element defines a named set of attributes. The attribute set can be applied to any output element.
`xsl:variable`	Provides a way to declare a variable in an XSLT document for the later use while transforming or rendering the given XML document.
`xsl:sort`	Provides a way to define sorting structure in an XSLT document for sorting the elements and nodes of any given XML document.
`xsl:otherwise`	Indicates what should happen when none of the `xsl:when` elements inside an `xsl:choose` element is satisfied. The `xsl:when` defines a condition to be tested and performs an action if the condition is true. This element is always a child element of `xsl:choose`. The `xsl:choose` element provides a way to choose between a number of alternatives based on conditions.

Element Name	Element Description
`xsl:preserve-space`	Preserves white space in a document.
`xsl:param`	Provides a way to define parameters in XSLT document.
`xsl:call-template`	Provides a way to call a named template. It contains zero or more `xsl:with-param` child elements.
`xsl:comment`	Creates an comment in the resulting XML document after transforming a given XML document.
`xsl:message`	Sends a text message to either the message buffer or a message dialog box.
`xsl:text`	Generates text in the resulting document.
`xsl:copy`	Copies the current node without childnodes and attributes to the resulting document.
`xsl:copy-of`	Copies the current node with childnodes and attributes to the output.
`xsl:element`	Creates an element with the specified name in the resulting document.
`xsl:number`	Inserts a formatted number into the result tree.
`xsl:decimal-format`	Defines default decimal format for number-to-text conversion.
`xsl:key`	Declares a key for subsequent use with the `key()` function.
`xsl:fallback`	This element provide a solution to handle the situation when an XML parser is not able to recognize some of newer or unknown XSLT elements while transforming any given XML document.
`xsl:include`	An XSLT document can include any other XSLT documents.
`xsl:processing-instruction`	This element generates a processing instruction for the output while transforming an XML document.
`xsl:namespace-alias`	Replaces the prefix associated with a given namespace with another prefix in the resulting output.

Work with XSLT Elements

In this section, we will investigate XSLT elements more closely.

xsl:stylesheet

The `xsl:stylesheet` element defines the root element of the style sheet or `xsl:transform`. The syntax is as follows:

```
<xsl:stylesheet
id = "name"
version = "version"
extension-element-prefixes = "list"
exclude-result-prefixes = "list">
TOP-LEVEL-ELEMENTS
</xsl:stylesheet>
```

The syntax for `xsl:transform` is as follows:

```
<xsl:transform
id = "name"
version = "version"
extension-element-prefixes = "list"
exclude-result-prefixes = "list">
TOP-LEVEL-ELEMENTS
</xsl:transform>
```

The following table describes the attributes in more detail.

Attribute Name	Value	Attribute Description
Version	version	Required. This attribute defines the version of the style sheet set by W3C. Value 1.0 is defined for the current recommended version of the style sheet by the W3c.
extension-element-prefixes	list	Optional. Defines a namespace as an extension namespace List (separated by white space) of namespace prefixes used for extension elements.
exclude-result-prefixes	list	Optional. List (separated by white space) of namespace prefixes that should not be included in the resulting document.
Id	name	Optional. Used to provide a unique identifier for the style sheet.

xsl:template

The xsl:template element defines what should be the output for elements matching a given pattern. The syntax of xsl:template is shown in the following code and table:

```
<xsl:template
name = "name"
match = "pattern"
mode = "mode"
priority = "number">
  <xsl:param>
  template-body
</xsl:template>
```

Attribute Name	Value	Attribute Description
name	name	Optional. The name of the template. If this attribute is omitted, there must be a match attribute.
match	pattern	Optional. The match pattern for the template. If this attribute is omitted, there must be a name attribute.
mode	mode	Optional. The mode of this template. This attribute allows multiple-time processing of an element with a different result every time.
priority	number	Optional. A number that indicates the priority of the template. The value of this should be between 0-9. Negative and positive signs are also included.

xsl:value-of

The xsl:value-of element inserts the value of the selected node as text. The syntax of xsl:value is shown in the following code and table:

```
<xsl:value-of select = "expression" disable-output-escaping = "yes|no"/>
```

Attribute Name	Value	Attribute Description
`Select`	expression	Required. Selects the value of the given node, and the selected value is converted into a string.
`disable-output-escaping`	Yes no	Optional. A `yes` value indicates that special characters (such as <) should be output as is. A `no` value indicates that special characters (such as <) should be output as `%lt`. Default is `no`.

xsl:apply-templates

The `xsl:apply-templates` element applies a template to the current element. The syntax of this element is shown in the following code and table:

```
<xsl:apply-templates select = "expression" mode = "name">
  <xsl:with-param>
  <xsl:sort>
</xsl:apply-templates>
```

Attribute Name	Value	Attribute Description
`select`	expression	Optional. The nodeset to be processed. If omitted, all childnodes under the current node will be processed by default. Nodes are selected and processed in order of document in case no sorting specification is defined.
`mode`	name	Optional. The processing mode for the elements.

Example

Listing 1-10, `example.xml`, contains the details of three students in a particular school.

Listing 1-10: example.xml

```
<?xml version = "1.0"?>
<?xml:stylesheet type = "text/xsl" href = "example2.xsl"?>
<students>
 <student>
  <name>ABC</name>
  <class>XIIth</class>
  <section>B</section>
  <gender>Male</gender>
  <age>20</age>
  <address>P-2,New Delhi</address>
  <phone>7777777</phone>
 </student>

 <student>
  <name>DEF</name>
  <class>IXth</class>
  <section>A</section>
  <gender>Male</gender>
  <age>17</age>
  <address>S-2,New Delhi</address>
  <phone>8888888</phone>
 </student>
 <student>
  <name>XYZ</name>
```

```
     <class>Ist</class>
     <section>A</section>
     <gender>Female</gender>
     <age>5</age>
     <address>C-3,New Delhi</address>
     <phone>9999999</phone>
    </student>
</students>
```

The corresponding XSLT in Listing 1-11, example2.xsl, transforms the XML document into HTML output, using the elements xsl:stylesheet, xsl:template, xsl:value-of, and xsl:apply-templates. Each row represents a student, and the columns represent the student's name, class, section, gender, age, address, and phone number.

The xsl:apply-templates element first selects a set of nodes, using the query specified in the select attribute. If this attribute is left unspecified, all the children of the current node are selected. For each of the selected nodes, xsl:apply-templates directs the XSLT processor to find an appropriate xsl:template to apply.

Listing 1-11: example2.xsl

© 2001 Dreamtech Software India Inc.
All Rights Reserved

```
<xsl:stylesheet version  = "1.0"
  xmlns:xsl = "http://www.w3.org/1999/XSL/Transform">
 <xsl:template match = "students">
 <html>
   <head>
<title>Demonstrating xsl:apply-templates</title>
</head>
 <body>
   <h1><center>Student Details</center></h1><hr/><br/>
   <table border = "5" align = "center">
     <tr>
      <th>Name</th>
<th>Class</th>
<th>Section</th>
<th>Gender</th>
<th>Age</th>
<th>Address</th>
<th>Phone</th>
     </tr>
     <xsl:apply-templates/>
   </table>
 </body>
 </html>
 </xsl:template>
 <xsl:template match = "students/student">
  <tr>
   <xsl:apply-templates/>
  </tr>
 </xsl:template>
 <xsl:template match = "name | class | section | gender | age | address |
phone">
  <td>
   <xsl:value-of select = "."/>
```

```
    </td>
  </xsl:template>
</xsl:stylesheet>
```

Figure 1-7 is the resulting output as rendered in Internet Explorer 5.5.

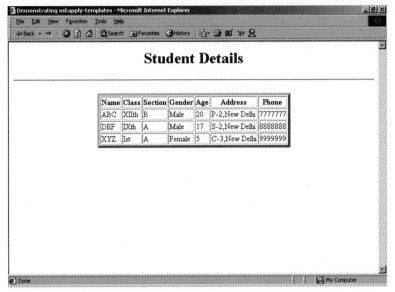

Figure 1-7: Output of example2.xsl

xsl:import

The xsl:import element imports one style sheet into another style sheet while performing the transformation process. The syntax of xsl:import is shown in the following code and table:

```
<xsl:import href = "uri"/>
```

Attribute Name	Value	Attribute Description
href	Uri	Required. URI of the external style sheet to be imported while executing the host style sheet.

xsl:apply-imports

The xsl:apply-imports element applies a template rule from an imported style sheet into the existing style sheet while performing the transformation process. The syntax of xsl:apply-imports is as follows:

```
<xsl:apply-imports>
<xsl:with-param>
</xsl:apply-imports>
```

This element contains zero or more xsl: with-param child elements.

xsl:with-param

This element is used to pass a parameter to a template. The value of the parameter can be accessed throughout the template. The syntax of xsl:with-param is as shown in the following code and table:

```
<xsl:with-param
```

```
  name = qname
  select = expression>
  <!-- Content: template -->
</xsl:with-param>
```

Attribute Name	Value	Attribute Description
name	Name	Required. This will define the name of the parameter used to pass some values to the template.
select	Expression	Optional. The value of the parameter. In case no content is specified, an empty string is generated.

xsl:output

The xsl:output element controls the format of the style sheet output. The syntax of xsl:output is shown in the following code and table:

```
<xsl:output
method = "xml|html|text|name"
version = "version"
encoding = "text"
omit-xml-declaration = "yes|no"
standalone = "yes|no"
doctype-public = "text"
doctype-system = "text"
cdata-section-elements = "namelist"
indent = "yes|no"
media-type = "mimetype"/>
```

Attribute Name	Value	Attribute Description
Method	text name	Optional.Defines the output ethod for generating the result document after the transforming process.
Version	version	Optional. The XML version of the output document if the method is defined as xml.
Encoding	text	Optional. Defines the character encoding, so the parser encodes the characters in sequences of bytes.
omit-xml-declaration	yes no	Optional. A yes values indicates that the XML declaration (<?xml...?>) should be omitted in the resulting document. A no value indicates that the XML declaration should be included in the resulting document.
Standalone	yes no	Optional. A yes value indicates that the result should be a standalone document. A no value indicates that the result should not be a standalone document.
doctype-public	text	Optional. The public identifier to be used in the <!doctype> declaration in the resulting document after the transformation process.
doctype-system	text	Optional. The system identifier to be used in the <!doctype> declaration in the resulting document after the transformation process.

Attribute Name	Value	Attribute Description
cdata-section-elements	namelist	Optional. A list (separated by white space) of elements whose content is to be output in CDATA sections.
Indent	yes no	Optional. A yes value indicates that the output should be indented to indicate the hierarchic structure (for readability). A no value indicates that the output should not be indented to indicate the hierarchic structure.
media-type	mimetype	Optional. Specifies the media type (MIME TYPE) of the outputted resulting document.

Example

The initial instructions in the XML document in Listing 1-10 need to be modified, as follows:

```
<?xml:stylesheet type = "text/xsl" href = "example3.xsl"?>
```

This example uses two style sheets: example3.xsl (see Listing 1-12), which imports the style sheet, and example3-import.xsl (see Listing 1-13), which uses the xsl:import element. In the example3.xsl style sheet, when a student element is found, the Arial font is applied to the elements in the node. Then the xsl:apply-imports element applies the matching rules in the imported example3-import.xsl style sheet.

Listing 1-12: example3.xsl

© 2001 Dreamtech Software India Inc.
All Rights Reserved

```
<xsl:stylesheet version = "1.0"
  xmlns:xsl = "http://www.w3.org/1999/XSL/Transform">
  <xsl:import href = "example3-import.xsl"/>
  <xsl:output method = "html"/>
  <xsl:template match = "student">
    <font face = "Arial">
          <xsl:apply-imports/>
    </font>
  </xsl:template>
</xsl:stylesheet>
```

Listing 1-13: example3-import.xsl

© 2001 Dreamtech Software India Inc.
All Rights Reserved

```
<xsl:stylesheet version = "1.0"
  xmlns:xsl = "http://www.w3.org/1999/XSL/Transform">
  <xsl:template match = "text()"/>
    <xsl:template match = "/">
      <html>
          <head>
<title>
                    Demonstrating xsl:import &amp; xsl:apply-imports
          </title>
</head>
          <body>
                <xsl:apply-templates/>
```

```
            </body>
        </html>
    </xsl:template>
    <xsl:template match = "student">
        <b>Student </b>
        <i><xsl:apply-templates select = "name"/></i>
        <xsl:text> is in class </xsl:text>
        <xsl:apply-templates select = "class"/>
        <xsl:text> section </xsl:text>
        <xsl:apply-templates select = "section"/><br/>
    </xsl:template>
    <xsl:template match = "name">
        <xsl:value-of select = "."/>
    </xsl:template>
    <xsl:template match = "class">
        <font color = "red">
<xsl:value-of select = "."/>
</font>
    </xsl:template>
    <xsl:template match = "section">
        <xsl:value-of select = "."/>
    </xsl:template>
</xsl:stylesheet>
```

Figure 1-8 is the resulting output as rendered in Internet Explorer 5.5.

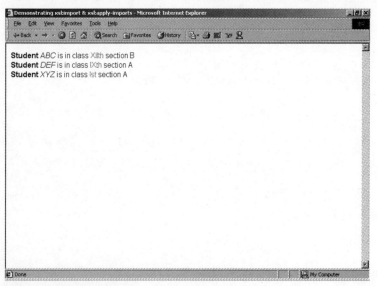

Figure 1-8: Output of example 3

xsl:for-each

The `xsl:for-each` element provides a way to create a looping structure for selecting or fetching the values of all childnodes under the given root node The following code and table show the syntax of `xsl:for-each`:

```
<xsl:for-each select = "expression">
  <xsl:sort>
template-body
```

```
</xsl:for-each>
where:
```

Attribute Name	Value	Attribute Description
select	expression	Required. Specify the name of the root node from which the loop is begun for selecting or fetching the values under it.

xsl:if

This element provides a way to write a conditional statement. The following code and table show the syntax of xsl:if:

```
<xsl:if test = "expression">
   template-body
</xsl:if>
```

Attribute Name	Value	Attribute Description
test	expression	Required. The condition to be tested from the source data.

xsl:attribute

This element creates an attribute to the nearest containing element. The syntax of this element is shown in the following code and table:

```
<xsl:attribute name = "attributename" namespace = "uri">
template-body
</xsl:attribute>
```

Attribute Name	Value	Attribute Description
name	attributename	Required. The name of the attribute to be added in the resulting document.
namespace	Uri	Optional. Defines the namespace URI of the added attribute by using the attributename attribute.

This element can be used within a template-body or within an xsl:attribute-set element.

xsl:attribute-set

The xsl:attribute-set element defines a named set of attributes. The attribute set can be applied to any output element. The syntax of this element is shown in the following code and table:

```
<xsl:attribute-set
name = "name"
   use-attribute-sets = "namelist">
<!-- Content: xsl:attribute -->
</xsl:attribute-set>
```

Attribute Name	Value	Attribute Description
name	Name	Required. The name of the attribute set.
use-attribute-	Name-list	Optional. A List (separated by white space) of other attribute sets to use in the attribute set.

sets		

Examples

The initial instructions in the XML document example.xml need to be modified as follows:

```
<?xml:stylesheet type = "text/xsl" href = "example4.xsl"?>
```

The corresponding XSLT, example4.xsl (see Listing 1-14), specifies a template that defines the structure of the overall output document, with rows for students whose age is less than 18 years. It uses templates and xsl:for-each and xsl:if elements to create td elements for the name, class, and section source elements.

Listing 1-14: example4.xsl

© 2001 Dreamtech Software India Inc.
All Rights Reserved

```
<xsl:stylesheet version = "1.0"
  xmlns:xsl = "http://www.w3.org/1999/XSL/Transform">
  <xsl:output method = "html"/>
  <xsl:template match = "/">
  <html>
  <head>
<title>Demonstrating xsl:for-each &amp; xsl:if</title>
</head>
      <body>
          <h1><center>Students with age less than 18 years.</center></h1>
          <xsl:for-each select = "students/student">
              <xsl:if test = "age &lt; 18">
          <table border = "2" align = "center">
   <xsl:attribute name = "bgcolor"> yellow</xsl:attribute>
                      <tr>
                                 <td>Name</td>
<td><xsl:value-of select = "name"/></td>
              </tr>
                      <tr>
                                 <td>Class</td>
<td><xsl:value-of select = "class"/></td>
              </tr>
                      <tr>
                                 <td>Section</td>
<td><xsl:value-of select = "section"/></td>
              </tr>
                  </table><br/>
          </xsl:if>
          </xsl:for-each>
      </body>
      </html>
  </xsl:template>
</xsl:stylesheet>
```

Figure 1-9 shows the output as rendered in Internet Explorer 5.5.

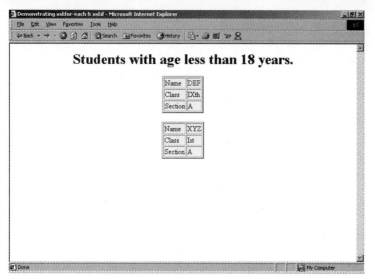

Figure 1-9: Output of Listing 1-14

Consider the XML document `audio.xml`, which contains information relevant to the audio collection.

```
<?xml version = "1.0"?>

<?xml:stylesheet type = "text/xsl" href = "audio.xsl"?>

<LIST>

<AUDIO>

    <ARTIST>Bryan Adams</ARTIST>

    <ALBUM>So far so good</ALBUM>

    <COMPANY>Virgin Records</COMPANY>

    <AMOUNT>$11.50</AMOUNT>

    <RELEASE>1994</RELEASE>

 </AUDIO>

 <AUDIO>

    <ARTIST>George Michael</ARTIST>

    <ALBUM>Faith</ALBUM>

    <COMPANY>CBS Records</COMPANY>

    <AMOUNT>$9.90</AMOUNT>

    <RELEASE>1988</RELEASE>

 </AUDIO>

 <AUDIO>

    <ARTIST>U2</ARTIST>
```

```
   <ALBUM>The Joshua Tree</ALBUM>

   <COMPANY>Atlantic</COMPANY>

   <AMOUNT>$8.50</AMOUNT>

   <RELEASE>1988</RELEASE>

 </AUDIO>

 <AUDIO>

   <ARTIST>Michael Jackson</ARTIST>

   <ALBUM>History</ALBUM>

   <COMPANY>SONY</COMPANY>

   <AMOUNT>$20.50</AMOUNT>

   <RELEASE>1995</RELEASE>

 </AUDIO>

</LIST>
```

The corresponding XSLT, `audio.xsl` (see Listing 1-15) transforms the XML document to display the audio collection by using the `xsl:for-each` element. This style sheet creates a table and uses the values of selected XML elements to fill the table with information.

Listing 1-15: audio.xsl

© 2001 Dreamtech Software India Inc.
All Rights Reserved

```
<xsl:stylesheet version = "1.0"
    xmlns:xsl = "http://www.w3.org/1999/XSL/Transform">
<xsl:template match = "/">
<html>
<head>
<title>Audio Selections...</title>
</head>
<body>
 <table border = "5" bgcolor = "aqua" width = "100%">
<tr>
        <th>Artist</th>
        <th>Album</th>
        <th>Amount</th>
</tr>
          <xsl:for-each select = "LIST/AUDIO">
<tr align = "center">
        <td><xsl:value-of select = "ARTIST"/></td>
        <td><xsl:value-of select = "ALBUM"/></td>
        <td><xsl:value-of select = "AMOUNT"/></td>
</tr>
</xsl:for-each>
    </table>
  </body>
  </html>
</xsl:template>
</xsl:stylesheet>
```

Figure 1-10 is the output as rendered in Internet Explorer.

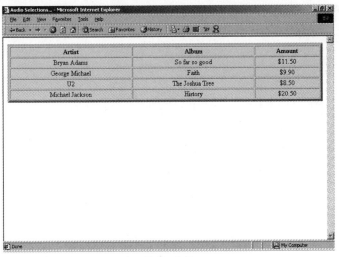

Figure 1-10: Output of preceding Listing 1-15

To put a conditional `if` test against the content of the file, you add an `xsl:if` element to the XSL document `audio1.xsl`, as shown in Listing 1-16.

Listing 1-16: audio1.xsl

```
<xsl:stylesheet version = "1.0"
    xmlns:xsl = "http://www.w3.org/1999/XSL/Transform">
<xsl:template match = "/">
<html>
<head>
<title>Audio Selections...</title>
</head>
<body>
 <table border = "5" bgcolor = "aqua" width = "100%">
<tr>
        <th>Artist</th>
        <th>Album</th>
        <th>Amount</th>
</tr>
            <xsl:for-each select = "LIST/AUDIO">
  <xsl:if match = ".[ARTIST = 'Bryan Adams']">
<tr align = "center">
        <td><xsl:value-of select = "ARTIST"/></td>
        <td><xsl:value-of select = "ALBUM"/></td>
        <td><xsl:value-of select = "AMOUNT"/></td>
</tr>
</xsl:if>
            </xsl:for-each>
    </table>
  </body>
  </html>
</xsl:template>
</xsl:stylesheet>
```

Figure 1-11 is the resulting output as rendered in Internet Explorer 5.5.

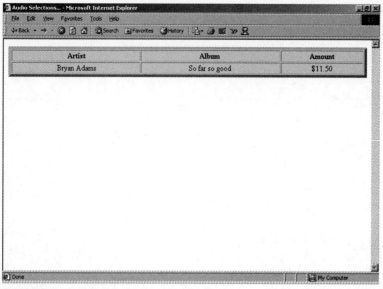

Figure 1-11: Output of Listing 1-16

The example in Listing 1-17 creates an attribute set that you can apply to any output element.

Listing 1-17: attrib.xsl

© 2001 Dreamtech Software India Inc.
All Rights Reserved

```
<xsl:attribute-set name = "font">
  <xsl:attribute name = "font-name">Arial</xsl:attribute>
  <xsl:attribute name = "font-size">16px</xsl:attribute>
  <xsl:attribute name="font-weight">bold</xsl:attribute>
</xsl:attribute-set>
```

The attribute set is used in an element in the following manner:

```
<p xsl:use-attribute-set = "font">This is a paragraph.</p>
```

xsl:variable

This element provides a way to declare a variable in an XSLT document. The following code and table show the syntax:

```
<xsl:variable name = "name" select = "expression">
  template-body
</xsl:variable>
```

Attribute Name	Value	Attribute Description
name	name	Required. The name of the variable.
select	expression	Optional. The variable value.

Example

The initial instructions in the XML document `example.xml` need to be modified, as follows:

```
<?xml:stylesheet type = "text/xsl" href = "example5.xsl"?>
```

The corresponding XSLT, example5.xsl (see Listing 1-18), transforms the XML document and calculates the number of students, the sum of the ages of the students, and the average age of the students and stores them in the following variables: studentCount, ageTotal, and avg, respectively, using the xsl:variable element.

Listing 1-18: example5.xsl

© 2001 Dreamtech Software India Inc.
All Rights Reserved

```
<xsl:stylesheet version = "1.0"
  xmlns:xsl = "http://www.w3.org/1999/XSL/Transform">
 <xsl:template match = "/">
 <html>
<head>
<title>Demonstrating xsl:variable</title>
</head>
 <body>
 <xsl:variable name = "studentCount" select = "count(students/student)"/>
 <xsl:variable name = "ageTotal" select = "sum(students/student/age)"/>
 <xsl:variable name = "avg" select = "$ageTotal div $studentCount"/>
 <table border = "5" bgcolor = "yellow" align = "center">
  <caption><b>STUDENT INFORMATION</b></caption>
  <tr>
   <td>Number of Students</td>
   <td><xsl:value-of select = "format-number($studentCount, '#.00')"/></td>
  </tr>
  <tr>
   <td>Sum (Age)</td>
   <td><xsl:value-of select = "format-number($ageTotal, '#.00')"/></td>
  </tr>
  <tr>
   <td>Average (Age)</td>
   <td><xsl:value-of select = "format-number($avg, '#.00')"/></td>
  </tr>
 </table>
 </body>
 </html>
 </xsl:template>
</xsl:stylesheet>
```

Figure 1-12 is the resulting output as rendered in Internet Explorer 5.5.

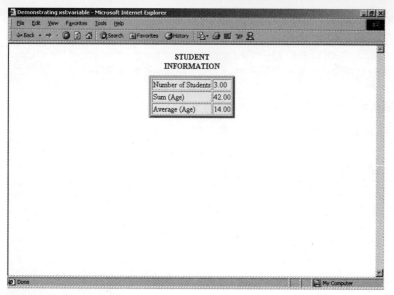

Figure 1-12: Output of Listing 1-18

xsl:sort

The `xsl:sort` element provides a way to define a sorting method for the elements outputted in the resulting documents. The following code and table show the syntax:

```
<xsl:sort
select = "expression"
order = "ascending|descending"
case-order = "upper-first|lower-first"
lang = "language-code"
data-type = "text|number|qname"/>
```

Attribute Name	Value	Attribute Description
select	expression	Optional. The sort expression.
order	ascending descending	Optional. The sorting order. Default sorting option is `ascending`.
case-order	upper-first lower-first	Optional. Do uppercase letters come before lowercase letters, or vice-versa? The default is uppercase first.
Lang	language-code	Optional. The language alphabet to use for sorting purposes. By default, the system-defined language is used
data-type	Text number qname	Optional. The data type of the data to be sorted. Is it a number, a text, or a user-defined data type? Default is `text`.

Examples

The initial instructions in the XML document `example.xml` need to be modified as follows:

```
<?xml:stylesheet type = "text/xsl" href = "example6.xsl"?>
```

The corresponding XSLT, `example6.xsl` (see Listing 1-19), transforms the XML document and generates a list of students sorted by name, using the `xsl:sort` element.

Listing 1-19: example6.xml

```
<xsl:stylesheet version = "1.0"
  xmlns:xsl = "http://www.w3.org/1999/XSL/Transform">
 <xsl:template match = "students">
  <html>
  <head>
<title>Demonstrating xsl:sort</title>
</head>
        <body>
  <h1><center>Sorted List...</center></h1>
  <table border = "5" align = "center">
   <tr>
    <th>Name</th>
<th>Class</th>
<th>Section</th>
<th>Gender</th>
     <th>Age</th>
<th>Address</th>
<th>Phone</th>
   </tr>
   <xsl:apply-templates select = "student">
    <xsl:sort select = "name"/>
   </xsl:apply-templates>
  </table>
 </body>
 </html>
 </xsl:template>
 <xsl:template match = "students/student">
  <tr>
   <td><xsl:value-of select = "name"/></td>
   <td><xsl:value-of select = "class"/></td>
   <td><xsl:value-of select = "section"/></td>
   <td><xsl:value-of select = "gender"/></td>
   <td><xsl:value-of select = "age"/></td>
   <td><xsl:value-of select = "address"/></td>
   <td><xsl:value-of select = "phone"/></td>
  </tr>
 </xsl:template>
</xsl:stylesheet>
```

Figure 1-13 is the resulting output as rendered in Internet Explorer 5.5.

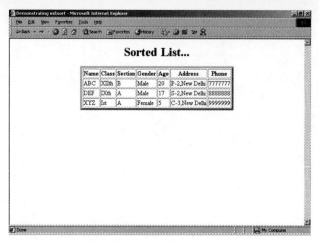

Figure 1-13: Output of Listing 1-19

You also can sort by using the `order-by` attribute of the `xsl:for-each` element.

Alter the document `audio.xsl` to produce a sorted output in ascending order by album (see Listing 1-20).

Listing 1-20: audio_order.xsl

© 2001 Dreamtech Software India Inc.
All Rights Reserved

```
<xsl:stylesheet version = "1.0"
     xmlns:xsl="http://www.w3.org/TR/WD-xsl">
<xsl:template match = "/">
<html>
<head>
<title>Sorting</title>
</head>
<body>
 <table border = "5" bgcolor = "aqua" width = "100%">
<tr>
        <th>Artist</th>
        <th>Album</th>
        <th>Amount</th>
</tr>
          <xsl:for-each select = "LIST/AUDIO" order-by="+ ALBUM">
<tr align = "center">
        <td><xsl:value-of select = "ARTIST"/></td>
        <td><xsl:value-of select = "ALBUM"/></td>
        <td><xsl:value-of select = "AMOUNT"/></td>
</tr>
          </xsl:for-each>
   </table>
  </body>
  </html>
</xsl:template>
</xsl:stylesheet>
```

Figure 1-14 is the resulting output as rendered in Internet Explorer 5.5.

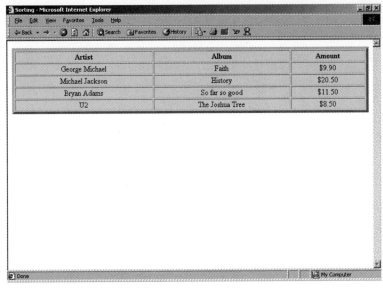

Figure 1-14: Output of Listing 1-20

xsl:otherwise

This element indicates what should happen when none of the xsl:when elements inside an xsl:choose element is satisfied. The following code shows the syntax of this element:

```
<xsl:otherwise>
  template-body
</xsl:otherwise>
```

xsl:when

The xsl:when element defines a condition to be tested and performs an action if the condition is true. This element is always a child element of xsl:choose. The following code and table show the syntax of xsl:when

```
<xsl:when test = "expression">
  template-body
</xsl:when>
```

Attribute Name	Value	Attribute Description
test	expression	Required. The condition to be tested from the source data.

xsl:choose

This element provides a way to choose between a numbers of alternatives, based on conditions. The following code shows the syntax of xsl:choose. This works with the conjunction of the <xsl:otherwise> and <xsl:when> elements.

```
<xsl:choose>
  <xsl:when>..
  <xsl:otherwise>..
</xsl:choose>
```

Example

The initial instructions in the XML document `example.xml` need to be modified, as follows:

```
<?xml:stylesheet type = "text/xsl" href = "example7.xsl"?>
```

In the corresponding XSLT, `example7.xsl` (see Listing 1-21), XSLT changes the XML document by "either/or" processing. It first tries to locate students whose age is less than 10 years, and then it looks for students whose age is less than 18 years. Finally, an `xsl:otherwise` element inserts a template if no match is found; that is, it performs the final "else" part.

Listing 1-21: example7.xsl

```
<xsl:stylesheet version = "1.0"
  xmlns:xsl = "http://www.w3.org/1999/XSL/Transform">
<xsl:output method = "html"/>
 <xsl:template match = "/">
 <html>
 <head>
<title>Demonstrating xsl:otherwise</title>
</head>
            <body>
             <xsl:for-each select = "//student">
                 <div>
                 <xsl:choose>
                 <xsl:when test = "self::*[age &lt; '10']">
                        <xsl:attribute name = "style">
                             background-color: yellow
                        </xsl:attribute>
                        <h2><xsl:value-of select = "name" /> -
Child.</h2>
                 </xsl:when>
                 <xsl:when test = "self::*[age &lt; '18']">
                        <xsl:attribute name = "style">
                             background-color: red
                        </xsl:attribute>
                        <h2><xsl:value-of select = "name" /> -
Teenager.</h2>
                 </xsl:when>
                 <xsl:otherwise>
                        <xsl:attribute name = "style">
                             background-color: orange
                        </xsl:attribute>
                        <h2><xsl:value-of select = "name" /> -
Adult.</h2>
                 </xsl:otherwise>
                 </xsl:choose>
                 <h2>Age <xsl:value-of select = "age"/></h2>
                 </div>
                 </xsl:for-each>
        </body>
        </html>
    </xsl:template>
</xsl:stylesheet>
```

Figure 1-15 is the resulting output as rendered in Internet Explorer 5.5.

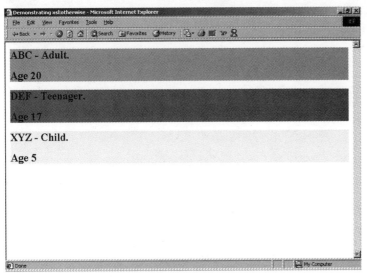

Figure 1-15: Output of Listing 1-21

xsl:strip-space/xsl:preserve-space

The `xsl:strip-space` element provides a way to define the handling of white space, and the `xsl:preserve-space` element preserves white space in a document. The following code and table show the syntax of these elements:

```
<xsl:preserve-space elements = "list"/>
<xsl:strip-space elements = "list"/>
```

Attribute Name	Value	Attribute Description
elemen ts	list	Required. List (separated by white space) of elements where white space should be preserved/removed. *Note:* The list can also contain * and `prefix:` * so that all elements or all elements from a particular namespace can be joined.

Example

The initial instructions in the XML document `example.xml` need to be modified, as follows:

```
<?xml:stylesheet type = "text/xsl" href = "example8.xsl"?>
```

The corresponding XSLT, `example8.xsl` (see Listing 1-22), demonstrates the use of an `xsl:strip-space` element that has an `elements` attribute whose value is a white-space–separated list of name tests.

Listing 1-22: example8.xsl

© 2001 Dreamtech Software India Inc.
All Rights Reserved

```
<xsl:stylesheet version = "1.0"
  xmlns:xsl = "http://www.w3.org/1999/XSL/Transform">
<xsl:output method = "html"/>
  <xsl:strip-space elements = "student"/>
```

```
    <xsl:template match = "/">
     <html>
        <head>
 <title>Demonstrating xsl:strip-space</title>
        </head>
        <body>
          <div>
     <xsl:attribute name="style">background-color:  aqua</xsl:attribute>
     <xsl:value-of select="count(//text())"/>
     nodes of type text were found.
          </div>
        </body>
     </html>
  </xsl:template>
</xsl:stylesheet>
```

Figure 1-16 is the resulting output as rendered in Internet Explorer 5.5.

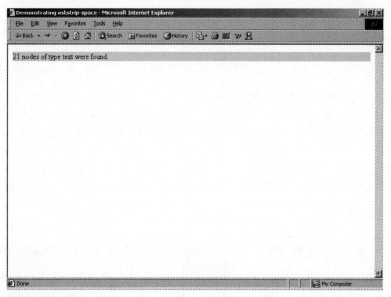

Figure 1-16: Output of Listing 1-22

xsl:param

This element provides a way to define parameters in XSLT documents. The following code and table show the syntax:

```
<xsl:param name = "name" select = "expression">
 template-body
</xsl:param>
```

Attribute Name	Value	Attribute Description
name	name	Required. The name of the parameter in an XSLT document.
select	expression	Optional. The value of the defined parameter.

Example

The initial instructions in the XML document `example.xml` need to be modified. You need to add the attribute `rollno` to each student element, as seen in Listing 1-23.

Listing 1-23: example2.xml

```xml
<?xml version = "1.0"?>
<?xml:stylesheet type = "text/xsl" href = "example9.xsl"?>
<students>
 <student rollno = "1">
  <name>ABC</name>
  <class>XIIth</class>
  <section>B</section>
  <gender>Male</gender>
  <age>20</age>
  <address>P-2,New Delhi</address>
  <phone>7777777</phone>
 </student>
 <student rollno = "2">
  <name>DEF</name>
  <class>IXth</class>
  <section>A</section>
  <gender>Male</gender>
<age>17</age>
  <address>S-2,New Delhi</address>
  <phone>8888888</phone>
 </student>
 <student rollno = "3">
  <name>XYZ</name>
  <class>Ist</class>
  <section>A</section>
  <gender>Female</gender>
  <age>5</age>
  <address>C-3,New Delhi</address>
  <phone>9999999</phone>
 </student>
</students>
```

The corresponding XSLT, `example9.xsl` (see Listing 1-24), transforms the XML document and uses the `xsl:param` element to list the details of the student whose roll number is 3.

Listing 1-24: example9.xsl

```xml
<xsl:stylesheet version = "1.0"
  xmlns:xsl = "http://www.w3.org/1999/XSL/Transform">
 <xsl:param name = "rollno">3</xsl:param>
  <xsl:output method = "html"/>
    <xsl:template match = "/">
      <html>
        <head>
<title>Demonstrating xsl:param</title>
        </head>
```

```
            <body>
                <xsl:apply-templates select = "//student[@rollno = $rollno]"/>
            </body>
        </html>
    </xsl:template>
    <xsl:template match = "student">
     <table border = "5" bgcolor = "yellow" align = "center">
        <caption><b>STUDENT DETAILS ROLL NO.3</b></caption>
        <tr>
<td>Name</td>
<td><xsl:value-of select = "name"/></td>
</tr>
            <tr>
<td>Age</td>
<td><xsl:value-of select = "age"/></td>
</tr>
<tr>
<td>Gender</td>
<td><xsl:value-of select = "gender"/></td>
</tr>
                    <tr>
<td>Class</td>
<td><xsl:value-of select = "class"/></td>
</tr>
            <tr>
<td>Section</td>
<td><xsl:value-of select = "section"/></td>
</tr>
            <tr>
<td>Address</td>
<td><xsl:value-of select = "address"/></td>
</tr>
<tr>
<td>Phone</td>
<td><xsl:value-of select = "phone"/></td>
</tr>
</table>
    </xsl:template>
</xsl:stylesheet>
```

Figure 1-17 is the resulting output as rendered in Internet Explorer 5.5.

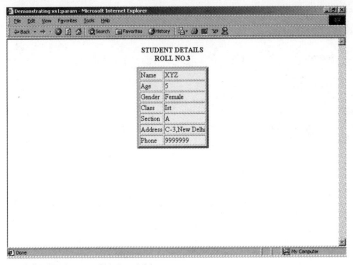

Figure 1-17: Output of Listing 1-24

xsl:call-template

This element provides a way to call a named template. It contains zero or more `xsl:with-param` child elements. The following code and table show the syntax for `xsl:call-template`:

```
<xsl:call-template name = "templatename">
  <xsl:with-param>
</xsl:call-template>
```

Attribute Name	Value	Attribute Description
name	template name	Required. The name of the template to be called.

Example

The initial instructions in the XML document `example.xml`, need to be modified, as follows:

```
<?xml:stylesheet type = "text/xsl" href = "example10.xsl"?>
```

The corresponding XSLT, `example10.xsl` (see Listing 1-25), demonstrates how the `xsl:call-template` element invokes a named template, an `xsl:template` element with an assigned-name attribute.

Listing 1-25 : example10.xsl

```
<xsl:stylesheet version = "1.0"
  xmlns:xsl = "http://www.w3.org/1999/XSL/Transform">
  <xsl:template match = "/">
  <html>
    <head>
<title>Demonstrating xsl:call-template &amp; xsl:with-param</title>
    </head>
    <body>
      <table align = "center" border = "1" bgcolor = "yellow">
        <tr>
```

```
<th>Name</th>
<th>Age</th>
        </tr>
        <xsl:for-each select = "students/student">
        <tr>
            <td><xsl:value-of select = "name"/></td>
            <td>
              <xsl:call-template name = "name">
                 <xsl:with-param name = "par_age" select = "age"/>
              </xsl:call-template>
            </td>
        </tr>
        </xsl:for-each>
    </table>
  </body>
</html>
</xsl:template>
<xsl:template name = "name">
  <xsl:param name = "par_age"/>
  <xsl:value-of select = "$par_age"/>
</xsl:template>
</xsl:stylesheet>
```

Figure 1-18 is the resulting output as rendered in Internet Explorer 5.5.

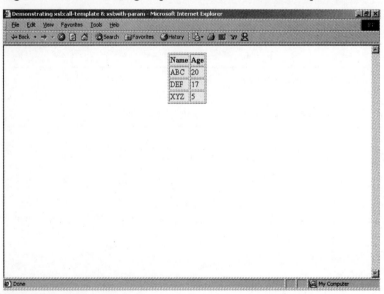

Figure 1-18: Output of Listing 1-25

xsl:comment

This element creates an XML comment in the resulting XML document after the transforming process.
Its syntax is as follows:

```
<xsl:comment>
  <!-- Content: template -->
</xsl:comment>
```

xsl:message

The `xsl:message` element sends a text message to either the message buffer or a message dialog box. The following code and table show the syntax:

```
<xsl:message
  terminate = "yes" | "no">
  <!-- Content: template -->
</xsl:message>
```

Attribute Name	Value	Attribute Description
terminate	yes no	Optional. A yes value terminates the processing after the message is written to the output. A no value continues the processing after the message is written to the output. Default is no.

Example

The following example declares a variable, `color`, and sets its value to the `color` attribute of the current element. If the current element has no `color` attribute, a message is written to the output.

```
<xsl:variable name = "color">
  <xsl:choose>
    <xsl:when test = "@color">
     <xsl:value-of select = "@color"/>
    </xsl:when>
    <xsl:otherwise>
     <xsl:message>
       <xsl:text>The element has no color attribute!</xsl:text>
     </xsl:message>
    </xsl:otherwise>
  </xsl:choose>
</xsl:variable>
```

xsl:text

This element generates text in the output. The following code and table show the syntax for `xsl:text`:

```
<xsl:text
  disable-output-escaping = "yes" | "no">
  <!-- Content: #PCDATA -->
</xsl:text>
```

Attribute Name	Value	Attribute Description
disable-output-escaping	yes no	Optional. A yes value indicates that special characters (such as <) should be output as is. A no value indicates that special characters (such as <) should be output as <. Default is no.

xsl:copy

This element copies the current node, *without* childnodes and attributes, to the output. The following code and table show the syntax for `xsl:copy`:

```
<xsl:copy use-attribute-sets = "name-list">
<!-- Content: template -->
```

```
</xsl:copy>
```

Attribute Name	Value	Attribute Description
use-attribute-sets	name-list	Optional. List (separated by white space) of other attribute sets to use in the node. This attribute is used only while copying element nodes.

Example

In the following example, the p node in the XML document is copied to the current output:

```
<xsl:template match = "p">
  <xsl:copy>
    <xsl:apply-templates/>
  </xsl:copy>
</xsl:template>
```

xsl:copy-of

This element copies the current node, *with* childnodes and attributes, to the output. The following code and table show the syntax for this element:

```
<xsl:copy-of select = "expression" />
```

Attribute Name	Value	Attribute Description
select	expression	Required. The element addresses the node to be copied to the output in a given XML document.

Example

In the following example, the p node in the XML document is copied, with all its b childnodes, to the current output:

```
<xsl:template match = "p">
  <xsl:copy-of select = "b"/>
  <xsl:apply-templates/>
</xsl:template>
```

xsl:element

This element creates an element with the specified name in the output. The following code and table illustrate the syntax for this element:

```
<xsl:element
   name = "name"
namespace = "uri"
   use-attribute-sets = "namelist">
   <!-- Content: template -->
</xsl:element>
```

Attribute Name	Value	Attribute Description
name	name	Required. Name of the element to be created in resulting XML document after completion of the transforming process.
namespace	uri	Optional. Namespace URI of the created element in resulting XML document after the transforming process..
use-attribute-sets	namelist	Optional. List (separated by white space) of attribute sets containing attributes to be added to the created element the in resulting XML document after the transforming process.

Example

This example selects the format attributes of each element in the message node tree; and for each one, it creates an element whose name is equal to the value of that attribute. Listing 1-26 shows the XML document, and Listing 1-27 shows the corresponding XSL document.

Listing 1-26: XML document

© 2001 Dreamtech Software India Inc.
All Rights Reserved

```
<?xml version = "1.0"?>
<message>
  <to format = "b">Ashish</to>
  <from format = "i">Madan</from>
  <text format = "u">Hello</text>
</message>
```

Listing 1-27: XSL document

© 2001 Dreamtech Software India Inc.
All Rights Reserved

```
<xsl:template match = "/">
  <xsl:for-each select = "message">
  <xsl:element name = "{@format}">
  <xsl:value-of select="."/><br />
  </xsl:element>
  </xsl:for-each>
</xsl:template>
```

The output is as follows:

```
Ashish
Madan
Hello
```

xsl:number

This element inserts a formatted number into the result tree. The following code and table show the correct syntax for xsl:number:

```
<xsl:number
  level = "single" | "multiple" | "any"
  count = "pattern"
  from = "pattern"
  value = "expression"
  format =  "{formatstring}"
  lang = "{languagecode}"
```

```
letter-value = "{alphabetic | traditional}"
grouping-separator = "{char}"
grouping-size = "{number}" />
```

Attribute Name	Value	Attribute Description
Level	single multiple any	Optional. Specifying the counting level in the tree of a given XML document.
Count	Pattern	Optional. The elements to count.
From	Pattern	Optional. Where to begin the counting.
Value	expression	Optional. A number to be converted to text. If not specfied, <xsl:element> inserts a number automatically based on the position of the node in the tree.
Format	{formatstring}	Optional. The output format of the number.
Lang	{languagecode}	Optional. The language to use for conversion.If not specified, it uses the default-system language
letter-value	{alphabetic} {traditional}	Optional. The letter value for numbering schemes.
grouping-separator	{character}	Optional. The character for separating groups of digits.
grouping-size	{number}	Optional. The number of digits in each group, indicating where the grouping-separator should be inserted.

Example

This example generates a sample digit with the grouping size of 3 and using the grouping separator "-" for generating the output

```
<xsl:number value = '250000' grouping-size = '3' grouping-separator = '.'/>
```

The output is as follows.

```
250.000
```

xsl:decimal-format

This element defines the default decimal format for number-to-text conversion. The following code and table show the syntax for xsl:decimal-format:

```
<xsl:decimal-format
name = "name"
decimal-separator = "char"
grouping-separator = "char"
infinity = "string"
minus-sign = "char"
NaN = "string"
percent = "char"
per-mille = "char"
zero-digit = "char"
digit = "char"
```

```
pattern-separator = "char"/>
```

Attribute Name	Value	Attribute Description
name	Name	Optional. Name of the decimal format.
decimal-separator	Char	Optional. A character to separate the integer and the fraction part of a number. Default is `.`.
grouping-separator	Char	Optional. A character to separate groups of digits. Default is `,`.
infinity	String	Optional. A string to represent infinity. Default is `infinity`.
minus-sign	Char	Optional. A character to represent negative numbers. Default is `-`.
NaN	String	Optional. A string to represent "Not a Number". Default is `NaN`.
percent	Char	Optional. A character to represent a percentage sign. Default is `%`.
per-mille	Char	Optional. A character to represent a per-mille sign. Default is `‰`.
zero-digit	Char	Optional. A character to indicate a place where a leading zero-digit is required. Default is `0`.
digit	Char	Optional. A character to indicate a place where a digit is required. Default is `#`.
pattern-separator	Char	Optional. A character to separate format strings. Default is `;`.

Example

The third argument in the format-number function refers to the format specified by the `name` attribute in the `<xsl:decimal-format>`.

```
<xsl:decimal-format> element.
<xsl:decimal-format name = "format"
decimal-separator = "," grouping-separator = "."/>
<xsl:template match = "/">
<xsl:value-of  select = "format-number(77777.7, '#.###,00', 'format')"/>
</xsl:template>
```

The output is as follows:

```
77.777,70
```

xsl:key

This element declares a key for subsequent use with the `key()` function. The following code and table show the syntax for `xsl:key`:

```
<xsl:key
  name = "name"
  match = "pattern"
  use = "expression"/>
```

Attribute Name	Value	Attribute Description
Name	name	Required. The name of the key.
Match	pattern	Required. Defines a pattern to be matched for the XSLT processor during the processing of the elements, while transforming the document
Use	expression	Required. The value of the key for each of the applicable nodes.

The `xsl:key` element is only useful in conjunction with the `key()` XSLT function node-set `key(name, object)`, which returns the nodes that match the object in the specified key.

Example

This sample demonstrates a simple use of `<xsl:key>` element for searching the XML document.

Listing 1-28: XML document

```
<students>
  <student name = "Ravinder" rollno = "777"/>
  <student name = "Ashish" rollno = "888"/>
</students>
```

Listing 1-29: XSL document

```
<xsl:key name = "stud" match = "student" use = "@id"/>
<xsl:apply-templates select = "key('stud','777')"/>
```

The first line of Listing 1-29 defines a key. In the second line, a student with id value `777` is found.

xsl:fallback

This element calls template content that can provide a reasonable substitute for the behavior of a new element when one is encountered. The syntax of this element is as follows:

```
<xsl:fallback>
  <!-- Content: template -->
</xsl:fallback>
```

Example

Listing 1-30 loops through each message element with a made-up XSL element `xsl:loop`. The XSL processor does not support this element and will use the `xsl:for-each` element instead.

Listing 1-30: ex_loop.xsl

```
<xsl:template match = "message">
  <xsl:loop select = "message">
<xsl:fallback>
    <xsl:for-each select = "message">
```

```
            <xsl:value-of select = "."/>
        <xsl:/for-each>
            </xsl:fallback>
            <xsl:value-of select = "."/>
</xsl:loop>
</xsl:template>
```

xsl:include

This listing specifies another XSLT style sheet to include. The following code and table illustrate the syntax for `xsl:include`:

```
<xsl:include href = "uri"/>
```

Attribute Name	Value	Attribute Description
href	uri	Required. Reference of the URI of style sheet to be included into existing style sheet.

Example

This following style sheet, `first.xsl`, contains the following `attribute-set`:

```
<xsl:attribute-set name = "font">
  <xsl:attribute name = "font-name">Arial</xsl:attribute>
  <xsl:attribute name = "font-size">16px</xsl:attribute>
  <xsl:attribute name = "font-weight">italic</xsl:attribute>
</xsl:attribute-set>
```

Another style sheet, `second.xsl`, uses the `attribute-set` (`"font"`) from `first.xsl`.

```
<xsl:include href = "first.xsl"/>
<p xsl:use-attribute-set = "font">This is a paragraph.</p>
```

xsl:processing-instruction

This element generates a processing instruction in the output. The following code and table show the syntax for `xsl:processing-instruction`:

```
<xsl:processing-instruction
  name = "name">
    <!-- Content: template -->
</xsl:processing-instruction>
```

Attribute Name	Value	Attribute Description
name	name	Required. The name of the processing instruction for the transformation of the given XML document.

xsl:namespace-alias

This element replaces the prefix associated with a given namespace with another prefix. The following code and table show the syntax for `xsl:namespace-alias`:

```
<xsl:namespace-alias
stylesheet-prefix = "ncname"
```

```
result-prefix = "ncname"/>
```

Attribute Name	Value	Attribute Description
stylesheet-prefix	ncname #default	Required. The namespace prefix used in the style sheet.
result-prefix	ncname #default	Required. The namespace prefix to be used in the output.

Summary

XML, or Extensible Markup Language, is the latest W3C standard for exchanging structured data. Like HTML, XML is a tagged language. XML is gaining respect very quickly, and it is likely that it will soon dominate HTML as the "language of choice," because you can customize it for any device, whether a laptop, a browser, or a cell phone.

XSL is also a W3C standard. Basically, it is a transformation and formatting language, mainly used to transform XML documents. This chapter has covered the extensive use of XSLT elements on XML documents.

Chapter 2

Designing a Cross-Platform–Based Application

This chapter discusses the design and development process of the application you build throughout the rest of the book. The following are some of the vital points covered in this chapter:

- ◆ Current architecture of Web applications
- ◆ Non-XML approach for the applications
- ◆ XML- and XSLT-based approach for the applications
- ◆ Architecture of the project

Introduction to Web-Application Architectures

A *Web application* is an application accessible to a client via the Web. The user of a Web application is never far away from the application and can use an application over the Internet by using any Web browser. A Web server, which is always connected to the Internet, hosts the application, and the user is able to access the application by typing a URL in the Web browser. The browser sends a request to the server to provide the application's contents, and the server responds by sending the content to the browser.

In recent years, several new kinds of wireless devices have been developed that are powerful enough to connect to the Internet or capable of displaying data on a small screen and interacting with the Internet. In addition, many new standards have appeared on the market that enable data transmission over different kinds of networks. These standards include wireless application protocol (WAP), I-mode, and so on. Consequently, in today's unpredictable scenario, if you are planning to build an application and host it over the Internet, it should be compatible with all the new kinds of devices and types of media; this capability not only increases the user base but also allows for wider accessibility over different networks.

Because it's a typical client-server application, a Web application can have different categories of clients accessing it. A database server contains a large amount of data and comes in numerous varieties (such as SQL Server, Access, Oracle, DB2, and so forth). It is the function of the Web server to coordinate with the database server for all types of database manipulations. In addition, server-side processing can occur in different languages (such as ASP, Java servlets, PHP, Cold Fusion, and so on). When writing an application, however, a programmer chooses one specific database and a compatible server-side programming language. Despite various clients accessing the applications, the server is stationary and remains in the background. It is the job of the server-side programming language to detect the client and, accordingly, run the client application. The following two approaches exist for this situation:

- ◆ Non-XML approach
- ◆ XML and XSLT approach

Non-XML Approach

A typical Web-based application uses SQL Server or Access as the database; ASP or some other server language for the server-side processing; and HTML, WML, or cHTML for the client-side application. The ASP detects the browser and, depending upon its type, calls and executes a client application written specifically for it. In this approach, programmers write different applications for different clients. Considering the heavy cost of application development, as well as the time constraint, this approach is not very useful. Even so, this approach remains popular even today.

Figure 2-1 illustrates this approach.

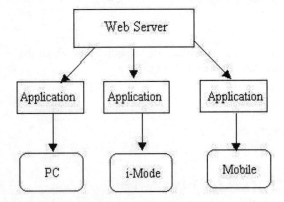

Figure 2-1: Non-XML approach to a Web application

The major problems with the non-XML approach are as follows:

◆ Because each application is written for a specific device, it is strongly bound to that device, which means that the entire application must be modified when the device is modified.

◆ A separate set of code is defined for each device. To make the application run on a new device, someone needs to write a new set of code.

◆ The application doesn't have a layered structure; thus, the entire application must be modified if only a small part must be changed.

After looking at these problems, programmers have begun using the new XML- and XSLT-based approach to develop applications.

XML and XSLT Approach

XML is similar to HTML, the main difference being that XML enables you to define and use tags as needed. This capability means that you easily can develop platform-independent and easily customizable Web applications. XML support is generally available on all platforms in the form of editors and parsers, so it is easy for the applications to run on various platforms. Developing an application in XML makes logic more accessible. XML also minimizes dependence on a particular device for the implementation of a new tag. Figure 2-2 outlines this approach.

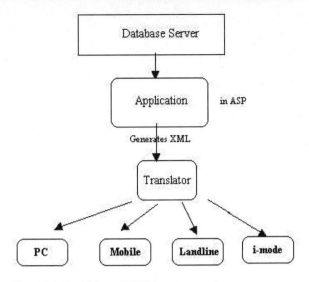

Figure 2-2: The XML- and XSLT-based approach

Compared to the non-XML approach, the XML- and XSLT-based approach has the following advantages:

♦ You write only one application for various devices, and the various client applications are developed through the XSL transformation.

♦ In order for the application to run on a new device, you need only insert a new module in the formatter layer of the application. The rest of the application remains untouched.

♦ Using this approach, you can make a particular application compatible with various devices.

XML as a database of applications

Using XML as a database for your application has many advantages. For example, you can choose any database server in the world for saving your data on the back end. When your application requires some kind of data from the server, the business logic layer in the XML- and XSLT-based approach fetches the data from the database server and then converts it into XML format. After you have all the data in XML format, you can transfer or relay this data across networks. The formatting layer is applied to the data on different servers to make it easily readable.

If your clients use different devices to access your application, it becomes crucial to have the data in XML format, because only then can you perform the different transformations for each kind of client by changing the data a little.

Use XML to build dynamic applications

The scenario under consideration is one in which the application must run on different platforms. To make such a cross-platform application, you must keep in mind the following requirements:

♦ **User agent:** The *user agent* is the specified client accessing the application. The user agent also can be specified or defined as a class of Web clients accessing the application. Depending upon the user agent, XSL transformations might take place on the intermediate XSL.

♦ **Document profile:** The document profile contains the information about the output formatting structure for the document to be displayed – that is, it describes how the page should appear for each output device.

♦ **Device profile:** The device profile is an RDF or other format document that describes the capabilities of the device in use (that is, the user agent). It contains various applications and functions supported by a specified device and that XSL can use for the final transformations.

The device profile and document profile work together in the creation of the XSL style sheet. This XSL style sheet is created according to the specific requirements of the client. This style sheet is then further used to transform the XML content into a Web page according to the device specifications.

Reusability of XSL style sheets

The intermediate XSLT are stored on the server and can be used to generate output documents for devices of the same category. It is not always necessary to create XSL for every operation.

Application Architecture

The application you are building throughout this book is a cross-platform application designed and developed for all types of clients that can access the Internet. Even if you don't have Internet connectivity, you can still use the application by dialing a telephone number on a landline phone and then using the interactive voice response system module of this application.

Application Clients

The various types of clients are as follows:

♦ Personal computers (PCs)

♦ Mobile phones with Wireless Markup Language (WML) browsers supporting ver 1.1

♦ Handheld Device Markup Language (HDML) browsers

♦ I-mode devices

♦ Landline phones

Personal computers

In a typical Web-based application, the user accesses a Web site from a personal computer. The client face of the application is written in HTML. In Chapter 4, you undertake your first transformation process for PC browsers. Using a PC, someone can connect to the Internet and access the application (see Figure 2-3).

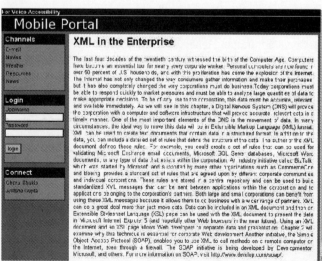

Figure 2-3: The home page of the Mobile Web application running on a PC browser

Mobile phones

People initially used mobile phones for wireless sound communications; but now, with the help of WML, they can use mobile phones to access the Internet, despite problems like a small screen and low data transfer rates (see Figure 2-4). In Chapter 5, you transform your application into one for WAP clients. On the server-side of the application, you add a new formatting layer to provide services to Wireless Application Protocol (WAP) devices for performing this task.

Figure 2-4: The home page of the Mobile Web application running on a cell-phone emulator

HDML browsers

Wireless Internet services are accessible by means of the PCS phone, which is equipped with a built-in HDML browser (see Figure 2-5). In Chapter 7, you perform the transformation task for the HDML market. Thus, by the end of Chapter 7, your application becomes accessible to three different platforms, including PCs, WAP devices, and HDML clients.

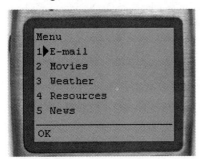

Figure 2-5: The home page of the Mobile Web application running on an HDML simulator

i-mode

i-mode is served by compact HTML (cHTML), which can enable users to access desktop HTML sites, although the sites look better if they are written in cHTML (see Figure 2-6). In Chapter 9, you transform your complete Mobile Web application for i-mode clients by adding a new set of XSLT documents to the formatting layer of the application.

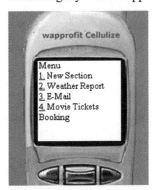

Figure 2-6: The home page of the Mobile Web application running on an i-mode emulator

Landline

If a user doesn't have any of the preceding four devices, the only alternative left for communication is the landline. The application you build in this book is also accessible by landline. It gives an audio response to the query of the user as an interactive voice response system. In Chapter 11, you transform the site to work with landline phone clients or voice-enabled clients.

Application Features

The features covered in the application are:

♦ **E-mail:** This feature enables the user to create and use e-mail services through the application. The new user can create an account, and the old user can use the account to receive or send e-mail messages.

♦ **Movies:** Through this feature, the various users can get a listing of movie theatres and show times. The important highlighted feature is the online booking of tickets.

♦ **Weather:** This option enables users to see the maximum and minimum temperature of various cities.

♦ **News:** This section provides updates on current happenings.

Application Database Structure

SQL Server 2000 is the database server for the application you build in this book. SQL Server provides extensive support for and interacts with XML data. The database has six tables, four of which are used for the movie information and online reservations. One table is for news updates, and one is for weather forecasting. A detailed description of each table follows.

Movie Table

The movie table stores the information about the movies currently running in the user's geographic area. The table structure is as follows:

Name	Type
Movie_id	numeric
Movie_name	character

Hall Table

This table stores the names and addresses of the halls in the town. The table structure is as follows:

Name	Type
Hall_id	numeric
Hall_name	character
Address	character

Status Table

This table sets the relationship between hall and movie. It also contains the information about show times, as well as the number of seats available in the balcony, the rear sections, and the upper sections. The date of the movie release is also stored in this table. The table structure is as follows:

Name	Type
Hall_id	numeric
Movie_id	numeric
Showtime	character
Balcony	numeric
Rear_stall	numeric
Upper_stall	numeric
Middle_stall	numeric
Movie_date	char

User Table

This table stores the client's reservation information. This information includes the name and other data, such as telephone number, mail ID, and so on. The number of total tickets bought, along with the description of the movie, hall, showtime, date, and section are also stored here. The table structure is as follows:

Name	Type
Firstname	character
Lastname	character
Telno	Char
Movie_id	numeric
Hall_id	numeric
Stall	Char
Movie_date	Char
Showtime	Char
Totaltickets	numeric
Emailadd	Char

The next two tables are for weather forecasts and news information.

Weather Table

This table stores the weather information from various states. The related information consists of temperature, rainfall, sunset, sunrise, and so on. This table's structure is as follows:

Name	Type
State	Char
Weather	datetime
Sunrise	Char
Sunset	Char
Moonrise	Char

Name	Type
Moonset	Char
Dayhumidity	Char
Nighthumidity	Char
Lowerwinddirection	Char
Higherwinddirection	Char
Lowerwindspeed	Char
Higherwindspeed	Char
High_temp	Char
Low_temp	Char
Rainfall	Char
Figure	char

News Table

This table stores news items under different categories, such as sports, business, politics, and so forth, along with their dates and headlines. The structure of this table is as follows:

Name	Type
Category	Char
Heading	Char
Description	Char
News_date	Date
Heading_id	Int

Application Work Flow

The Mobile Web Application is a database-driven application. All the data is stored on the server in SQL Server. When a request is made, the data is first converted into XML format and then the XSLT is applied to the data, based on the type of client that is accessing the application.

The business logic layer is written in ASP and the data is stored in SQL Server on the Web server. To make the application run on various platforms, ASP detects the browser sending the request and performs all the tasks that contain some kind of business logic inside them. Then, depending upon the request, the client application is developed and reflected in readable form for that specific browser.

The site opens up with an ASP file, which checks for the user agent (the browser sending the request). The main function of this ASP file is to detect the browser and device type. ASP redirects the client to the appropriate page on which the modules for the specific client categories are specified. Consider the creation of the URL that points to the file requested. A file with an extension of .html, for example, is called for a PC browser, and a WML file is called for a WAP browser. The device capabilities are checked so that the XSL transformations can create the appropriate client page. After the device capabilities are verified, the intermediate XSL style sheet is chosen, depending upon the value of the device and document profiles. Finally, from this XSL style sheet, the final output is created and displayed on the respective browser.

The preceding procedure consists of the following steps:

1. A browser sends a request.

2. The ASP file reads the request and detects the corresponding user agent.

3. The application reads the XML content from the content file.

4. The selection process of the intermediate XSL takes place, depending upon the data read from the device and document profiles.

5. The transformation of XML content on the basis of the intermediate XSL occurs, creating the output document.

6. The application sends the response to the corresponding device.

Summary

This chapter gave a brief introduction to Web application development and examined both XML and non-XML approaches. The chapter also introduced the Mobile Web Application, which you develop throughout this book. This application illustrates details of cross-platform application development. In addition, this chapter briefly described the application's purpose, target clients, structures, and development. The remaining chapters of this book address the finer nuances of the development of the application.

Chapter 3

Database Interaction Techniques Using XML and SQL Server 2000

Perhaps the most significant and practical feature in SQL Server 2000 is the support that it provides for XML. XML has the most viable solution for separating data from presentation information, thereby providing a standard way to define and exchange data between applications and databases. This chapter emphasizes the following topics:

♦ XML support in SQL Server 2000

♦ SQL statements using HTTP

♦ Using template files over HTTP

♦ Writing XPath queries for SQL Server 2000

♦ Executing stored procedures using XML

♦ Methods of retrieving data using XML

♦ Bulk Load and Updategrams

Overview of XML Support in SQL Server 2000

Because SQL Server 2000 is a relational database management system (RDBMS), it has added XML support in its new release for inserting, updating, and deleting values in the database. XML provides a way for transferring data on the Internet. Consequently, it is possible for a database to be accessible from any application program on any platform. XML is integrated with the SQL Server RDBMS, which helps in creating efficient applications for the Internet and the corporate world.

The main features and functionalities include provision for URLs to retrieve data from SQL Server using the HTTP protocol, functionality of system-stored procedures for manipulating XML data, and the capability to store and generate XML data using the FOR XML clause in the SELECT statement. Other important features include Updategrams and Bulk Load, which are discussed elsewhere in this chapter.

SQL Statements Using HTTP

SQL statements executed at the URL in Microsoft Internet Explorer (IE) can access the database in SQL Server through HTTP. The data retrieved in the browser is in XML format.

Now look at a few SELECT statements that make use of the virtual directory named virtual, which is used to access the testDB database created in SQL Server 2000. The testDB database includes department and employee tables. The structure of the department table is as follows.

```
dept_no            varchar(10) primary key,
dept_name          varchar(45) not null
```

Insert the following data in the department table:

```
'100', 'PURCHASE'
'101', 'SALES'
'102', 'MARKETING'
'103', 'FINANCE'
'104', 'HRD'
'105', 'INFORMATION TECHNOLOGY'
'106', 'PERSONAL'
```

The structure of the employee table is as follows:

```
emp_code              varchar(10) primary key,
emp_name              varchar(45) not null,
emp_d_o_b             datetime,
emp_sex     char(1),
emp_f_name            varchar(45),
emp_d_o_j             datetime,
emp_marital_status varchar(10),
emp_qualification  varchar(20),
emp_desig             varchar(20) not null,
emp_dept_no                     varchar(10) foreign key references department
(dept_no),
emp_r_addr            varchar(40) not null,
emp_r_city                      varchar(25),
emp_r_state          varchar(25),
emp_r_pincode        varchar(10),
emp_r_phone          varchar(21),
emp_e_mail_id        varchar(50),
emp_pager_no                    varchar(15),
emp_c_phone          varchar(21),
emp_basic_sal                   integer not null,
emp_prob_period      varchar(2),
remarks     varchar(200)
```

Insert the following data in the employee table. Note that the IIS Virtual Directory Management for SQL Server is used for creating virtual names. The virtual name `virtual` is used throughout this chapter.

```
'1001', 'Ashish Baveja', '10/10/76', 'M', 'MR.S. K. Baveja', '06/25/00',
'Single', 'M.C.A', 'Senior Programmer', '105', '7 Golf Links', 'New Delhi',
'Delhi', '110002', '7777777', 'ashish@quest.com', '98110-77777', '96280-777777',
15000, '6', 'Hard Working'

'1002', 'Madan Kumar Dhyani', '01/07/74', 'M', 'MR. J. P. Dhyani', '07/01/99',
'Single', 'M.C.A', 'System Analyst', '105', '8 Green Park', 'New Delhi',
'Delhi', '110001', '8888888', 'madan@quest.com', '98110-88888', '96280-888888',
20000, '3', 'Sincere'

'1003', 'Ravinder Bhushan Srivastava', '05/18/72', 'M', 'MR. S. B. Srivastava',
'07/01/99', 'Single', 'M.C.A', 'EDP Manager', '103', '9 South Extension I', 'New
Delhi', 'Delhi', '110003', '9999999', 'ravinder@quest.com', '98110-99999',
'96280-999999', 10000, '12', 'Diligent'

'1004', 'Sunil Kumar', '02/29/68', 'M', 'MR. R. N. Kumar', '06/15/96', 'Single',
'L.L.B.', 'Legal Advisor', '104', '9 National Park', 'New Delhi', 'Delhi',
'110005', '6666666', 'sunil@quest.com', '98110-66666', '96280-666666', 25000,
'5', 'Meticulous'
```

```
'1005', 'Jaya Srivastava', '01/30/74', 'F', 'MR. S. C. Srivastava', '02/01/01',
'Single', 'P.G.D.C.A.', 'Junior Programmer', '104', '5 Central Area', 'New
Delhi', 'Delhi', '110005', '5555555', 'jaya@quest.com', '98110-55555', '96280-
555555', 10000, '5', 'Intelligent'

'1006', 'Ekta Baveja', '07/30/74', 'F', 'MR. S. K. Baveja', '02/09/00',
'Single', 'M.B.A.', 'Marketing Manager', '102', '6 Southern Avenue', 'New
Delhi', 'Delhi', '110006', '3333333', 'ekta@quest.com', '98110-33333', '96280-
333333', 10000, '3', 'Intelligent'

'1007', 'Sandeep Dhyani', '12/06/77', 'M', 'MR. Harish Dhyani', '07/15/01',
'Single', 'B.E.(Mech)', 'Senior Supervisor', '100', '10 Central Links', 'New
Delhi', 'Delhi', '110007', '4444444', 'sandeep@quest.com', '98110-44444',
'96280-444444', 15000, '8', 'Hard Working'

'1008', 'Aparna Srivastava', '12/29/78', 'F', 'MR. S. K. Srivastava',
'08/01/01', 'Single', 'M.B.A.', 'Sales Executive', '101', '15 South Extension
II', 'New Delhi', 'Delhi', '110009', '1111111', 'aparna@quest.com', '98110-
11111', '96280-111111', 12000, '5', 'Average'

'1009', 'Gaurav Sharma', '09/12/74', 'M', 'MR. R. K. Sharma', '06/30/00',
'Single', 'P.G.D.B.A.', 'Personal Manager', '106', '16 Defence Colony', 'New
Delhi', 'Delhi', '110009', '2222222', 'aparna@quest.com', '98110-22222', '96280-
222222', 12000, '5', 'Meticulous'
```

The following SELECT statement returns the records in the employee table in the testDB database. The XML mode is set to RAW.

```
http://quest/virtual?sql=SELECT+emp_code,emp_name,emp_desig,emp_dept_no+FROM+emp
loyee+FOR+XML+RAW&root=ROOT
```

In the above code quest is representing the Web server name of our machine. This can be change to localhost or the name of your Web server.

```
http://localhost/virtual?sql=SELECT+emp_code+FROM+employee+FOR+XML+RAW&root=ROOT
```

Now understand how this process actually works on the server side. The request is to be routed to a server named quest, as specified by the URL that issues a request to the virtual directory named virtual. The parameter root with value ROOT is used when the query returns multiple elements. The three modes, namely RAW, AUTO, and EXPLICIT, are discussed later in this chapter.

You can retrieve more than one element by using the SELECT '<ROOT>' opening tag and SELECT '</ROOT>' closing tag, as in the following example:

```
http://quest/virtual?sql=SELECT+'<ROOT>';SELECT+emp_code,emp_name,emp_desig,emp_
dept_no+FROM+employee+FOR+XML+AUTO;SELECT+'</ROOT>'
```

The output of the preceding queries would be the same and are rendered by Internet Explorer as shown in Figure 3-1.

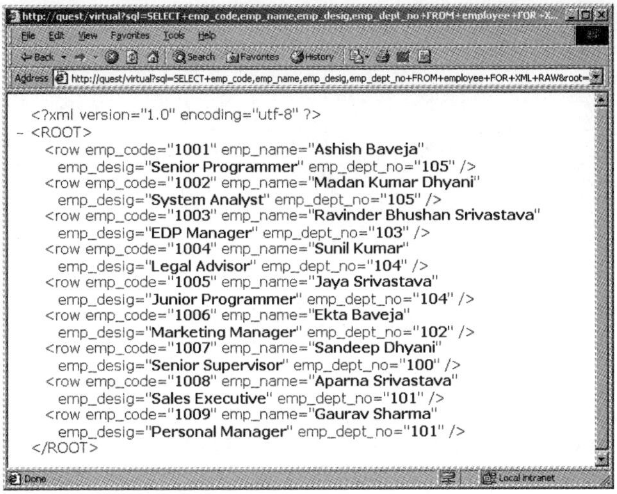

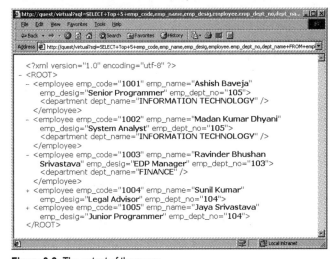

Figure 3-1: Output of the preceding query

The next `SELECT` statement makes use of multiple employee and department tables in the testDB database.

```
http://quest/virtual?sql=SELECT+Top+5+emp_code,emp_name,emp_desig,employee.emp_d
ept_no,dept_name+FROM+employee,department+WHERE+employee.emp_dept_no=department.
dept_no+Order+by+emp_code+FOR+XML+AUTO&root=ROOT
```

Figure 3-2 shows the output of the query as rendered by Internet Explorer.

Figure 3-2: The output of the query

The following `SELECT` statement returns the name of employees working in the department, where the department name begins with `INFO`. Employee names are selected for the department whose name starts with `INFO`; you achieve this result by using the `LIKE` clause and the special character %, represented by %25 in the URL (the ASCII value of % is 37, and the hexadecimal equivalent of 37 is 25). The % symbol signifies 0 or more characters. The output is retrieved as a concatenated string because the query is specified without the `FOR XML` clause. Without the `FOR XML` clause, only a single column can be specified in the `SELECT` statement.

```
http://quest/virtual?sql=SELECT+emp_name+FROM+employee,department+WHERE+dept_nam
e+LIKE+'INFO%25'+and+employee.emp_dept_no=department.dept_no
```

Figure 3-3 shows the output of the query as rendered by Internet Explorer.

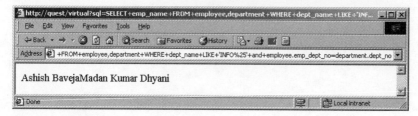

Figure 3-3: The output of the preceding query

The next SELECT statement makes use of the special character ?, represented by %3F in the URL (the ASCII value of ? is 63, and the hexadecimal equivalent of 63 is 3F), to retrieve the details about the employee whose employee code is passed as the argument.

```
http://quest/virtual?sql=SELECT+emp_code,emp_name,emp_desig,dept_name+FROM+emplo
yee,department+WHERE+employee.emp_dept_no=department.dept_no+AND+emp_code=%3F+FO
R+XML+AUTO&emp_code=1001&root=ROOT
```
Figure 3-4 shows the output of the query as rendered by Internet Explorer.

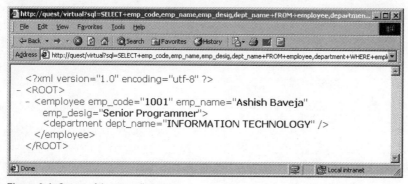

Figure 3-4: Output of the preceding query

Call Stored Procedures over HTTP

To format and further manipulate the data in XML format, you can execute the stored procedures at the URL through the HTTP protocol. Following is a discussion of a few stored procedures that make use of the virtual directory `virtual` that is used to access the testDB database created in SQL Server 2000. The stored procedure, `employeeDetail`, retrieves the employee code, name, designation, and department name from employee and department tables in the testDB database. The stored procedure is as follows:

```
CREATE PROCEDURE employeeDetail
AS
    SELECT '<ROOT>';
    SELECT emp_code, emp_name, emp_desig, dept_name
    FROM employee, department
    WHERE employee.emp_dept_no = department.dept_no
    FOR XML AUTO;
    SELECT '</ROOT>'
GO
```

The URL that executes the stored procedure is the following:

```
http://quest/virtual?sql=EXECUTE+employeeDetail
```

Figure 3-5 shows the output of the query as rendered by Internet Explorer.

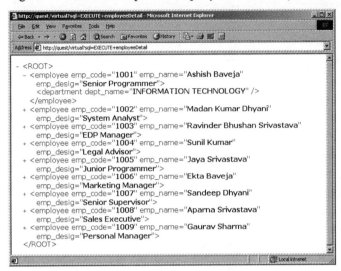

Figure 3-5: The output of executing stored procedures

Modify the `employeeDetail` stored procedure by adding a parameter to it that helps retrieve employee information according to the passed argument. To remove the existing stored procedure, `employeeDetail`, add the following statements:

```
IF EXISTS (SELECT name FROM sysobjects
    WHERE name = 'employeeDetail' AND type = 'P')
    DROP PROCEDURE employeeDetail
GO
```

The modified `employeeDetail` stored procedure now becomes the following:

```
IF EXISTS (SELECT name FROM sysobjects
    WHERE name = 'employeeDetail' AND type = 'P')
    DROP PROCEDURE employeeDetail
GO
CREATE PROCEDURE employeeDetail @param int
AS
    SELECT '<ROOT>';
    SELECT emp_code, emp_name, emp_desig, dept_name
    FROM employee, department
    WHERE employee.emp_dept_no = department.dept_no
                    AND emp_basic_sal > @param
    FOR XML AUTO;
    SELECT '</ROOT>'
GO
```

The URL that executes the stored procedure is as follows:

```
http://quest/virtual?sql=EXECUTE+employeeDetail+@param=10000
```

Figure 3-6 shows the output of the query as rendered by Internet Explorer.

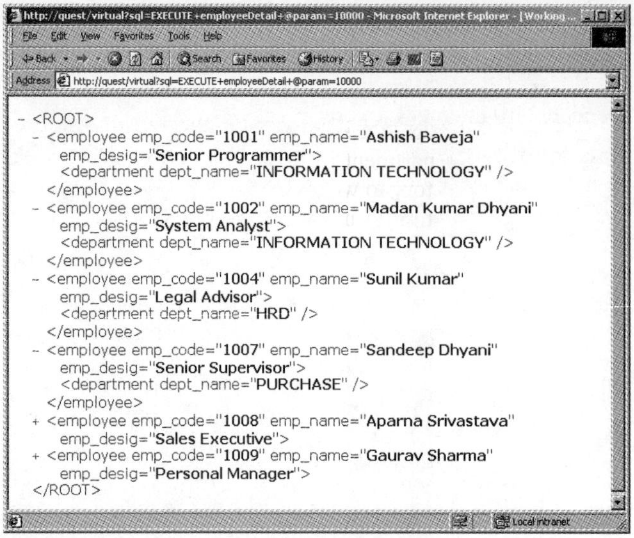

Figure 3-6: Executing the stored procedure with parameters

Using XML Template Files over HTTP

Templates are well-formed XML documents used to specify SQL statements or XPath (XML Path) queries that are long enough to be executed at the URL. (XPath queries are covered later in this chapter.) This ensures that the security measures for important information, such as SQL queries, are hidden and hence cannot be manipulated at the URL.

Templates are well-formed XML documents, so they have the following syntax:

```
<ROOT xmlns:sql="urn:schemas-microsoft-com:xml-sql" sql:xsl='XSL Style-
sheetName'>
  <sql:header>
      <sql:param>..</sql:param>
      <sql:param>..</sql:param>...n
  </sql:header>
  <sql:query>
      sql statement(s)
  </sql:query>
  <sql:xpath-query mapping-schema="SchemaFile.xml">
      XPath query
  </sql:xpath-query>
</ROOT>
```

The namespace declaration `xmlns:sql="urn:schemas-microsoft-com:xml-sql"` is mandatory, because an SQL namespace defines various elements in the template. The attribute `sql:xsl` identifies an XSL style sheet to be applied to the final XML document. `<ROOT>` signifies the top-level element in the XML hierarchy. The `<sql:header>` element is a container of single or multiple parameter values, which are specified in `<sql:param>` elements. The elements `<sql:query>` and `<sql:xpath-query>` specify the SQL statements and XPath queries, respectively. The `<sql:xpath-query>` element has an attribute-mapping schema that indicates the schema that is mapped to the template. Two important concepts to remember are that all the elements in a template are optional, and any name can be assigned to the namespace.

Now look at a few templates that make use of the virtual directory `virtual` that is used to access the testDB database created in SQL Server 2000. The virtual name assigned to the template type is a template that is created through IIS Virtual Directory Management for SQL Server.

The template in Listing 3-1, `sample1.xml`, makes use of a `SELECT` statement for retrieving employee information (namely, employee code, employee name, and department name) from employee and department tables. Note that this .xml file is saved in the directory to which the virtual name, `virtual`, has been assigned by using the configuring SQL XML support in IIS utility provided with SQL server 2000.

Listing 3-1: sample1.xml

```
<ROOT xmlns:sql="urn:schemas-microsoft-com:xml-sql">
  <sql:query>
      SELECT emp_code, emp_name, dept_name
      FROM employee, department
      WHERE employee.emp_dept_no = department.dept_no
      FOR XML AUTO
  </sql:query>
</ROOT>
```

You can execute this template by using the following URL:

```
http://quest/virtual/template/sample1.xml
```

Figure 3-7 shows the output of the template as rendered by Internet Explorer.

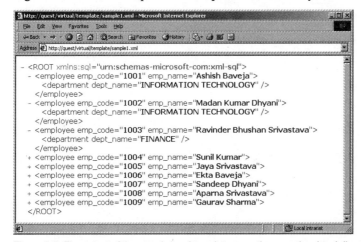

Figure 3-7: The output of the sample1.xml template executing over the virtual directory

You also can obtain the output in Figure 3-7 if the template is directly specified in the URL, as in the following example:

```
http://quest/virtual?template=<ROOT+xmlns:sql="urn:schemas-microsoft-com:xml-
sql"><sql:query>SELECT+emp_code,emp_name,dept_name+FROM+employee,department+
WHERE+employee.emp_dept_no=department.dept_no+FOR+XML+AUTO</sql:query>
</ROOT>
```

The template in Listing 3-2, `sample2.xml`, executes the preceding stored procedure, `employeeDetail`, inside the `<sql:query>` tag.

Listing 3-2: sample2.xml

```
<ROOT xmlns:sql="urn:schemas-microsoft-com:xml-sql">
  <sql:query>
    exec employeeDetail 15000
  </sql:query>
</ROOT>
```

You can execute this template by using the following URL:

```
http://quest/virtual/template/sample2.xml
```

Figure 3-8 shows the output of the template as rendered by Internet Explorer.

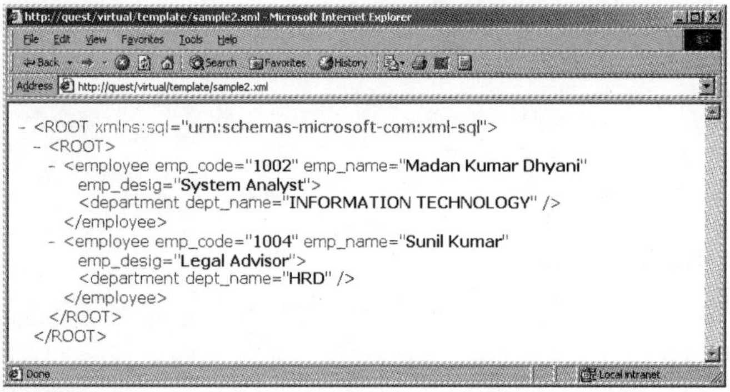

Figure 3-8: Executing the stored procedure from the sample2.xml template

You also can add parameters to SQL statements specified in templates. The `<sql:header>` element, being a container of single or multiple parameter values, specifies `<sql:param>` elements for adding parameters. The template in Listing 3-3, `sample3.xml`, executes the modified stored procedure, `employeeDetail`, which requires a single parameter, `@param`.

Listing 3-3: sample3.xml

```
<ROOT xmlns:sql='urn:schemas-microsoft-com:xml-sql'>
  <sql:header>
      <sql:param name='param'>12000</sql:param>
  </sql:header>
  <sql:query>
      exec employeeDetail @param
  </sql:query>
</ROOT>
```

The following statements represent the default value (here, `12000`) for the parameter `param`:

```
<sql:header>
    <sql:param name='param'>12000</sql:param>
</sql:header>
```

The modified stored procedure `employeeDetail` is as follows:

```
IF EXISTS (SELECT name FROM sysobjects
   WHERE name = 'employeeDetail' AND type = 'P')
   DROP PROCEDURE employeeDetail
GO
CREATE PROCEDURE employeeDetail @param int
AS
    SELECT emp_code, emp_name, emp_desig, dept_name
```

```
    FROM employee, department
    WHERE employee.emp_dept_no = department.dept_no
                AND emp_basic_sal > @param
    FOR XML AUTO
GO
```

You can execute this template by using the following URL:

```
http://quest/virtual/template/sample3.xml
```

Figure 3-9 shows the output of the template as rendered by Internet Explorer.

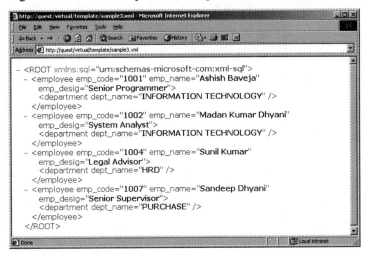

Figure 3-9: Output of the preceding query

If the parameter is passed along with the URL, the default value is ignored while retrieving the data, as in the following example:

```
http://quest/virtual/template/sample3.xml?param=15000
```

Figure 3-10 shows the output of the template as rendered by Internet Explorer.

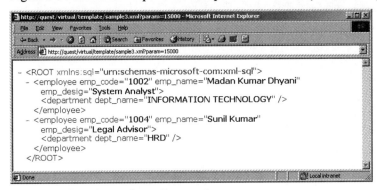

Figure 3-10: Executing templates with parameters

The template in Listing 3-4, `sample4.xml`, makes use of more than one parameter.

Listing 3-4: sample4.xml

```
<ROOT xmlns:sql='urn:schemas-microsoft-com:xml-sql'>
  <sql:header>
```

```
        <sql:param name='param1'>15000</sql:param>
        <sql:param name='param2'>25000</sql:param>
    </sql:header>
    <sql:query>
        exec employeeDetail @param1, @param2
    </sql:query>
</ROOT>
```

The modified stored procedure `employeeDetail` is as follows:

```
IF EXISTS (SELECT name FROM sysobjects
    WHERE name = 'employeeDetail' AND type = 'P')
    DROP PROCEDURE employeeDetail
GO
CREATE PROCEDURE employeeDetail @param1 int, @param2 int
AS
    SELECT emp_code, emp_name, emp_desig, dept_name
    FROM employee, department
    WHERE employee.emp_dept_no = department.dept_no
                AND emp_basic_sal BETWEEN @param1 AND @param2
    FOR XML AUTO
GO
```

You can execute this template by using the following URL:

```
http://quest/virtual/template/sample4.xml
```

Figure 3-11 shows the output of the template as rendered by Internet Explorer.

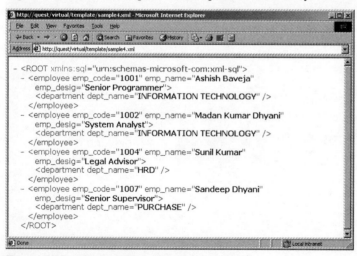

Figure 3-11: Executing templates with multiple parameters

Write XPath Queries for SQL Server 2000

XPath queries help retrieve data from an XML document and provide an efficient source of navigation within an XML document. You can specify the XPath query within a template or directly at the URL. The latest XPath language specification is available online on the W3C Web site:
`http://www.w3c.org/TR/xpath`.

Specify Nodes and Axes in XPath Queries

XPath queries are represented by a location path, which specifies expressions for selecting nodes from the context node. The location path is either absolute or relative and consists of location steps that contain axes and node tests.

Axis

The axis represents the hierarchical relationship between the selected nodes from the location step and the context node. The axes `child`, `parent`, `attribute`, and `self` are supported in XPath queries.

Node test

The node test represents the type of node selected by the location step. There is a principal node type for the axes `child`, `parent`, `attribute`, and `self`. The `attribute` axis has the `<attribute>` principal node type, whereas other axes have the `<element>` principal node type.

Now consider how the XPath queries are executed using the mapping schema in Listing 3-5, `testSchema.xml`.

Listing 3-5: testSchema.xml

```xml
<?xml version="1.0" ?>
<Schema xmlns="urn:schemas-microsoft-com:xml-data"
        xmlns:dt="urn:schemas-microsoft-com:datatypes"
        xmlns:sql="urn:schemas-microsoft-com:xml-sql">

  <ElementType name="emp" sql:relation="employee">
    <AttributeType name="emp_code" />
    <AttributeType name="emp_name" />
    <AttributeType name="emp_d_o_b" />
    <AttributeType name="emp_sex" />
    <AttributeType name="emp_f_name" />
    <AttributeType name="emp_d_o_j" />
    <AttributeType name="emp_marital_status" />
    <AttributeType name="emp_qualification" />
    <AttributeType name="emp_desig" />
    <AttributeType name="emp_dept_no" />
    <AttributeType name="emp_r_addr" />
    <AttributeType name="emp_r_city" />
    <AttributeType name="emp_r_state" />
    <AttributeType name="emp_r_pincode" />
    <AttributeType name="emp_r_phone" />
    <AttributeType name="emp_e_mail_id" />
    <AttributeType name="emp_pager_no" />
    <AttributeType name="emp_c_phone" />
    <AttributeType name="emp_basic_sal" />
    <AttributeType name="emp_prob_period" />
    <AttributeType name="remarks" />

    <attribute type="emp_code" />
    <attribute type="emp_name" />
    <attribute type="emp_d_o_b" />
    <attribute type="emp_sex" />
    <attribute type="emp_f_name" />
    <attribute type="emp_d_o_j" />
    <attribute type="emp_marital_status" />
    <attribute type="emp_qualification" />
```

```
    <attribute type="emp_desig" />
    <attribute type="emp_dept_no" sql:relation="department" sql:field="dept_no">
      <sql:relationship
                key-relation="employee"
                key="emp_dept_no"
                foreign-relation="department"
                foreign-key="dept_no" />
    </attribute>
    <attribute type="emp_r_addr" />
    <attribute type="emp_r_city" />
    <attribute type="emp_r_state" />
    <attribute type="emp_r_pincode" />
    <attribute type="emp_r_phone" />
    <attribute type="emp_e_mail_id" />
    <attribute type="emp_pager_no" />
    <attribute type="emp_c_phone" />
    <attribute type="emp_basic_sal" />
    <attribute type="emp_prob_period" />
    <attribute type="remarks" />

    <element type="dept">
      <sql:relationship
                key-relation="employee"
                key="emp_dept_no"
                foreign-relation="department"
                foreign-key="dept_no" />
    </element>
  </ElementType>

  <ElementType name="dept" sql:relation="department">
    <AttributeType name="dept_no"/>
    <AttributeType name="dept_name" />
    <attribute type="dept_no"/>
    <attribute type="dept_name" />
  </ElementType>
</Schema>
```

The template in Listing 3-6, example1.xml, executes XPath queries that make use of the testSchema.xml mapping schema. Here, the XPath query selects all the dept child elements of the context node, using /dept.

Listing 3-6: example1.xml

```
<ROOT xmlns:sql="urn:schemas-microsoft-com:xml-sql">
  <sql:xpath-query mapping-schema="testSchema.xml">
    /dept
  </sql:xpath-query>
</ROOT>
```

You can execute this template by using the following URL:

```
http://quest/virtual/template/example1.xml
```

Figure 3-12 shows the output of the template as rendered by Internet Explorer.

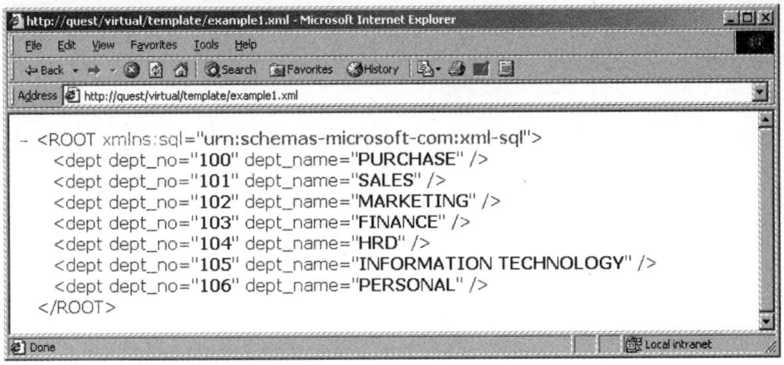

Figure 3-12: The output of executing the XPath query

The URL also can directly specify the XPath query, as follows:

```
http://quest/virtual/schema/testSchema.xml/child::dept?root=ROOT
```

Here, `schema` is a virtual name of type `schema`, `child` is the axis, and `dept` is the node test. Figure 3-13 shows the output as rendered by Internet Explorer.

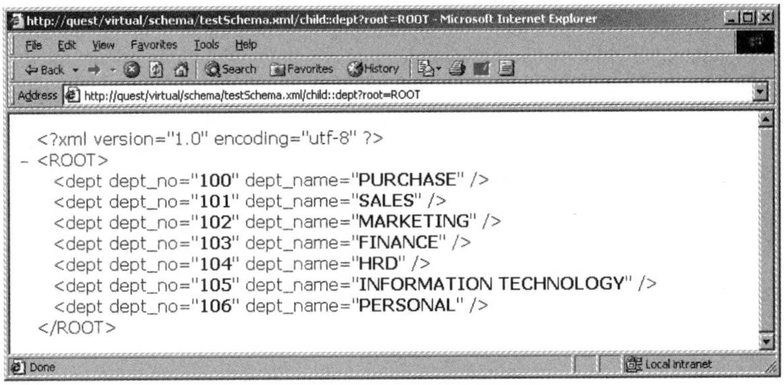

Figure 3-13: Executing the XPath query directly from the URL

Consider the template in Listing 3-7, `example2.xml`.

Listing 3-7: example2.xml

```
<ROOT xmlns:sql="urn:schemas-microsoft-com:xml-sql">
  <sql:xpath-query mapping-schema="testSchema.xml">
    /emp/dept
  </sql:xpath-query>
</ROOT>
```

Here, the XPath query selects all the `dept` child elements of the `emp` element children of the context node using `/emp/dept`.

You can execute this template by using the following URL:

```
http://quest/virtual/template/example2.xml
```

Figure 3-14 shows the output as rendered by Internet Explorer.

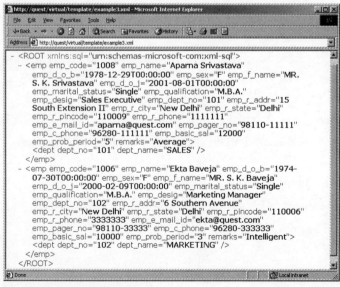

Figure 3-14: The output of example2.xml

The URL also can directly specify the XPath query, as follows:

```
http://quest/virtual/schema/testSchema.xml/child::emp/child::dept?root=ROOT
```

Here, `schema` is a virtual name of type `schema`, `child` is the axis, and `emp` and `dept` are the node tests.

The template in Listing 3-8, `example3.xml`, executes an XPath query that retrieves `emp` child elements of the context node with "`M.B.A`" as the `emp_qualification` attribute's value.

Listing 3-8: example3.xml

```
<ROOT xmlns:sql="urn:schemas-microsoft-com:xml-sql">
  <sql:xpath-query mapping-schema="testSchema.xml">
    /emp[attribute::emp_qualification="M.B.A."]
  </sql:xpath-query>
</ROOT>
```

Execute this template by using the following URL:

```
http://quest/virtual/template/example3.xml
```

Figure 3-15 shows the output as rendered by Internet Explorer.

Figure 3-15: Output of example3.xml

The URL also can directly specify the XPath query, as follows:

```
http://quest/virtual/schema/testSchema.xml/emp[attribute::emp_qualification="M.B
.A."]?root=ROOT
```

Here, `schema` is a virtual name of type `schema`, `attribute` is the axis, and `emp_qualification` is the node test.

Operators in XPath Queries

In XPath queries three main types of operators are available. They are

- ♦ Relational operators
- ♦ Arithmetic operators
- ♦ Boolean operators

In next few paragraphs, we discuss the three operators one by one.

Relational operators

The template in Listing 3-9, `example4.xml`, demonstrates the use of the relational operator `>` for retrieving all `emp` elements with `emp_basic_sal` greater than `15000`. Likewise, you can use other relational operators `(<, >=, <=, !=, =)`.

Listing 3-9: example4.xml

```
<ROOT xmlns:sql="urn:schemas-microsoft-com:xml-sql">
  <sql:xpath-query mapping-schema="testSchema.xml">
    /emp[@emp_basic_sal>15000]
  </sql:xpath-query>
</ROOT>
```

You can execute this template by using the following URL:

```
http://quest/virtual/template/example4.xml
```

Figure 3-16 shows the output as rendered by Internet Explorer.

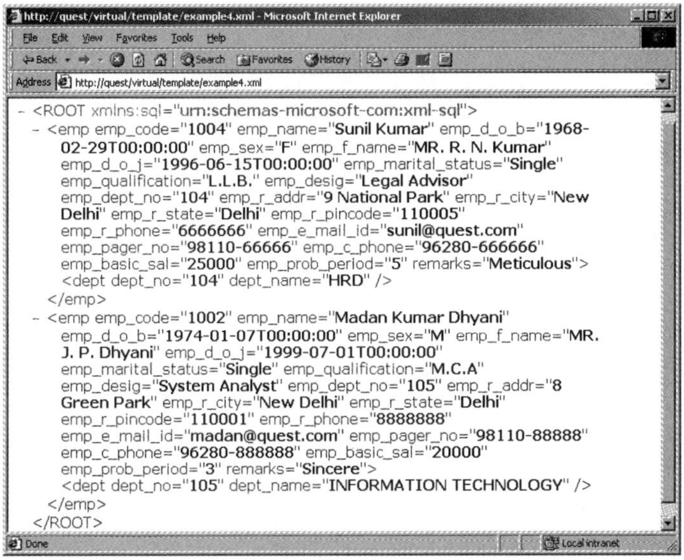

Figure 3-16: The output of example4.xml

The URL also can directly specify the XPath query, as follows:

```
http://quest/virtual/schema/testSchema.xml/child::emp[@emp_basic_sal>15000]?root
=ROOT
```

Arithmetic operators

The template in Listing 3-10, `example5.xml`, demonstrates the use of the arithmetic operator + for retrieving all `emp` elements with `emp_basic_sal` plus `3000` equal to `15000`. Likewise, you can use other arithmetic operators (`-`, `*`, `div`).

Listing 3-10: example5.xml

```
<ROOT xmlns:sql="urn:schemas-microsoft-com:xml-sql">
  <sql:xpath-query mapping-schema="testSchema.xml">
    /emp[@emp_basic_sal+3000=15000]
  </sql:xpath-query>
</ROOT>
```

Execute this template by using the following URL:

```
http://quest/virtual/template/example5.xml
```

Figure 3-17 shows the output as rendered by Internet Explorer.

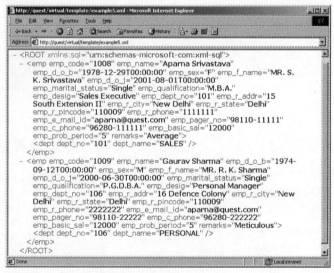

Figure 3-17: The output of example5.xml

The URL also can directly specify the XPath query, as follows:

```
http://quest/virtual/schema/testSchema.xml/child::emp[@emp_basic_sal+3000=15000]
?root=ROOT
```

Boolean operators

The template in Listing 3-11, `example6.xml`, demonstrates the use of the Boolean operator and for retrieving the `emp` element with `emp_name` as "Madan Kumar Dhyani" and `emp_dept_no` as "105." Likewise, you can use some other Boolean operators (such as `or`).

Listing 3-11: example6.xml

```
<ROOT xmlns:sql="urn:schemas-microsoft-com:xml-sql">
  <sql:xpath-query mapping-schema="testSchema.xml">
```

```
        /emp[@emp_name="Madan Kumar Dhyani" and @emp_dept_no="105"]
    </sql:xpath-query>
</ROOT>
```

You can execute this template by using the following URL:

```
http://quest/virtual/template/example6.xml
```

Figure 3-18 shows the output as rendered by Internet Explorer.

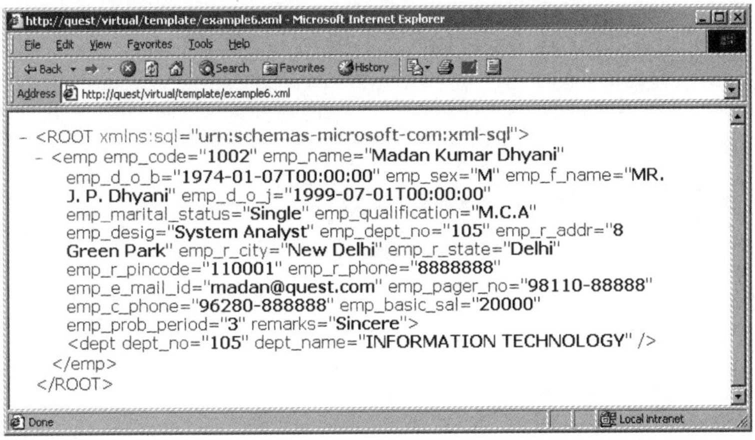

Figure 3-18: Output of example6.xml

The URL also can directly specify the XPath query, as follows:

```
http://quest/virtual/schema/testSchema.xml/child::emp[@emp_name="Madan  Kumar
Dhyani" and @emp_dept_no="105"]?root=ROOT
```

Variables in XPath Queries

The template in Listing 3-12, `example7.xml`, demonstrates the usage of XPath variables in XPath queries. The template contains an XPath query that takes a single parameter. In case no parameter value is passed, the default value for this particular parameter, indicated in `<sql:header>`, is used.

Listing 3-12: example7.xml

```
<ROOT xmlns:sql="urn:schemas-microsoft-com:xml-sql">
  <sql:header>
    <sql:param name='dept_no'>105</sql:param>
  </sql:header>
  <sql:xpath-query mapping-schema="testSchema.xml">
    dept[@dept_no=$dept_no]
  </sql:xpath-query >
</ROOT>
```

Execute this template by using the following URL:

```
http://quest/virtual/template/example7.xml
```

Note that the default value is taken because no parameter is passed to the URL.

Figure 3-19 shows the output as rendered by Internet Explorer.

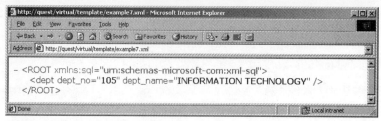

Figure 3-19: The output of example7.xml

Now pass a parameter to the URL, as follows:

```
http://quest/virtual/template/example7.xml?dept_no=101
```

Figure 3-20 shows the output as rendered by Internet Explorer.

Figure 3-20: Output of example7.XML with parameters

The URL also can directly specify the XPath query, as follows:

```
http://quest/virtual/schema/testSchema.xml/dept[@dept_no=$dept_no]?dept_no=101&r
oot=ROOT
```

Boolean Functions in XPath Queries

To demonstrate some Boolean functions (namely, not(), true(), and false()), add the following record to the employee table. Note that this employee has not been assigned a department number.

```
'1010', 'Dhruv Khanna', '06/08/74', 'M', 'MR. Anil Khanna', '05/25/98',
'Single', 'M.B.A.', 'Research Manager', null, '17 Defence Enclave', 'New Delhi',
'Delhi', '110024', '2233445', 'dhruv@quest.com', '98110-22334', '96280-223344',
10000, '4', 'Brilliant'
```

The template in Listing 3-13, example8.xml, demonstrates the use of the not() Boolean function for retrieving the emp element lacking department information.

Listing 3-13: example8.xml

```
<ROOT xmlns:sql="urn:schemas-microsoft-com:xml-sql">
  <sql:xpath-query mapping-schema="testSchema.xml">
    /emp[not(dept)]
  </sql:xpath-query>
</ROOT>
```

Execute this template by using the following URL:

```
http://quest/virtual/template/example8.xml
```

Figure 3-21 shows the output as rendered by Internet Explorer.

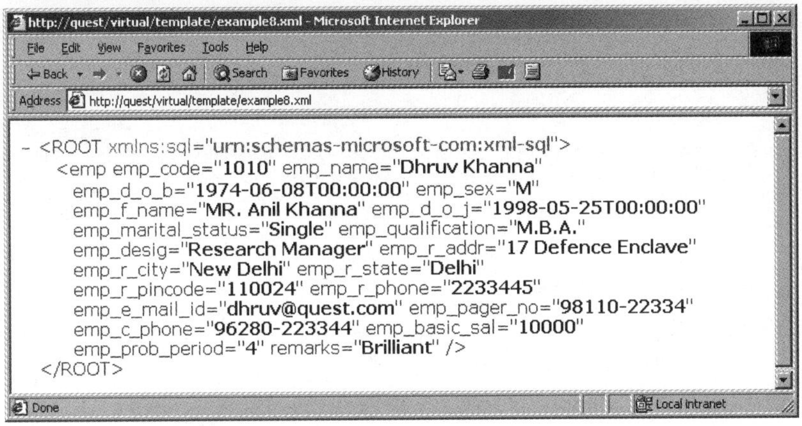

Figure 3-21: Output of example8.XML

The URL also can directly specify the XPath query, as follows:

```
http://quest/virtual/schema/testSchema.xml/child::emp[not(child::dept)]?root=ROO
T
```

Likewise, you can use other Boolean functions, such as `true()` and `false()`, as follows:

```
http://quest/virtual/schema/testSchema.xml/child::emp[child::dept=false()]?root=
ROOT
```

The preceding URL displays the employee record that is missing the department information.

```
http://quest/virtual/schema/testSchema.xml/child::emp[child::dept=true()]?root=R
OOT
```

The preceding URL displays all employee information, excluding the record with missing department information.

Methods of Retrieving Data Using XML

The `SELECT` statement's `FOR XML` clause retrieves query data as an XML document. You can use the `FOR XML` clause both in queries and in stored procedures. The `FOR XML` clause has the following syntax:

```
FOR XML mode [, XMLDATA] [, ELEMENTS][, BINARY BASE64]
```

Here, `XML mode` indicates one of three modes — `RAW`, `AUTO`, or `EXPLICIT` — which are used to retrieve data. `XMLDATA` attaches the XML schema to the XML document. `ELEMENTS` results in retrieving the columns as subelements. `BINARY BASE64` causes any binary data that the query retrieves to be specified in base64-encoded format.

The three modes of data retrieval are briefly described in this section. The following sections present a more detailed look at each mode.

The `FOR XML` clause of the `SELECT` statement can include the following three modes for retrieving data:

◆ `RAW` — Using this mode, you can convert every row in the query output into an XML element.

◆ `AUTO` — This mode converts every row in the query output in the XML hierarchy.

◆ `EXPLICIT` — This mode enables you to convert the query result set output into an XML document.

RAW Mode

The following URL retrieves the employee information, using the RAW mode specified in the FOR   XML clause:

```
http://quest/virtual?sql=SELECT+emp_code,emp_name,emp_desig,dept_name+FROM+emplo
yee,department+WHERE+employee.emp_dept_no=department.dept_no+ORDER+BY+emp_code+F
OR+XML+RAW&root=ROOT
```

Figure 3-22 shows the output as rendered by Internet Explorer.

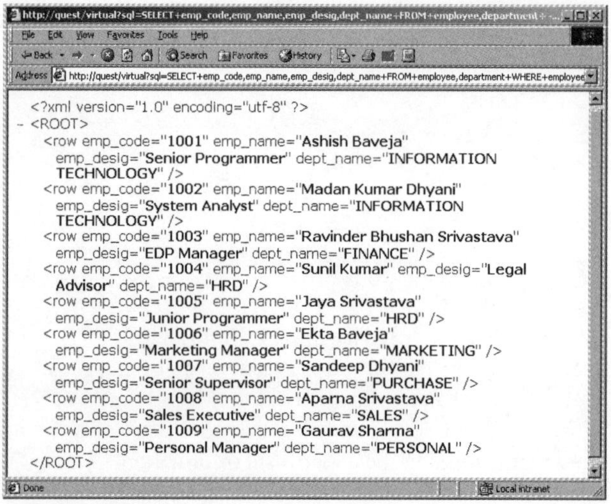

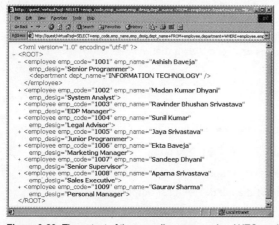

Figure 3-22: The output of the preceding query, using RAW mode

AUTO Mode

The following URL retrieves the employee information, using the AUTO mode specified in the FOR   XML clause:

```
http://quest/virtual?sql=SELECT+emp_code,emp_name,emp_desig,dept_name+FROM+emplo
yee,department+WHERE+employee.emp_dept_no=department.dept_no+ORDER+BY+emp_code+F
OR+XML+AUTO&root=ROOT
```

Figure 3-23 shows the output as rendered by Internet Explorer.

Figure 3-23: The output of the preceding query, using AUTO mode

EXPLICIT Mode

Now consider how to use EXPLICIT mode to retrieve the XML hierarchy. First, take a look at the following code:

```
SELECT 1                     as Tag,
       NULL                  as Parent,
       employee.emp_code as [emp!1!emp_code],
       NULL                  as [dept!2!dept_name]
FROM employee
UNION ALL
SELECT 2,
       1,
       employee.emp_code,
       department.dept_name
FROM employee, department
WHERE employee.emp_dept_no = department.dept_no
ORDER BY [emp!1!emp_code]
FOR XML EXPLICIT
```

The SELECT statement must specify the metadata columns, Tag and Parent, for creating the XML hierarchy. The top-level element of emp with a Tag value of 1 and Parent tag of emp is NULL. The child element of emp is dept, which is assigned a Tag value of 2 and Parent tag value of 1. Using the UNION ALL statement, the first SELECT statement retrieves the emp elements and the next SELECT statement retrieves the dept elements.

The output generated by using Query Analyzer in SQL Server 2000 is as follows:

```
<emp emp_code="1001">
   <dept dept_no="INFORMATION TECHNOLOGY"/>
</emp>
<emp emp_code="1002">
   <dept dept_no="INFORMATION TECHNOLOGY"/>
</emp>
<emp emp_code="1003">
   <dept dept_no="FINANCE"/>
</emp>
<emp emp_code="1004">
   <dept dept_no="HRD"/>
</emp>
<emp emp_code="1005">
   <dept dept_no="HRD"/>
</emp>
<emp emp_code="1006">
   <dept dept_no="MARKETING"/>
</emp>
<emp emp_code="1007">
   <dept dept_no="PURCHASE"/>
</emp>
<emp emp_code="1008">
   <dept dept_no="SALES"/>
</emp>
<emp emp_code="1009">
   <dept dept_no="PERSONAL"/>
</emp>
```

Add a Schema in Your Query with XMLDATA

The following SELECT statement makes use of the XMLDATA schema to retrieve employee and related department information:

```
SELECT emp_code, emp_name, emp_desig, dept_name
FROM employee, department
WHERE employee.emp_dept_no=department.dept_no
ORDER BY emp_code
FOR XML RAW, XMLDATA
```

The XMLData schema generated by using Query Analyzer in SQL Server 2000 is as follows:

```
<Schema name="Schema2"      xmlns="urn:schemas-microsoft-com:xml-data"
xmlns:dt="urn:schemas-microsoft-com:datatypes">
   <ElementType name="row"  content="empty" model="closed">
      <AttributeType name="emp_code" dt:type="string"/>
      <AttributeType name="emp_name" dt:type="string"/>
      <AttributeType name="emp_desig" dt:type="string"/>
      <AttributeType name="dept_name" dt:type="string"/>
      <attribute type="emp_code"/>
      <attribute type="emp_name"/>
      <attribute type="emp_desig"/>
      <attribute type="dept_name"/>
   </ElementType>
</Schema>
<row xmlns="x-schema:#Schema2" emp_code="1001" emp_name="Ashish Baveja"
  emp_desig="Senior Programmer" dept_name="INFORMATION TECHNOLOGY"/>
<row xmlns="x-schema:#Schema2" emp_code="1002" emp_name="Madan Kumar Dhyani"
  emp_desig="System Analyst" dept_name="INFORMATION TECHNOLOGY"/>
<row xmlns="x-schema:#Schema2" emp_code="1003" emp_name="Ravinder Bhushan
  Srivastava" emp_desig="EDP Manager" dept_name="FINANCE"/>
<row xmlns="x-schema:#Schema2" emp_code="1004" emp_name="Sunil Kumar"
  emp_desig="Legal Advisor" dept_name="HRD"/>
<row xmlns="x-schema:#Schema2" emp_code="1005" emp_name="Jaya Srivastava"
  emp_desig="Junior Programmer" dept_name="HRD"/>
<row xmlns="x-schema:#Schema2" emp_code="1006" emp_name="Ekta Baveja"
  emp_desig="Marketing Manager" dept_name="MARKETING"/>
<row xmlns="x-schema:#Schema2" emp_code="1007" emp_name="Sandeep Dhyani"
  emp_desig="Senior Supervisor" dept_name="PURCHASE"/>
<row xmlns="x-schema:#Schema2" emp_code="1008" emp_name="Aparna Srivastava"
  emp_desig="Sales Executive" dept_name="SALES"/>
<row xmlns="x-schema:#Schema2" emp_code="1009" emp_name="Gaurav Sharma"
  emp_desig="Personal Manager" dept_name="PERSONAL"/>
```

Specify Style Sheets While Retrieving Data

The following URL contains a SELECT statement that uses the XSL style sheet department.xsl, which retrieves the department information for all departments in the department table:

```
http://quest/virtual?sql=SELECT+dept_no,dept_name+from+department+FOR+XML+AUTO&x
sl=department.xsl&root=ROOT
```

The XSL style sheet department.xsl, saved in the virtual directory, is as follows:

```
<?xml version='1.0' encoding='UTF-8'?>
 <xsl:stylesheet xmlns:xsl="http://www.w3.org/1999/XSL/Transform" version="1.0">
    <xsl:template match = '*'>
```

```
        <xsl:apply-templates />
    </xsl:template>
    <xsl:template match = 'department'>
       <TR>
  <TD ALIGN='center'><xsl:value-of select = '@dept_no' /></TD>
        <TD ALIGN='center'><xsl:value-of select = '@dept_name' /></TD>
       </TR>
    </xsl:template>
    <xsl:template match = '/'>
      <HTML>
        <HEAD>
           <STYLE>th{background-color: rgb(100,200,150)}</STYLE>
        </HEAD>
        <BODY>
         <TABLE border='1' style='width:500;' align='center'>
           <TR><TH colspan='2'>Department Information</TH></TR>
           <TR>
       <TH>Department Number</TH>
       <TH>Department Name</TH>
    </TR>
           <xsl:apply-templates select = 'ROOT' />
         </TABLE>
        </BODY>
      </HTML>
    </xsl:template>
</xsl:stylesheet>
```

Figure 3-24 shows the output of the query as rendered by Internet Explorer.

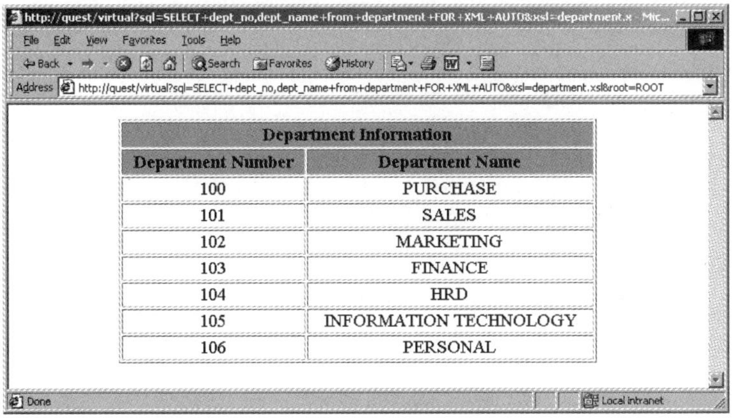

Department Information	
Department Number	Department Name
100	PURCHASE
101	SALES
102	MARKETING
103	FINANCE
104	HRD
105	INFORMATION TECHNOLOGY
106	PERSONAL

Figure 3-24: The output of the query with style sheet included

Bulk Load Method

XML for SQL Server 2000 includes XML Bulk Load, which enables high-speed bulk loads of data within the XML tags. A mapping schema is necessary for XML Bulk Load to identify database tables and, accordingly, insert the data in the database tables.

To demonstrate Bulk Load, create two tables, `Student_info` and `Exam_info`, as follows:

```
Student_info (Reg_no, Level, Name, Address)
Exam_info (Reg_no, Roll_no)
```

The Student_info table is defined as follows:

```
Reg_no int primary key,
Level  char(1) NOT NULL,
Name  varchar(20) NOT NULL,
Addressvarchar(20)
```

The Exam_info table is defined as follows:

```
Roll_no        int primary key,
Reg_no         int FOREIGN KEY REFERENCES Student_info(Reg_no)
```

The foreign key `Reg_no` in the `Exam_info` table refers to the primary key `Reg_no` in the `Student_info` table.

Now, consider the mapping schema in Listing 3-14, `Stud-Exam-Schema.xml`, which uses `sql:relationship` to represent the relationship between the `Student_info` and `Exam_info` tables.

Listing 3-14: Stud-Exam-Schema.xml

```
<?xml version="1.0" ?>
<Schema xmlns="urn:schemas-microsoft-com:xml-data"
        xmlns:dt="urn:schemas-microsoft-com:xml:datatypes"
        xmlns:sql="urn:schemas-microsoft-com:xml-sql" >
   <ElementType name="Reg_no" content="textOnly"/>
   <ElementType name="Level" content="textOnly"/>
   <ElementType name="Name" content="textOnly"/>
   <ElementType name="Address" content="textOnly"/>
   <ElementType name="ROOT" sql:is-constant="1">
      <element type="Stud_info" />
   </ElementType>
   <ElementType name="Stud_info" sql:relation="Student_info" >
      <element type="Reg_no" />
      <element type="Level" />
      <element type="Name" />
      <element type="Address"/>
      <element type="Exam" >
          <sql:relationship
                  key-relation     ="Student_info"
                  key              ="Reg_no"
                  foreign-relation="Exam_info"
                  foreign-key      ="Reg_no" />
      </element>
   </ElementType>
    <ElementType name="Exam" sql:relation="Exam_info" >
      <AttributeType name="Roll_no" />
      <attribute type="Roll_no" />
    </ElementType>
</Schema>
```

A record is generated for the `Student_info` table when a `Stud_info` element comes into scope, and the values of the `Stud_info` attributes (`Reg_no`, `Level`, `Name`, and `Address`) are copied to the record. Likewise, when the `Exam` element comes into scope, a record for the `Exam_info` table is generated and the value of the `Roll_no` attribute is copied to the record.

The value of the `Reg_no` attribute of the `Exam` element is retrieved from the `Reg_no` attribute of the `Stud_info` element. The information specified in `sql:relationship` is used to obtain the `Reg_no` foreign key value. As the `Exam` element goes out of scope, XML Bulk Load sends the record to the database. Finally, the `Stud_info` element node goes out of scope. XML Bulk Load sends the student record to the database. Similarly, all the subsequent students in the XML document are processed.

Now, consider the XML input data in Listing 3-15, `Stud-Exam-XMLData.xml`.

Listing 3-15: Stud-Exam-XMLData.xml

```xml
<ROOT>
 <Stud_info>
   <Reg_no>76680</Reg_no>
   <Level>B</Level>
   <Name>Madan Kumar Dhyani</Name>
   <Address>666 Golf Links</Address>
   <Exam Roll_no="700800" />
 </Stud_info>
 <Stud_info>
   <Reg_no>43059</Reg_no>
   <Level>B</Level>
   <Name>Ravinder Bhushan</Name>
   <Address>777 Green Park</Address>
   <Exam Roll_no="600900" />
 </Stud_info>
 <Stud_info>
   <Reg_no>778899</Reg_no>
   <Level>A</Level>
   <Name>Ashish Baveja</Name>
   <Address>888 South Extn.</Address>
   <Exam Roll_no="200700" />
 </Stud_info>
</ROOT>
```

You execute Bulk Load by using the HTML file in Listing 3-16, `BulkLoad.html`.

Listing 3-16: BulkLoad.html

```html
<html>
 <head>
  <title>Insertion using Bulk Load</title>
  <script language="vbscript">
    sub insert()
    set BL_object = CreateObject("SQLXMLBulkLoad.SQLXMLBulkLoad")
    BL_object.ConnectionString =
"provider=SQLOLEDB.1;datasource=quest;database=Northwind;uid=sa;pwd="
         BL_object.ErrorLogFile = "c:\error.log"
         BL_object.CheckConstraints = True
         BL_object.Execute "c:\main\Stud-Exam-Schema.xml", "c:\main\Stud-
Exam-XMLData.xml"
         set BL_object=Nothing
       end sub
    </script>
  </head>
  <body bgcolor=rgb(100,100,100)>
    <center><h1><font color=yellow>Insertion using Bulk
Load</font></h1></center>
    <hr>
    <form>
       <center>
         <input type=button value="Insert" onclick="insert()">
```

```
        </center>
      </form>
    </body>
</html>
```

Figure 3-25 shows the output as rendered by Internet Explorer.

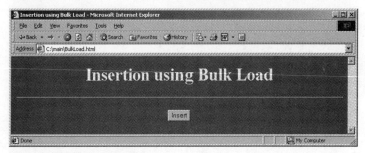

Figure 3-25: Insertion using Bulk Load

Updategrams

Updategrams are XML documents used for performing database manipulation. An Updategram supports multiple insert, update, and delete tags for database operation.

Use Updategrams to Insert Data

To demonstrate Updategrams, execute the URL that has the XML template `createstudent.xml` (Listing 3-17), which creates a table named student in testDB database:
`http://quest/virtual/template/createstudent.xml`.

Listing 3-17: createstudent.xml

```
<ROOT xmlns:sql="urn:schemas-microsoft-com:xml-sql">
  <sql:query>
    CREATE TABLE student
    (Roll_no int NOT NULL,
     Name varchar(25),
     Age int,
     primary key(Roll_no))
  </sql:query>
</ROOT>
```

Now execute the Updategram in Listing 3-18, `insertstudent.xml`, to insert data into the student table.

Listing 3-18: insertstudent.xml

```
<root xmlns:sql="urn:schemas-microsoft-com:xml-sql"
 xmlns:updg="urn:schemas-microsoft-com:xml-updategram" >
  <updg:sync>
    <updg:after>
      <student Roll_no='101' Name='Ravinder Bhushan' Age='29'/>
    </updg:after>
  </updg:sync>
</root>
```

The transaction starts with an opening updg:sync tag and ends with a closing updg:sync tag. The attributes of the student element (Roll_no, Name, and Age) between the opening and closing updg:after tags represent the data to be inserted into the student table.

The following URL executes this Updategram:
http://quest/virtual/template/insertstudent.xml.

Now execute the Updategram in Listing 3-19, updatestudent.xml, to update data in the student table.

Listing 3-19: updatestudent.xml

```
<root xmlns:sql="urn:schemas-microsoft-com:xml-sql"
 xmlns:updg="urn:schemas-microsoft-com:xml-updategram" >
  <updg:sync>
    <updg:before >
  <student Roll_no='101'/>
</updg:before>
<updg:after>
   <student Roll_no='101' Name='Jaya Shrivastava' Age='27'/>
</updg:after>
 </updg:sync>
</root>
```

The attribute Roll_No of the student element between the opening and closing updg:before tags represents the row in the student table that is to be updated. The attribute of the student element between the opening and closing updg:after tags represents the values with which contents of the fields are updated.

The following URL executes this Updategram:

http://quest/virtual/template/updatestudent.xml

Now execute the Updategram in Listing 3-20, deletestudent.xml, to delete data from the student table.

Listing 3-20: deletestudent.xml

```
<root xmlns:sql="urn:schemas-microsoft-com:xml-sql"
 xmlns:updg="urn:schemas-microsoft-com:xml-updategram" >
  <updg:sync>
    <updg:before >
 <student Roll_no='101'/>
    </updg:before>
   </updg:sync>
</root>
```

The Roll_No attribute of the student element between the opening and closing updg:before tags represents the data to be deleted from the student table.

The following URL executes this Updategram:

http://quest/virtual/template/deletestudent.xml

Parameters in Updategrams

Parameters can be passed to Updategrams. The Updategram in Listing 3-21, parameter.xml, uses parameters that update the data in the student table.

Listing 3-21: parameter.xml

```
ROOT xmlns:updg="urn:schemas-microsoft-com:xml-updategram">
<updg:header>
  <updg:param name='Roll_no'/>
  <updg:param name='Name' />
  <updg:param name='Age' />
</updg:header>
  <updg:sync >
    <updg:before>
        <student Roll_no='$Roll_no' />
    </updg:before>
    <updg:after>
        <student Name='$Name' Age='$Age'/>
    </updg:after>
  </updg:sync>
</ROOT>
```

The following URL executes this Updategram using parameters:

```
http://quest/virtual/template/parameter.xml?Roll_no=101&Name=Aparna&Age=10
```

Also note that, by assigning a value to the `nullvalue` attribute, NULL can be passed as a parameter value in Updategrams.

Post Data from an HTML Form

The file in Listing 3-22, update.html, passes Roll_No, Name, and Age values to an Updategram that updates a record in the student table in the testDB database. The POST method has been used to transmit student data to the server.

Listing 3-22: update.html

```
<head>
<TITLE>Sample Form </TITLE>
</head>
<body>
<center>
<font color=rgb(255,100,100) size=5>
For a given Rollno, Student Name and Age is updated.
</font>
</center>
<hr>
<p align=right>
<form action="http://quest/virtual" method="POST">
<B>Student Rollno</B>
<input type=text name=Roll_no>
<br>
<B>Name</B>
<input type=text name=Name>
<br>
<B>Age</B>
<input type=text name=Age>
</p>
<input type=hidden name=contenttype value=text/xml>
<input type=hidden name=template value='
<ROOT xmlns:updg="urn:schemas-microsoft-com:xml-updategram">
<updg:header>
```

```
  <updg:param name="Roll_no"/>
  <updg:param name="Name" />
  <updg:param name="Age" />
</updg:header>
  <updg:sync>
    <updg:before>
       <student Roll_no="$Roll_no" />
    </updg:before>
    <updg:after>
      <student Name="$Name" Age="$Age" />
    </updg:after>
  </updg:sync>
</ROOT>
'>
<p align=right><input type="submit"></p>
</form>
</body>
```

Figure 3-26 shows the input screen as rendered by Internet Explorer.

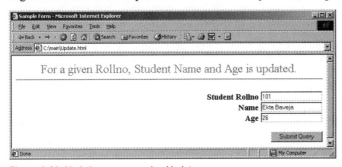

Figure 3-26: Updating process using Updategram

Use ADO to Post an Updategram

Any Visual Basic application can efficiently use ADO for setting up a SQL Server connection and for successful execution of Updategrams. `ADODB.Connection` creates a connection with SQL Server, and `ADODB.Command` executes the connection. `DBGUID_MSSQLXML` sets the command dialect. `ADODB.Stream` is used for commanding the I/O stream.

The Visual Basic code in Listing 3-23 updates the name and age of a particular student in the student table.

Listing 3-23: Sample VB Code

```
Private Sub Form_Load()
    dim connection1 As New ADODB.Connection
    dim command1 command1 As New ADODB.Command
    dim streamInput As New ADODB.Stream
    dim streamOutput As New ADODB.Stream

    connection1.Provider = "SQLOLEDB"
    connection1.Open "server=(local); database=testDB; uid=sa; "

    Set command1.ActiveConnection = connection1

    SQLxml1 = "<ROOT xmlns:updg='urn:schemas-microsoft-com:xml-updategram' >"
    SQLxml1 = SQLxml1 & " <updg:sync >"
```

```
SQLxml = SQLxml & "  <updg:before>"
SQLxml = SQLxml & "    <student  Roll_no='101' />"
SQLxml = SQLxml & "  </updg:before>"
SQLxml = SQLxml & "  <updg:after>"
SQLxml = SQLxml & "    <student  Name='M.K.Dhyani' Age='27'/>"
SQLxml = SQLxml & "  </updg:after>"
SQLxml = SQLxml & " </updg:sync>"
SQLxml = SQLxml & "</ROOT>"


command1.Dialect = "{5d531cb2-e6ed-11d2-b252-00c04f681b71}"


streamInput.Open
streamInput.WriteText SQLxml
streamInput.Position = 0

Set command1.CommandStream = streamInput


streamOutput.Open
streamOutput.LineSeparator = adCRLF
command1.Properties("Output Stream").Value = streamOutput
command1.Properties("Output Encoding").Value = "UTF-8"
command1.Execute , , adExecuteStream
streamOutput.Position = 0
Debug.Print streamOutput.ReadText
End Sub
```

Consider how you can use a mapping schema with an Updategram to update the database. The mapping schema in Listing 3-24, StudSchema.xml, declares the stud element with the Rollno, Name, and Age attributes for mapping with columns of the student table in the testDB database.

Listing 3-24: StudSchema.xml

```
<?xml version="1.0" ?>
  <Schema xmlns="urn:schemas-microsoft-com:xml-data"
       xmlns:dt="urn:schemas-microsoft-com:datatypes"
       xmlns:sql="urn:schemas-microsoft-com:xml-sql">
    <ElementType name="stud" sql:relation="student" >
      <AttributeType name="Rollno" />
      <AttributeType name="Name" />
      <AttributeType name="Age" />
      <attribute type="Rollno" sql:field="Roll_no" />
      <attribute type="Name" sql:field="Name" />
      <attribute type="Age" sql:field="Age" />
    </ElementType>
  </Schema>
```

The Visual Basic code in Listing 3-25 has the mapping schema StudSchema.xml for executing the Updategram that updates student records.

Listing 3-25: Sample VB Code

```
Private Sub Form_Load()
  dim connection1 As New ADODB.Connection
  dim command1 As New ADODB.Command
  dim streamInput As New ADODB.Stream
  dim streamOutput As New ADODB.Stream

  connection1.Provider = "SQLOLEDB"
```

```
connection1.Open "server=(local); database=testDB; uid=sa;"
Set command1.ActiveConnection = connection1

streamInput.Open
streamInput.WriteText "<ROOT xmlns:updg='urn:schemas-microsoft-com:xml-
updategram' >"
streamInput.WriteText "   <updg:header>"
streamInput.WriteText "         <updg:param name='Roll_no'/>"
streamInput.WriteText "         <updg:param name='Name' />"
streamInput.WriteText "         <updg:param name='Age' />"
streamInput.WriteText "   </updg:header>"
streamInput.WriteText "   <updg:sync >"
streamInput.WriteText "         <updg:before>"
streamInput.WriteText "            <student Roll_no='101' />"
streamInput.WriteText "         </updg:before>"
streamInput.WriteText "         <updg:after>"
streamInput.WriteText "            <student Name='Jaya Srivastava' Age='27' />"
streamInput.WriteText "         </updg:after>"
streamInput.WriteText "   </updg:sync>"
streamInput.WriteText "</ROOT>"

command1.Dialect = "{5d531cb2-e6ed-11d2-b252-00c04f681b71}"
command1.Properties("Mapping Schema") = "c:\main\StudSchema.xml"
streamInput.Position = 0
Set command1.CommandStream = streamInput

streamOutput.Open
streamOutput.LineSeparator = adCRLF
command1.Properties("Output Stream").Value = streamOutput
command1.Execute , , adExecuteStream
streamOutput.Position = 0
Debug.Print streamOutput.ReadText
End Sub
```

Summary

SQL Server 2000 provides integrated XML support to the RDBMS environment. SQL Server 2000's integrated XML functionality offers useful features, such as XML Updategrams, Bulk Load, an automation object model, and functions (such as XPath queries, T-SQL function, the FOR XML extensions to the SELECT statement, and two new system-stored procedures).

Chapter 4

Developing the Wireless Web Application Using XML and XSLT

In this chapter, we develop a portal specifically for PC browsers based on the XML and XSLT approach. The portal is divided into the following four sections:

♦ Weather

♦ News

♦ E-mail

♦ Movie-ticket buying

Write Interfaces Using XML and XSLT

This portal is targeted for PC browsers. Previous chapters discussed the details of the project design and its architecture as a whole application, as well as the database design for the complete project. This chapter takes a look at the first implementation of the project on a PC browser.

Write the Project Model

Before actually settling down to start the project, you should design a skeleton or a working model of the project. That way, when you start writing the code, you save valuable time that you can use for programming.

In the portal, you develop four basic services for the user that highlight the essence of the XML and XSLT approach being employed for content transformation. Here, you are using SQL Server 2000 as a database for the portal because it provides extensive support for XML. Figure 4-1 is a work flow diagram of the portal.

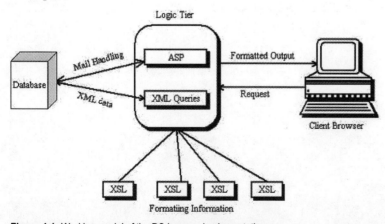

Figure 4-1: Working model of the PC browser implementation

Figure 4-1 illustrates the portals modularized into different parts. At the back end is a database server — SQL Server 2000 — that contains all the data for the portal. In the middle tier are two different technologies for handling the request and all other operations in the portal. These technologies are as follows:

- **Active Server Pages (ASP):** You use Active Server Pages for writing the logic part of the application. One of the strongest reasons for using ASP is to provide mail services. For mailing services, you use the I-Mail server at the back end and the Dimac J-Mail component.

- **XML queries:** To date, there are two standard ways of serving XML data in an application. The first method is to generate XML content on-the-fly by using ASP or any other server-side language; the second and the most commonly used method is to produce static XML files by using some custom application or by authoring it. Now, after the launch of SQL Server 2000, developers can write SQL queries to fetch contents from the database directly in XML format. Developers can now write their SQL queries in XML files and execute these over the HTTP protocol and get the data in XML format.

- **XSL documents:** After using queries and receiving the data in XML format, you will need to transform this data into a PC- or other browser-compliant format, such as HTML, WML, HDML, or something else. For this purpose, you will write different XSL documents for different platforms. By using this approach, you separate the content from its markup information.

Create XML Files for Static Data

In this section, you develop the home page section of the portal. To begin, create an XML file containing static data. Listing 4-1 is the XML code for the home page section.

Listing 4-1: main.xml

© 2001 Dreamtech Software India Inc.
All Rights Reserved

```xml
<?xml version="1.0" encoding="utf-8"?>
<?xml:stylesheet type="text/xsl" href="main.xsl"?>
<main>
   <title>Mobile Portal</title>
   <voicenumber>For Voice Accessibility</voicenumber>
  <linklist>
    <set>
      <link>E-mail</link>
      <url>http://192.168.1.100/login.xml</url>
    </set>
    <set>
      <link>Movies</link>
      <url>http://192.168.1.100/main/template/movies.xml</url>
    </set>
    <set>
      <link>Weather</link>
 <url>http://192.168.1.100/main/template/weather.xml</url>
    </set>
    <set>
      <link>Resources</link>
      <url>http://192.168.1.100/resources.xml</url>
    </set>
    <set>
      <link>News</link>
      <url>http://192.168.1.100/main/template/News.xml</url>
    </set>
```

```
    </linklist>
    <article>
  <headline>XML in the Enterprise</headline>
  <information> Welcome to the xml and  xsl based application, This application
provides you four main facilities to check your mails, browse the current
weather of main cities. It also provides a news section and one movie ticket
booking system. You can book tickets and check the current listing of the movies
in theatres using this application.</information>
     <authorname>Charul Shukla</authorname>
     </article>
  <developerlist>
     <record>
       <name>Charul Shukla</name>
       <mail>charul_shukla@dreamtechsoftware.com</mail>
     </record>
  <record>
       <name>Jyotsna Gupta</name>
       <mail>jyotsna@dreamtechsoftware.com</mail>
     </record>
  </developerlist>
</main>
```

When you load this XML file into Internet Explorer without including the XSL document, you get a result like that shown in Figure 4-2.

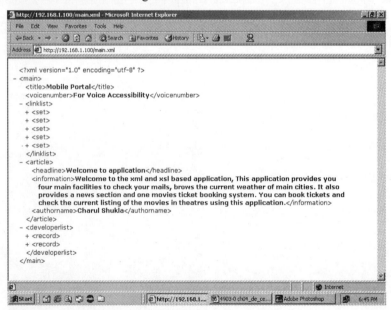

Figure 4-2: Loading main.xml in Internet Explorer 5.5

Before proceeding further into the subtle nuances, some essential points invite a discussion. When you load `main.xml` in the browser, it applies a default style sheet and also makes a tree structure of all the parent nodes and subnodes. If the MSXML parser finds some kind of problem in the document, it displays an error message and also tells you the line number where the error occurred so that you can rectify it.

For example: In the aforementioned code, if the title tag has not been closed by adding the `/` in the title tag on the right side, the MSXML parser is unable to parse the file and displays an error message (see Figure 4-3).

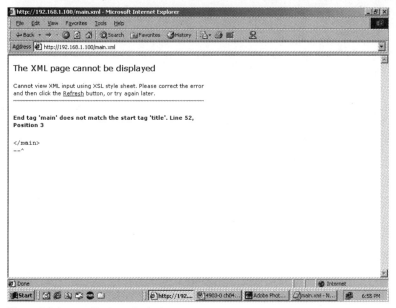

Figure 4-3: Internet Explorer displaying an error message

Consider the architecture of the document and the process by which you store information in it. You first start the document with the XML declaration and then create the first and root element to the document and name it `<main>`. After this, in line 2, you create a tag named `<title>` and store the information for the title of the Web page. Further, you add another element, named `<voicenumber>`, and store some additional information in it.

Now it's time to design a structure for the next part of information, that is, the links information on the home page. In this part of the code, you store the names and URLs for the various links on the portal. The structure of this section looks something like the following:

```
<linklist>
    <set>
       <link>E-mail</link>
       <url>http://192.168.1.100/login.xml</url>
    </set>
</linklist>
```

Here, you declare a root element called `<linklist>` and, under that, you write the different subsets of the `<set>` element. In the `<set>` element, you create a tag named `<link>`. This tag enables you to store the name of that particular link; and in the next line, you create another element, called `<url>`, and store the URL for that particular link in it. In the preceding example, you store the links information for e-mail; then in the next line, you store the URL for this link, as follows: `http://192.168.1.100/login.xml`. If you are going to deploy this code on your machine, please change the server address.

By following the same structure, you can store information about all other links and their URLs in the respective elements.

After finishing the link section, you can write the main information part for the portal home page, which contains one article and the name of the author of this article. The structure of this section is as follows:

```
<article>
   <headline>Headline for the Article </headline>
   <information>our Article </information>
```

```
<authorname>name of the author </authorname>
</article>
```

Lastly, you finish by employing a set of tags in the main.xml document. You use this section for storing the names of the developers of this portal. For this section, use the following design:

```
<developerlist>
    <record>
        <name>name of the developer</name>
        <mail>mailing address of developer</mail>
    </record>
</developerlist>
```

In this code, you first declare a tag named <developerlist>. Under that tag, store the information under a <record> tag so that you can repeat this <record> tag if the number of developers is more than one. The names of the developers are placed between the <name> and </name> tags and the mailing address of the developer(s) between the <mail> and </mail> tags.

In the last part of the document, close the root element by typing </main> in the XML document. By thus closing the <main> tag, you finish creating the home page section of the XML data for the portal.

Check How Well-Formed the XML Document Is

You can determine how well-formed the XML document is by clicking the right button of your mouse and then selecting the Validate XML options from the resulting menu. A window appears that indicates that your document is valid and well-formed, which means you can work with it (see Figure 4-4).

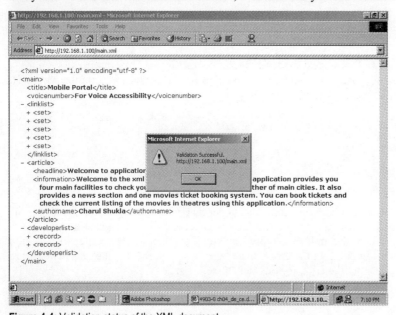

Figure 4-4: Validation status of the XML document

Generate HTML from XML Content Using XSLT

In this section, you prepare an XSLT and CSS document for the home-page section of the portal. You need to transform the main.xml document into HTML on-the-fly by using the XSLT transformation logic. Preparing the XSLT document for the home page is definitely a challenging task. For convenience, this task is divided into the following two main components:

♦ Writing the Cascading Style Sheet (CSS) for the home page section

♦ Writing the XSLT document for the home page section

Write the CSS Document for the Home Page

First prepare the CSS document in Listing 4-2, named `style.css`. Here, you include all the styles for all your future needs.

Listing 4-2: style.css

```css
.MIDDLE {
    PADDING-RIGHT: 6px; PADDING-LEFT: 6px; PADDING-BOTTOM: 6px; PADDING-TOP: 6px
}
body {
    BACKGROUND: #999999; MARGIN: 0px; COLOR: #000000;
}
.mainheading {
    FONT-WEIGHT: bold; FONT-STYLE: italic
}
.heading2 {
    PADDING-LEFT: 30px; FONT-WEIGHT: normal; FONT-SIZE: 24pt; BACKGROUND:
#000000; MARGIN-BOTTOM: 0px; PADDING-BOTTOM: 0px; COLOR: #ffffff; PADDING-TOP:
0px; FONT-FAMILY: verdana, arial, sans-serif
}
.heading2 A {
    COLOR: #ffffff; TEXT-DECORATION: none
}
.heading2 A:visited {
    COLOR: #ffffff; TEXT-DECORATION: none
}
.heading2 A:hover {
    COLOR: #ffffff
}
td {
    FONT-SIZE: 10pt; FONT-FAMILY: arial, sans-serif
}

td.CONTENT {
    BACKGROUND: #999999
}
td.NAVIGATION {
    PADDING-RIGHT: 10px; PADDING-LEFT: 0px; BACKGROUND: #999999; PADDING-BOTTOM:
10px; PADDING-TOP: 0px
}
td.NAVIGATION TABLE {
    BACKGROUND: #000000; MARGIN: 0px 0px 10px
}
td.NAVHEADING {
    FONT-WEIGHT: bold; FONT-SIZE: 12pt; BACKGROUND: #000000; COLOR: #ffffff;
FONT-STYLE: normal; FONT-FAMILY: arial, sans-serif
}
td.NAVIGATION TABLE TR td.NAVHEADING {
    FONT-WEIGHT: bold; FONT-SIZE: 12pt; BACKGROUND: #000000; COLOR: #ffffff;
FONT-STYLE: normal; FONT-FAMILY: arial, sans-serif
}
td.NAVIGATION TABLE TR td {
```

```
       PADDING-RIGHT: 2px; PADDING-LEFT: 2px; FONT-SIZE: 9pt; BACKGROUND: #a9a9a9;
PADDING-BOTTOM: 2px; PADDING-TOP: 2px; FONT-FAMILY: arial, sans-serif
}
TABLE.CONTENT {
       BACKGROUND: #a9a9a9; xborder-width: 1px; xborder-style: solid; xborder-
color: #000000
}
.hand {  cursor: hand}

.button
{
       BACKGROUND-COLOR: #ffffff;
       BORDER-BOTTOM: #000000 2px solid;
       BORDER-LEFT: #000000 2px solid;
       BORDER-RIGHT: #000000 2px solid;
       BORDER-TOP: #000000 2px solid;
       FONT: 9pt Arial, Helvetica, sans-serif;
       HEIGHT: 20px;
       color:#000000;
       PADDING-BOTTOM: 3px;
       PADDING-LEFT: 3px;
       PADDING-RIGHT: 3px;
       PADDING-TOP: 3px;
       WIDTH: 143px
}

a {  color: #000000; text-decoration: none}

a:hover {  color: #FFFFFF; text-decoration: none}
```

You are now ready to write your XSLT document.

Write the XSLT Document for the Home Page

At first glance, the code for an XSLT document seems long. Remember, however, that it is an essential exercise because you want your pages to look good and stylish.

In the XSLT document in Listing 4-3, main.xsl, you use HTML elements extensively to increase the attractiveness of the home page.

Listing 4-3: main.xsl

```
<?xml version='1.0'?>
<xsl:stylesheet xmlns:xsl="http://www.w3.org/TR/WD-xsl">

<xsl:template match="/">

<html>
  <head>

 <link rel="stylesheet" href="style.css" />

 <script language="javascript">
  function Submitpage()
```

```
    {
      if (Validate())
      {
        //document.frm.txtAction.value = "SendMail";
        var name,pass;
        name=document.frm.uname.value;
        pass=document.frm.pass.value;
        document.location="http://192.168.1.100/interface.asp?uname="  + name +
"&amp;pass=" + pass;
      }
    }
    function Validate()
    {
      if(document.frm.uname.value.length == 0)
      {
       document.frm.uname.focus();
       document.frm.uname.select();
       alert("You must enter your login name.");
       return false;
      }

      if(document.frm.pass.value.length == 0)
      {
       document.frm.pass.focus();
       document.frm.pass.select();
       alert("You must enter your password.");
       return false;
      }

      return true;
    }



    function init()
    {
      document.frm.uname.focus();
    }

    function openwindow()
    {
      window.open('show.xml','abc','width=300,height=300');
    }
</script>


      <title><xsl:apply-templates select="main/title" /> </title>
  </head>

<body >
    <table cellspacing="0" cellpadding="0" width="100%" border="0">
      <tr>
        <td bgcolor="#ffffff">
          <table style="MARGIN: 0px 0px 1px" cellspacing="0"
          cellpadding="0" width="100%" bgcolor="#000000"
          border="0">
```

```
        <tr>
          <td align="left" height="15" >
          <b style="cursor: hand" onclick="openwindow()" ><xsl:apply-
templates select="main/voicenumber" /></b>
          </td>

          <td align="middle">

          </td>
        </tr>
      </table>

      <table cellspacing="0" cellpadding="0" width="100%"
      border="0">
        <tr>
          <td>
            <div class="heading2">
              <xsl:apply-templates select="main/title" />
            </div>
          </td>
        </tr>
      </table>
    </td>
  </tr>

  <tr>
    <td class="MIDDLE" valign="top">

      <table cellspacing="0" cellpadding="0" border="0">
        <tr>
          <td class="NAVIGATION" valign="top" width="180">
            <table cellspacing="1" cellpadding="0" width="100%"
            border="0">
              <tr>
                <td class="NAVHEADING">Channels</td>
              </tr>

              <tr>
                <td>
                <xsl:apply-templates select="main/linklist" />

                </td>
              </tr>
            </table>

            <table style="MARGIN: 0px 0px 10px"
            cellspacing="1">
              <tr>
                <td class="NAVHEADING">Login</td>
              </tr>

              <tr>
                <td>

                  <form name="frm">
                    <p>Username <input name="uname" class="BUTTON" />
```

```
                          Password<br /><input name="pass" class="BUTTON"
type="password" /><br />
                    <br />
                       <input style="FONT-WEIGHT: 550; BORDER-LEFT-COLOR:
#999999; BORDER-BOTTOM-COLOR: #999999; COLOR: #000000; BORDER-TOP-COLOR:
#999999; BACKGROUND-COLOR: #cccccc; BORDER-RIGHT-COLOR: #999999" width="25px"
type="button" name="submit" value="login" class="hand" onclick="Submitpage()"
/></p>
                 </form>


             </td>
           </tr>
         </table>

         <table cellspacing="1">
           <tr>
             <td class="NAVHEADING" valign="center"
             width="180">Connect</td>
           </tr>

           <tr>
             <td class="NAVIGATION" valign="bottom"
             width="180">
    <xsl:apply-templates select="main/developerlist" />
                 </td>
             </tr>
           </table>

         </td>

         <td valign="top" width="100%">
           <div class="INTRO">
             <table
             style="PADDING-RIGHT: 10px; PADDING-LEFT: 10px; PADDING-
BOTTOM: 2px; PADDING-TOP: 2px"
                 cellspacing="0" cellpadding="0" border="0">
               <tr>
                 <td valign="top" bgcolor="#ffffff"
                 width="50%">
                    <h2><b><xsl:apply-templates
select="main/article/headline" /></b></h2>

                    <p align="justify"><xsl:apply-templates
select="main/article/information" /></p>

      <p><xsl:apply-templates select="main/article/authorname" /></p>
                 </td>
               </tr>
             </table>
           </div>
         </td>
       </tr>
     </table>
   </td>
 </tr>
```

```
      </table>
    </body>
</html>




</xsl:template>



<xsl:template match="main/linklist">



<table cellspacing="0" cellpadding="0" width="100%" border="0">
  <xsl:for-each select="set">
<tr>
<td BORDER="0">
<a>
    <xsl:attribute name="HREF">
     <xsl:value-of select="url" />
    </xsl:attribute>
    <xsl:value-of select="link" />
   </a>
</td>
</tr>
  </xsl:for-each>
</table>
</xsl:template>

<xsl:template match="main/developerlist">
<table cellspacing="0" cellpadding="0" width="100%" border="0">
  <xsl:for-each select="record">
<tr>
<td BORDER="0">
<a>
    <xsl:attribute name="HREF">
     mailto:<xsl:value-of select="mail" />
    </xsl:attribute>
    <xsl:value-of select="name" />
   </a>
</td>
</tr>
  </xsl:for-each>
</table>
</xsl:template>

<xsl:template match="main/article/headline">
 <xsl:value-of select="." />
</xsl:template>



<xsl:template match="main/article/information">
 <xsl:value-of select="." />
</xsl:template>

<xsl:template match="main/article/authorname">
```

```
  <b><i><xsl:value-of select="." /></i></b>
</xsl:template>

<xsl:template match="main/title">
  <xsl:value-of select="." />
</xsl:template>

<xsl:template match="main/voicenumber">
  <font color="white"><xsl:value-of select="." /></font>
</xsl:template>


<xsl:template match="main/outside">
  <font color="white"><xsl:value-of select="." /></font>
</xsl:template>
</xsl:stylesheet>
```

Figure 4-5 shows what appears in Internet Explorer when you load `main.xsl` by using the following line of code:

```
<?xml:stylesheet type="text/xsl" href="main.xsl"?>
```

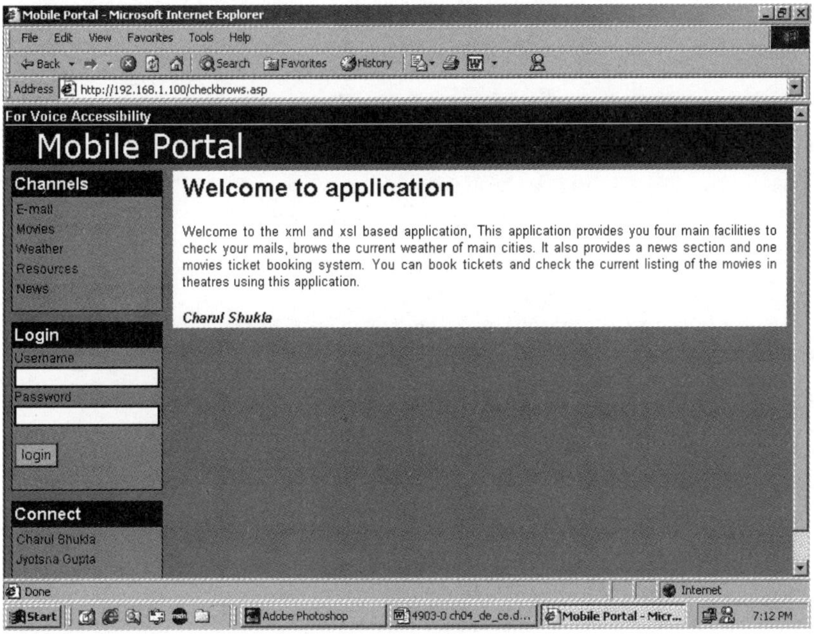

Figure 4-5: The output of the XML document after applying the XSL document

Take a closer look at the XSLT document so that you can understand how they work. In Figure 4-5 on uppermost part of the screen, you first construct a table, in which you display the text picked up from the `main.xml` document, using the `<xsl: value-of select >` statement in the end of the XSL document.

Now you construct a table for displaying the top bar of the portal from `main.xml`.

```
        <table style="MARGIN: 0px 0px 1px" cellspacing="0"
          cellpadding="0" width="100%" bgcolor="#000000"
          border="0">
            <tr>
              <td align="left" height="15" >
              <b style="cursor: hand" onclick="openwindow()" ><xsl:apply-
templates select="main/voicenumber" /></b>
              </td>


              <td align="middle">

              </td>
            </tr>
          </table>

<table cellspacing="0" cellpadding="0" width="100%"
          border="0">
            <tr>
              <td>
                <div class="heading2">
                  <xsl:apply-templates select="main/title" />
                </div>
              </td>
            </tr>
          </table>
```

This code constructs a table and displays the data in the < xsl:apply –templates select /> element.

Now consider the following code:

```
<xsl:template match="main/voicenumber">
 <font color="white"><xsl:value-of select="." /></font>
</xsl:template>

<xsl:template match="main/title">
 <xsl:value-of select="." />
</xsl:template>
```

Here, you first match the template by using the <xsl:template> element and then fetch the value from the XML document by using the <xsl: value-of select> element under the <xsl:template > element. By giving the ". " in the <xsl:value-of select> element, all the data resides under this tag in the XML document.

As you can see, here you are using only two <xsl:template > sections for picking up two different things from the XML document. The first fetches the value from the "main/voicenumber" section, while the second fetches the value from the "main/title" section.

When you load your XML document into the browser and apply the XSLT documents, this code generates the output shown in Figure 4-6.

For Voice Accessibility

Mobile Portal

Figure 4-6: Generating the top bar by using the XSLT document

After constructing the top bar for the home page, it's time to write the code for creating the left-hand side links and the channel bar for the home page. On the left side of the screen are the following three main tables:

- Channels
- Login
- Connect

Channels Section

Listing 4-4 contains the code for the Channels table.

Listing 4-4: Code for the Channels table

© 2001 Dreamtech Software India Inc.
All Rights Reserved

```
<table cellspacing="0" cellpadding="0" border="0">
        <tr>
          <td class="NAVIGATION" valign="top" width="180">
            <table cellspacing="1" cellpadding="0" width="100%"
            border="0">
              <tr>
                <td class="NAVHEADING">Channels</td>
              </tr>

              <tr>
                <td>
                <xsl:apply-templates select="main/linklist" />

                </td>
              </tr>
            </table>
```

In Listing 4-4, you construct a table and apply a template to it. Listing 4-5 contains the code for the template.

Listing 4-5: Code for template

© 2001 Dreamtech Software India Inc.
All Rights Reserved

```
<xsl:template match="main/linklist">
<table cellspacing="0" cellpadding="0" width="100%" border="0">
  <xsl:for-each select="set">
<tr>
<td BORDER="0">
<a>
    <xsl:attribute name="HREF">
     <xsl:value-of select="url" />
    </xsl:attribute>
    <xsl:value-of select="link" />
   </a>
</td>
</tr>
  </xsl:for-each>
</table>
</xsl:template>
```

In the above code, after matching a template to the path, you use `<xsl:for-each select>` to construct a logical loop condition for fetching all the `<set> </set>` recordsets from your XML document. Under the loop, you use the `<xsl:attribute>` statement to generate links for all the channels. Figure 4-7 depicts this structure.

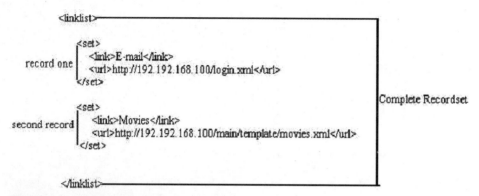

Figure 4-7: The structure of the records for the channels

Figure 4-8 shows the output of this XSLT document.

Figure 4-8: The output of the XSL code

Login Section

The XSLT code for generating the login section is present, as shown in Listing 4-6.

Listing 4-6: XSLT code for the login section

```
<table style="MARGIN: 0px 0px 10px"
            cellspacing="1">
        <tr>
          <td class="NAVHEADING">Login</td>
        </tr>

        <tr>
          <td>

            <form name="frm">
              <p>Username <input name="uname" class="BUTTON" />
              Password<br /><input name="pass" class="BUTTON"
type="password" /><br />
```

```
                        <br />
                        <input style="FONT-WEIGHT: 550; BORDER-LEFT-COLOR:
#999999; BORDER-BOTTOM-COLOR: #999999; COLOR: #000000; BORDER-TOP-COLOR:
#999999; BACKGROUND-COLOR: #cccccc; BORDER-RIGHT-COLOR: #999999" width="25px"
type="button" name="submit" value="login" class="hand" onclick="Submitpage()"
/></p>
                    </form>


                </td>
            </tr>
        </table>
```

In the preceding code, you create a table in which you design a form for accepting the user input. This section logs on to the mail services of the portal. A user can directly log on to the mail service by providing the USER-ID and PASSWORD.

Figure 4-9 shows the output of the preceding code.

Figure 4-9: Displaying the login form

Connect Section

In the connect section, you display the names of the developers so that if users have some queries or need some help, they can send an e-mail message and seek help from the developers. The user sends the message simply by clicking on the name of the developer.

The code for generating the table is listed from lines 136 to 148, as shown in Listing 4-7.

Listing 4-7: XSLT code for the connect section

```
<table cellspacing="1">
            <tr>
                <td class="NAVHEADING" valign="center"
                width="180">Connect</td>
            </tr>


            <tr>
                <td class="NAVIGATION" valign="bottom"
                width="180">
        <xsl:apply-templates select="main/developerlist" />
                </td>
            </tr>
```

```
                                      </table>
```

The code for fetching records from the XML file is shown in Listing 4-8.

Listing 4-8: XSLT code for fetching records from the XML file

```
<xsl:template match="main/developerlist">;
<table cellspacing="0" cellpadding="0" width="100%" border="0">
  <xsl:for-each select="record">
<tr>
<td BORDER="0">
<a>
   <xsl:attribute name="HREF">
     mailto:<xsl:value-of select="mail" />
   </xsl:attribute>
   <xsl:value-of select="name" />
  </a>
</td>
</tr>
  </xsl:for-each>
</table>
</xsl:template>
```

Figure 4-10 shows the output of the code.

Figure 4-10: Displaying the developer list

Article Section

In this section, you fetch an article from your XML document and display it on the right-hand side of the Mobile Portal home page (see Listing 4-9). Listing 4-10 shows the template code used to fetch the article.

Listing 4-9: XSLT code for transforming the fetched article

```
                <table
                style="PADDING-RIGHT: 10px; PADDING-LEFT: 10px; PADDING-
BOTTOM: 2px; PADDING-TOP: 2px"
                   cellspacing="0" cellpadding="0" border="0">
                 <tr>
                   <td valign="top" bgcolor="#ffffff"
                   width="50%">
                     <h2><b><xsl:apply-templates
select="main/article/headline" /></b></h2>

                     <p align="justify"><xsl:apply-templates
select="main/article/information" /></p>
```

```
        <p><xsl:apply-templates select="main/article/authorname" /></p>
                </td>
            </tr>
        </table>
```

Listing 4-10: Template code for fetching the article

```
<xsl:template match="main/article/headline">
 <xsl:value-of select="." />
</xsl:template>


<xsl:template match="main/article/information">
 <xsl:value-of select="." />
</xsl:template>

<xsl:template match="main/article/authorname">
 <b><i><xsl:value-of select="." /></i></b>
</xsl:template>
```

Here, you pick up the values for the article section by employing the same process used for the other sections. Now, finish reading the discussion of the coding for the home page and its content.

Apply XSLT to XML Documents

Applying the XSLT document to the XML file is an easy task. You can do this by adding the following line of code to your XML file:

```
<?xml:stylesheet type="text/xsl" href="main.xsl"?>
```

Add this line after the XML header declaration line and then save the file.

Display the XML Document in the Browser Using XSLT

Loading the content using XSL is similar to loading other files in browsers. To do this, you just have to type the location of your file on the address bar in the browser. When you type your home page location in the browser, it shows formatted output after rendering it by using the XSLT document, as shown in Figure 4-11.

Figure 4-11: Formatted output using XSLT

Develop the Weather Section

This section discusses the working of the weather section. The section also includes a discussion of how to use XSLT in it. For the first time, you take a look at the database interaction, as well as at the process for XSLT formatting while working with SQL Server 2000.

When you run the query in Listing 4-11, it returns the output in the form of XML. It contains the details on all cities.

Listing 4-11: XML query

```
<?xml version ='1.0' encoding='UTF-8'?>
<?xml:stylesheet type="text/xsl" href="weather.xsl"?>


<ROOT xmlns:sql="urn:schemas-microsoft-com:xml-sql">


  <sql:query>
   SELECT  * from weather


FOR XML AUTO,ELEMENTS


  </sql:query>


</ROOT>
```

After formatting the XML data, you want to display it in browser. In order to do this, you need to transform it by using the XSLT document. But before doing this, take a look on the pure XML data we are getting as a result of the query in Figure 4-12.

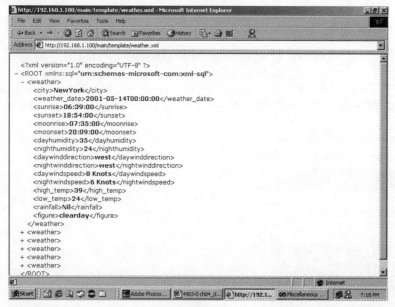

Figure 4-12: XML output of the query in weather.xml

Work with Advanced HTML Elements Using XSLT in the Weather Section

For the weather information, you make extensive use of tables for displaying data in a tabular form. You also use images with these.

In Listing 4-12, you construct the right-side area of the page for displaying the weather information in tabular form.

Listing 4-12: Code snippets from Weather.xsl

```
      <table>
                <tr>
                 <table width="100%" border="1" cellspacing="0"
cellpadding="0"  bordercolor="#000000">
                <tr>
        <td width="100%"  BGCOLOR="#000000" >

        <a href="http://192.168.1.100/main.xml" style="text-decoration : none
">Home</a> |
        <font color="white" >(For details click on Cityname)</font>
        </td>

      </tr>

      <tr>
        <td valign="top"><xsl:apply-templates select="ROOT" />
        <p align="justify"></p>
      </td>
      </tr>
    </table>
                </tr>
                </table>
```

Here you have only applied one template, named "ROOT". Using this template, you pick up all the data and then transform it.

Now take a look at the code of the template called ROOT, shown in Listing 4-13. This template fetches the records from weather.xml and then transforms all the data into HTML format.

Listing 4-13: ROOT template

```
<xsl:template match="ROOT">


 <table cellspacing="1" cellpadding="1" width="97%"  align="center" border="0"
bgcolor="#000000" style="MARGIN: 10px 2px 0px 2px">



                <tr class="dark">
                  <td valign="center" width="20%">
                  <b>City</b></td>
```

```
                <td valign="center" width="12%">Max</td>
                <td valign="center" width="12%">Min</td>
                <td valign="center" align="middle" width="15%">Forecast</td>
                <td valign="center" align="middle" width="19%">Humidity
(day)</td>
                <td valign="center" halign="middle" width="19%"> Rainfall
(mm)</td>
            </tr>

      <xsl:for-each select="weather">
            <tr class="medium">
      <td valign="center" align="middle" width="20%">
      <b>
            <a class="blacklinks">
      <xsl:attribute name="href" >
http://192.168.1.100/main/template/details.xml?city=<xsl:value-of  select="state"
/>
      </xsl:attribute>
      <xsl:value-of select=" city" />
      </a>

      </b>
    </td>

      <td class="text" valign="center" align="middle" width="12%"><xsl:value-of
select="high_temp" /> <sup>o</sup>C</td>
                <td valign="center" align="middle" width="12%"><xsl:value-of
select="low_temp" />
      <sup>o</sup>C
    </td>
                <td style="BORDER-RIGHT: black 1px groove; BORDER-TOP: black
1px groove; BORDER-LEFT: black 1px groove; BORDER-BOTTOM: black 1px groove"
valign="center" align="middle" bgcolor="#ffffff" width="15%">


          <xsl:if match=".[figure='clearday']">

              <img>
                  <xsl:attribute name="src">
                      http://192.168.1.100/images/clearday.gif
                  </xsl:attribute>
              </img>
          </xsl:if>
          <xsl:if match=".[figure='showerday']">
              <img>
                  <xsl:attribute name="src">
                      http://192.168.1.100/images/showerday.gif
                  </xsl:attribute>
              </img>
          </xsl:if>
          <xsl:if match=".[figure='uppercloud']">
              <img>
                  <xsl:attribute name="src">
                      http://192.168.1.100/images/uppercloud.gif
                  </xsl:attribute>
              </img>
```

```
                </xsl:if>
                <xsl:if match=".[figure='rainyday']">
                        <img>
                                <xsl:attribute name="src">
                                        http://192.168.1.100/images/rainyday.gif
                                </xsl:attribute>
                        </img>
                </xsl:if>
                <xsl:if match=".[figure='semicloud']">
                        <img>
                                <xsl:attribute name="src">
                                        http://192.168.1.100/images/semicloud.gif
                                </xsl:attribute>
                        </img>
                </xsl:if>


                </td>

                <td valign="center" align="middle" width="19%"><xsl:value-of
select="dayhumidity" />%</td>

                <td align="middle" width="19%"><xsl:value-of select="rainfall"
/></td>

            </tr>
                </xsl:for-each>
            </table>
</xsl:template>
```

The preceding code uses an `<xsl:if>` statement to choose the name of the figure on the basis of the value of the figure tag in the `weather.xml` document. Under the `<xsl:if statement>`, you use `<xsl:attribute>` to fill in the name of the image in the `src` attribute in the image tag. In the remaining part of the template code, you pick up different values for the different kinds of data that you need to display in the browser.

Figure 4-13 shows the output of the weather section.

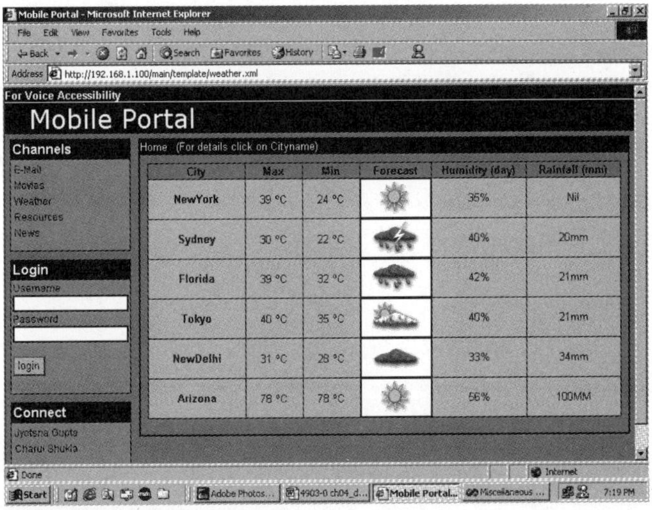

Figure 4-13: The weather section

Dynamic Links in XML

You can dynamically generate links to make XSL documents more logical. In the weather section of the portal, you create links dynamically for getting the detailed weather information of a particular city. In this section, then, each city is a link. After clicking on the city name, the user can see the details of the weather report for that city. Listing 4-14 shows the XSLT code for generating dynamic links.

Listing 4-14: XSLT code for generating a dynamic link

```
<a class="blacklinks">
<xsl:attribute name="href" >
http://192.168.1.100/main/template/details.xml?city=<xsl:value-of  select="city"
/>
</xsl:attribute>
<xsl:value-of select="city" />
</a>
```

In the preceding code, `<xsl:value-of select="city" />` is selecting the name of the state, and the path of the `detail.xml` file (Listing 4-15) is set in the `href` attribute of the `<a>` tag. The value of that state passes as a parameter in `detail.xml`. This `detail.xml` will generate the code for detailed information of that state, passed on as a parameter to this file.

Listing 4-15: detail.xml

© 2001 Dreamtech Software India Inc.
All Rights Reserved

```
<?xml version ='1.0' encoding='UTF-8'?>
<?xml:stylesheet type="text/xsl" href="details.xsl"?>
<ROOT xmlns:sql="urn:schemas-microsoft-com:xml-sql">

  <sql:header>
 <sql:param name='city'>NewYork</sql:param>
  </sql:header>

  <sql:query>
   SELECT  * from weather where city = @city
   FOR XML AUTO,ELEMENTS

  </sql:query>
</ROOT>
```

The next example shows us how to include a template for displaying the detailed weather information on the right-hand side of the home page. The detailed weather report displays only after a user selects the name of the state from the available list of options in the previous file.

In details.xsl, we include our template name "ROOT" as follows.

```
<xsl:apply-templates select="ROOT" />
```

Listing 4-16 shows the code for the ROOT template.

Listing 4-16: ROOT template in detail.xsl

© 2001 Dreamtech Software India Inc.
All Rights Reserved

```
<xsl:template match="ROOT">
 <table cellspacing="1" cellpadding="1" width="97%"  align="center" border="0"
bgcolor="#000000" style="MARGIN: 10px 2px 0px 2px">
```

```
        <xsl:for-each select="weather">
   <table cellspacing="1" cellpadding="1" width="97%"  align="center" border="0"
bgcolor="#000000" style="MARGIN: 10px 2px 0px 2px">
       <tr class="medium">
       <td valign="center" align="middle" width="100%"><b><xsl:value-of
select="city" /></b></td>
          </tr>
   </table>
    <table cellspacing="1" cellpadding="1" width="97%"  align="center"
border="0" bgcolor="#000000" style="MARGIN: 10px 2px 10px 2px">

              <tr class="medium">
       <td valign="center" align="middle" width="40%"><b>Maximum
Temprature</b></td>
       <td class="text" valign="center" align="middle" width="60%"><xsl:value-of
select="high_temp" /> <sup>o</sup>C</td>

           </tr>
          <tr class="medium">
       <td valign="center" align="middle" width="40%"><b>Minimum
Temprature</b></td>
       <td class="text" valign="center" align="middle" width="60%"><xsl:value-of
select="low_temp" /> <sup>o</sup>C</td>

           </tr>

          <tr class="medium">
       <td valign="center" align="middle" width="40%"><b>Sunrise</b></td>
       <td class="text" valign="center" align="middle" width="60%"><xsl:value-of
select="sunrise" /></td>

           </tr>
          <tr class="medium">
       <td valign="center" align="middle" width="40%"><b>Sunset</b></td>
       <td class="text" valign="center" align="middle" width="60%"><xsl:value-of
select="sunset" /></td>

           </tr>

          <tr class="medium">
       <td valign="center" align="middle" width="40%"><b>Moonrise</b></td>
       <td class="text" valign="center" align="middle" width="60%"><xsl:value-of
select="moonrise" /></td>

           </tr>
          <tr class="medium">
       <td valign="center" align="middle" width="40%"><b>Moonset</b></td>
       <td class="text" valign="center" align="middle" width="60%"><xsl:value-of
select="moonset" /></td>

           </tr>
          <tr class="medium">
       <td valign="center" align="middle" width="40%"><b>Day Time
Humidity</b></td>
       <td class="text" valign="center" align="middle" width="60%"><xsl:value-of
select="dayhumidity" /></td>
```

```
                </tr>
            <tr class="medium">
        <td valign="center" align="middle" width="40%"><b>Night Time
Humidity</b></td>
        <td class="text" valign="center" align="middle" width="60%"><xsl:value-of
select="nighthumidity" /></td>

                </tr>
            <tr class="medium">
        <td valign="center" align="middle" width="40%"><b>Daytime Wind
Direction</b></td>
        <td class="text" valign="center" align="middle" width="60%"><xsl:value-of
select="daywinddirection" /></td>

                </tr>
            <tr class="medium">
        <td valign="center" align="middle" width="40%"><b>Night Time Wind
Direction</b></td>
        <td class="text" valign="center" align="middle" width="60%"><xsl:value-of
select="nightwinddirection" /></td>

                </tr>
            <tr class="medium">
        <td valign="center" align="middle" width="40%"><b>Daytime Wind
Speed</b></td>
        <td class="text" valign="center" align="middle" width="60%"><xsl:value-of
select="daywindspeed" /></td>

                </tr>                 <tr class="medium">
        <td valign="center" align="middle" width="40%"><b>Night Time Wind
Speed</b></td>
        <td class="text" valign="center" align="middle" width="60%"><xsl:value-of
select="nightwindspeed" /></td>

                </tr>
                </table>
 </xsl:for-each>
 </table>
</xsl:template>
```

In the preceding code, we have used the `<xsl:for-each>` element for a table and then picked up data from `details.xml` using the `<xsl: value-of select >` statement from different tags. When we apply this XSLT document to our `details.xml`, it gives us the output shown in Figure 4-14.

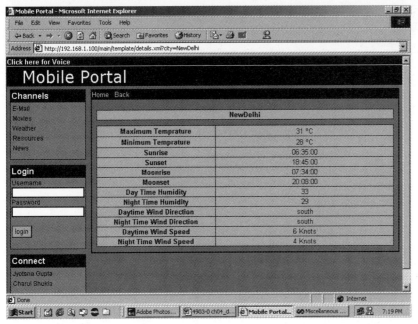

Figure 4-14: Screen-shot of a de7tailed weather report

Develop the News Section of the Portal

This section contains a discussion of the working of the Mobile Portal's news section. Begin by writing a query for fetching the records from the database and then transform them into HTML format using XSLT.

Listing 4-17 is an XML query for news.

Listing 4-17: news.xml

© 2001 Dreamtech Software India Inc.
All Rights Reserved

```
<?xml version ='1.0' encoding='UTF-8'?>
<?xml:stylesheet type="text/xsl" href="news.xsl"?>
<ROOT xmlns:sql="urn:schemas-microsoft-com:xml-sql">
<sql:header>
     <sql:param name='ctg'>highlight</sql:param>
</sql:header>
<sql:query>
  SELECT  * from news where category = @ctg
  FOR XML AUTO,ELEMENTS
</sql:query>
</ROOT>
```

When you run this query, it returns the XML output shown in Figure 4-15.

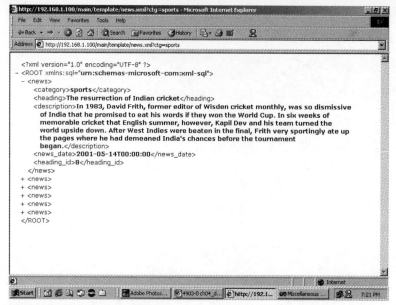

Figure 4-15: The output of news.xml

Pass Parameters Using the Query String

You can generate XML contents from the database on the basis of the value passed in the parameter. In the news section of the portal, when a user clicks on the different sections of the news (such as politics, economics, and so on), the name of that particular category is passed on as a parameter to `news.xml`. After this, the `news.xml` file generates the dynamic XML contents from the database for that particular category. The following code generates dynamic XML contents for the passed category:

```
<sql:header>
    <sql:param name='ctg'>highlight</sql:param>
</sql:header>
```

Using the `<sql:header>` element, you can pass the parameters for the query into a query string. It helps to use one single query for obtaining the different kinds of records on the basis of parameters passed to the query.

Apply XSLT to Dynamic XML Contents in the News Section

In this section, you learn how you can convert or format the contents dynamically generated in `news.xml` into readable and presentable form using XSLT. You use the `news.xsl` file in `news.xsl` to transform these contents into presentable form. Listing 4-18 shows the template code for the news section.

Listing 4-18: Template code for news section

© 2001 Dreamtech Software India Inc.
All Rights Reserved

```
<xsl:template match="ROOT">
<xsl:for-each select="news">
<table  border="1" bordercolor="#000000" cellpadding="5" cellspacing="5"
Class="MIDDILE" style="MARGIN: 10px 10px 10px 10px" >
 <tr>
  <td bgcolor="#dicece">
```

```
<b><xsl:value-of select="heading" /></b>
   <li><xsl:value-of select="description" /></li>
</td>
</tr>
</table>
</xsl:for-each>
</xsl:template>
```

In the preceding code, `<xsl:value-of select="heading"/>` selects the value of `heading` from the generated code of `news.xml`. In the same way, `<xsl:value-of select="description" />` selects the `description` of the news for the current selected heading.

Listing 4-19 contains the code for including the template.

Listing 4-19: Code snippets from news.xsl

```
            <table>
              <tr>
               <table width="100%" border="1" cellspacing="0"
cellpadding="0" height="458" bordercolor="#000000" BGCOLOR="#A9A9A9">
               <tr>
               <td height="45%" width="100%"  BGCOLOR="#000000" >
                     <a href="http://192.168.1.100/main.xml" style="text-
decoration : none ">Home</a> |
                     <a href="news.xml?ctg=highlight" style="text-
decoration : none">Highlight</a> |
                     <a href="news.xml?ctg=politics" style="text-decoration
: none">Politics</a> |
                     <a href="news.xml?ctg=sports" style="text-decoration :
none">Sports</a> |
                     <a href="news.xml?ctg=economics" style="text-
decoration : none">Economics</a>
               </td>

       </tr>
    <tr>
       <td colspan="3" valign="top"><xsl:apply-templates select="ROOT"
/></td>
    </tr>
    </table>



            </tr>
            </table>
```

In the preceding code, you have created four different links for the different news sections. These sections are as follows:

- Highlights
- Politics
- Sports
- Economics

When a user clicks on the section, you again call `news.xml` with a parameter attached to the link, which the user clicks. By doing this, you can display the news about the specific section.

After applying XSLT, the news screen look like the one shown in Figure 4-16.

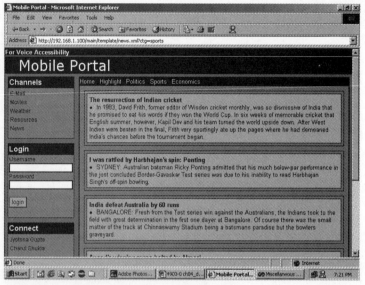

Figure 4-16: The formatted output of news.xml

Develop the E-Mail Section of the Mobile Portal

In this section, you build a simple e-mail section for the Mobile Portal. This section enables the user to send and receive e-mail. For the first time in the portal, you use Active Server Pages (ASP).

For this purpose, you use an I-Mail server as the mail server for the domain, and along with this you use a J-Mail component for sending and receiving e-mail messages through ASP. You can download I-Mail server from `http://www.ipswitch.com`; for the J-Mail component, just go to `http://wwww.dimac.net`. Trial editions for both pieces of software are available from these sites.

Design the Login Form for E-mail

The users of the Mobile Portal are able to log in to check e-mail messages from any page of the portal by using the left-hand side login form that is provided on every page of the portal. When a user clicks on the e-mail link on the left bar, the code calls the page named `login.xml` to display a login form for the user login.

Figure 4-17 shows the login form that is displayed on the left-hand side of the portal.

Figure 4-17: The login form on the left-side of every page

The user provides the username and password and then clicks the Login button. When a user clicks the Login button, the `interface.asp` file is called for authentication of the user and also to display the user's e-mail messages.

If the user clicks on the e-mail link on any page of the portal, it calls the `login.xml` file to display the login form and accept user input for logging in.

Now take a look at the code for the files `login.xml` (Listing 4-20) and `login.xsl`. In `login.xml`, there are no big changes in the document structure; it is similar to the home page document. However, the following two changes are carried out in the document:

♦ Removing the article information from the document because it's not needed now.

♦ Adding one new set of elements, starting from the `<logininfo>` tag.

Listing 4-20: login.xml

```
<logininfo>
 <username />
 <password />
 <login />
</logininfo>
```

In this listing, you just create some empty elements so that you can use them while designing the login form using XSLT.

Look at the `login.xsl` file. The first change in the document occurs in the code of the left-side login form, which you've already used. Now in this file, you disable its functionality by adding the one DHTML property named `DISABLED="TRUE"`. This code stops the working of both the input boxes and login button by disabling them. You also rename the form name as `FRM1`, instead of `FRM`, which you have been using so far in all the documents.

Listing 4-21: Code snippet from login.xsl

© 2001 Dreamtech Software India Inc.
All Rights Reserved

```
                    <form name="frm1">
                        <p>Username <input name="uname" class="BUTTON"
disabled="true"/>
                        Password<br /><input name="pass" class="BUTTON"
type="password" disabled="true"/><br />
                        <br />
                        <input style="FONT-WEIGHT: 550; BORDER-LEFT-COLOR:
#999999; BORDER-BOTTOM-COLOR: #999999; COLOR: #000000; BORDER-TOP-COLOR:
#999999; BACKGROUND-COLOR: #cccccc; BORDER-RIGHT-COLOR: #999999" width="25px"
type="button" name="submit" value="login" class="hand" onclick="Submitpage()"
disabled="true" /></p>
                    </form>
```

Second, you design the e-mail login form for the user, which will appear in the right portion of the screen, where the articles and news and other information appear.

Listing 4-22: Template code for login form
© 2001 Dreamtech Software India Inc.
All Rights Reserved

```
<xsl:template match="main/logininfo/username">
<table cellspacing="1" cellpadding="1" width="97%"  align="center" border="0"
bgcolor="#000000" style="MARGIN: 10px 2px 0px 2px">
```

```
 <tr>
 </tr>
</table>
<table border="0" cellspacing="0" cellpadding="0" height="252" width="300"
align="center">
 <tr>
  <td height="1" colspan="3" bgcolor="black"> <img src="blank.gif" /></td>
 </tr>
 <tr>
 <td width="1" bgcolor="black"><img src="blank.gif" /></td>
 <td align="center" valign="middle" width="400" height="250" bgcolor="silver">
    <form name="frm">
    <font face="Arial, Helvetica"><strong><big><big>E-mail
Login</big></big></strong><br />
    </font>
    <table border="0">
     <tr>
      <td align="right"><font size="2" color="black"
face="arial"><b>Login</b></font></td>
      <td align="center"><input type="text" name="uname" size="20"
class="button" title="Please Type your User-ID here" Autocomplete="off"/> </td>
     </tr>
     <tr>
      <td align="right"><font size="2" color="black"
face="arial"><b>Password</b></font></td>
      <td align="center"><input type="password" name="pass" size="20"
class="button" title="Please Type your Password here" Autocomplete="off"/></td>
     </tr>
     <tr>
      <td colspan="2" align="center"><font size="2" color="black"
face="arial"><input type="button" name="submit" value="Logon"
onclick="Submitpage()" title="Click here to login" /></font> </td>
     </tr>
    </table>
 </form>
 </td>
 <td width="1" bgcolor="black"><img src="blank.gif" /></td>
 </tr>
 <tr>
 <td height="1" colspan="3" bgcolor="black"><img src="blank.gif" /></td>
 </tr>
</table>
</xsl:template>
```

In the preceding code, you begin by first constructing a template, using one of the empty elements declared in the XML document. After matching the template, you build up some tables and a form for accepting the username and password from the user.

After clicking the login button, the same file is called again (`interface.asp`), to display the E-mail login form and its user authentication boxes. Figure 4-18 shows the output of the code.

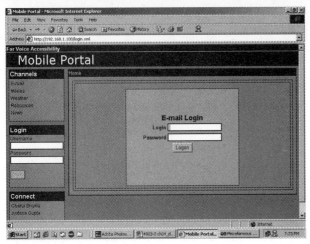

Figure 4-18: The formatted output of login.xml

In starting the `interface.asp` file, you declare two variables to store the value of the username and the password coming from `login.xml` in the form of a query string. The following is the code:

```
<%
Dim user,pass
user=Request.querystring ("uname")
pass=Request.querystring ("pass")
%>
```

After storing the values in the variables, the program retrieves all the e-mail messages from the user's inbox by means of the code in Listing 4-23.

Listing 4-23: Code snippet from interface.asp

© 2001 Dreamtech Software India Inc.
All Rights Reserved

```
<%
Set pop3 = Server.CreateObject( "Jmail.pop3" )
pop3.Connect user, pass,"http://192.168.1.100"
Response.Write( "You have " & pop3.count & " emails in your mailbox!<br><br>" )
c = 1

if (pop3.Count>0) then
while (c<=pop3.Count)
  Set msg = pop3.Messages.item(c)
  ReTo = ""
  ReCC = ""
  Set Recipients = msg.Recipients
  separator = ", "

  ' We now need to get all the recipients,
  ' both normal and Carbon Copy (CC) recipients
  ' and store them in a variabel
  For i = 0 To Recipients.Count - 1
   If i = Recipients.Count - 1 Then
    separator = ""

   End If
   Set re = Recipients.item(i)
```

```
  If re.ReType = 0 Then
    ReTo = ReTo & re.Name & " (" & re.EMail & ")" & separator

  else
    ReCC = ReTo & re.Name & " (" & re.EMail & ")" & separator

  End If
 Next
%>
```

In this code, you first create an object named `Jmail.pop3`. This object retrieves mail from the user inbox by using the J-Mail component. After creating an object, you connect the mail server by writing the following line of code:

```
pop3.Connect user, pass,http://192.168.1.100
```

In this line, you connect the mail server by providing the username, password, and the official `mailhost` address of the server. Look for the I-Mail server configuration or contact your administrator for the `mailhost` server address.

After connecting the mail server, the user sees how many e-mail messages are currently in his or her mailbox. Next, by adding some additional lines of code, you retrieve all the messages from the inbox of the user, one by one, and display them in a tabular format, as follows:

```
 <tr class="medium">
 <td valign="center" align="middle" width="20%"><b><%= msg.FromName %></b></td>
 <td class="text" valign="center" align="middle" width="55%"><a
Class="blacklinks"href="http://192.168.1.100/showmail.asp?user=<%=user%>&pass=<%
=pass%>&value=<%=c%>" ><%= msg.Subject %></a></td>
 <td valign="center" align="middle" width="25%"><%=msg.date %></td>
</tr>
```

Having displayed the data in a tabular format, you close the loop used to retrieve all the messages from the inbox by typing the following lines of code:

```
<%
c = c + 1
wend
end if
%>
```

When you execute this code, its gives you output similar to that shown in Figure 4-19.

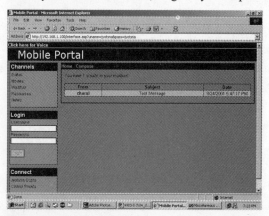

Figure 4-19: The user inbox

All the messages are now displayed in a tabular format. You can select any particular message to read by clicking on the subject of that message. Consequently, it calls another file, named `showmail.asp`, and passes some parameters to it, such as username, the password, and the messageid for that particular message.

At the beginning of `showmail.asp`, you declare some variables to store information, as follows:

```
<%
Dim megno,uname,pass
msgno=Request.querystring ("value")
uname=Request.querystring ("user")
pass=Request.querystring ("pass")
%>
```

In this code, you store three kinds of information: username, password, and the messageid. Here again you use the same code to retrieve the message from the inbox. In this case, however, you do not construct a loop for it; instead, you simply retrieve the single message in the basis of the messageid. This process is shown in Listing 4-24.

Listing 4-24: Code snippet from showmail.asp

```
<%

Set pop3 = Server.CreateObject( "Jmail.pop3" )
pop3.Connect uname, pass,"http://192.168.1.100"
 Set msg = pop3.Messages.item(msgno)
 ReTo = ""
 ReCC = ""
 Set Recipients = msg.Recipients
 separator = ", "

 ' We now need to get all the recipients,
 ' both normal and Carbon Copy (CC) recipients
 ' and store them in a variabel
 For i = 0 To Recipients.Count - 1
  If i = Recipients.Count - 1 Then
   separator = ""

  End If
  Set re = Recipients.item(i)
  If re.ReType = 0 Then
   ReTo = ReTo & re.Name & " (" & re.EMail & ")" & separator

  else
   ReCC = ReTo & re.Name & " (" & re.EMail & ")" & separator

  End If
 Next


%>
```

After retrieving the message, you build a tabular format in order to display the message, as shown in Listing 4-25.

Listing 4-25: Code for displaying the message

```
<tr class="medium">
<td valign="center" width="20%" align="left"><b>From</b></td>
 <td valign="center" align="left" width="80%"><%= msg.FromName %></td>
</tr>

<tr class="medium">
<td valign="center" width="20%" align="left"><b>Subject</b></td>
 <td class="text" valign="left" align="left" width="80%"><%= msg.Subject %></td>
</tr>

<tr class="medium">
<td valign="center" width="20%" align="left"><b>Receiving Date</b></td>
 <td valign="center" align="left" width="80%"><%=msg.date %></td>
</tr>

<tr class="medium">
<td valign="top" width="20%" align="left"><b>Message</b></td>
 <td valign="center" align="left" width="80%"><%=msg.body %></td>

</tr>
```

Figure 4-20 shows the output of `showmail.asp`.

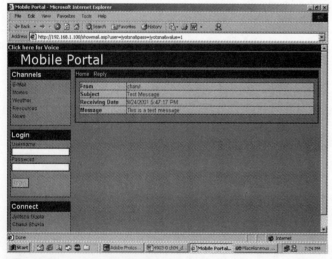

Figure 4-20: The output of showmail.asp

So far, you have completed only one portion of the e-mail section: the receiving and reading mail section. The following section includes a discussion on the remaining portions of the e-mail section, namely, sending e-mail messages to a user by means of the I-Mail server and the J-Mail component. The interface here is also based on the XML and XSLT approach.

Design a Form to Compose Mail

This section contains a discussion of how to accept information from the user for transmission to the destination. To understand this process, follow the construction of the send mail section of the e-mail system. The design of the compose mail section uses XML and XSLT.

First, take a look at the code of `send.xml` in Listing 4-26. This is almost the same code that you wrote for the home and the login sections of the Mobile Portal. In this file, however, you also add some empty elements, such as `login.xml`. Following are the changes incorporated in the coding:

Listing 4-26: send.xml

```
<jmail>
 <jmessage>
  <from />
  <to />
  <CC />
  <BCC />
  <subject />
  <body />
  <send />
 </jmessage>
</jmail>
```

Now look at the XSLT document named `send.xsl` (Listing 4-27), from which you generate your output in the shape of a formatted document containing a form for sending the e-mail message. Listing 74-28 shows templates for use with `send.xsl`.

Listing 4-27: code snippet from send.xsl

© 2001 Dreamtech Software India Inc.
All Rights Reserved

```
<table width="97%" border="1" bordercolor="#000000" cellpadding="5"
cellspacing="5" Class="MIDDILE" style="MARGIN: 10px 10px 10px 10px" >
<tr>
<td bgcolor="#dicece">
                                        <table style="PADDING-RIGHT: 10px;
PADDING-LEFT: 10px; PADDING-BOTTOM: 2px; PADDING-TOP: 2px" cellspacing="0"
cellpadding="0" border="0">
<tr>

<form action="http://192.168.1.100/send.asp" method="post" name="frm">
<input type="hidden" name="txtAction"></input>

<table width="400px" hight="300px" border="0" cellspacing="5" cellpadding="10" >
<tr>
<td>
<table width="100%" border="0" cellspacing="0" cellpadding="0">
<tr>
                                        <xsl:apply-templates
select="main/jmail/jmessage/from" />
</tr>
<tr>
                                        <xsl:apply-templates
select="main/jmail/jmessage/to" />
</tr>
<tr>
```

```
                                              <xsl:apply-templates
select="main/jmail/jmessage/CC" />
</tr>
<tr>
                                              <xsl:apply-templates
select="main/jmail/jmessage/BCC" />
</tr>
<tr>
                                              <xsl:apply-templates
select="main/jmail/jmessage/subject" />
</tr>
  <xsl:apply-templates select="main/jmail/jmessage/body" />
<tr>
                                              <xsl:apply-templates
select="main/jmail/jmessage/send" />
</tr>
</table>
</td>
</tr>
</table>
</form>

</tr>
</table>
```

Listing 4-28: Code listing for different templates of send.xsl

```
<xsl:template match="main/jmail/jmessage/from">

 <td>
  From:
 </td>
 <td>
  <input type="text" name="txtFrom" size="40">
  </input>
 </td>

</xsl:template>
```

In the above template, you match it first with the corresponding section and then construct two columns in a row – one for displaying the text "FROM" and the second for the input box – which you use for accepting the sender name data.

```
<xsl:template match="main/jmail/jmessage/to">

 <td>
  To:
 </td>

 <td>
  <input type="text" name="txtTo" size="40">
  </input>
 </td>
</xsl:template>
```

In the preceding template, you construct two columns in a row — one for displaying the text "TO" and the second for the input box — which you use for accepting data for the recipient's address.

```
<xsl:template match="main/jmail/jmessage/CC">
 <td>
  CC:
 </td>
 <td>
  <input type="text" name="cc" size="40">

  </input>
 </td>
</xsl:template>
```

In the preceding template, you construct two columns in a row – one for displaying the text "CC" and the second for the input box – which you use for accepting data for the carbon copy.

```
<xsl:template match="main/jmail/jmessage/BCC">
 <td>
  BCC:
 </td>
 <td>
  <input type="text" name="bcc" size="40">

  </input>
 </td>
</xsl:template>
```

You construct two columns in a row in the preceding template – one for displaying text "BCC" and the second for the input box – which you use for accepting data for a blind carbon copy.

```
<xsl:template match="main/jmail/jmessage/subject">
 <td>
  Subject:
 </td>
 <td>
  <input type="text" name="txtSubject" size="40">
  </input>
 </td>
</xsl:template>
```

You construct two columns in a row in the preceding template also – one for displaying the text "Subject" and the second for the input box – which accepts data for the subject.

```
<xsl:template match="main/jmail/jmessage/body">
 <tr>
  <td colspan="2">Message Body:
  </td>
 </tr>
 <tr>
  <td colspan="2">
   <textarea name="txtBody" rows="8" cols="65">
   </textarea>
  </td>
 </tr>

</xsl:template>
```

In the preceding template, you construct two columns in a row – one for displaying the text `"body"` and the second for the input box – which accepts data for the body of the message.

```
<xsl:template match="main/jmail/jmessage/send">
 <td align="left" colspan="2">
  <br/><input type="button" name="btnSend" value="Send Mail"
onclick="SubmitPage()">
  </input>
 </td>
</xsl:template>
```

Finally, in the preceding template, you build a submit button for sending the form.

Figure 4-21 shows the output of `send.xsl`.

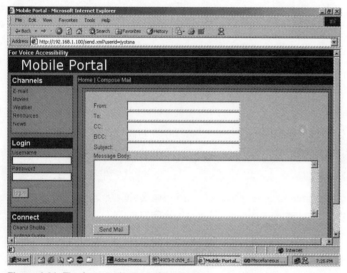

Figure 4-21: The formatted output of send.xml

Validation of Forms Using JavaScript

Many times users forget to enter all the requested available data or they enter the wrong data. To overcome these kinds of problems, you can use some client-side scripting technologies, such as VBscripts and JavaScripts, to validate the form before submitting it to the server.

You use JavaScript to validate the form before submitting it to the server and to display an error message if an error occurs (see Listing 4-29).

Listing 4-29: Code listing of JavaScript in send.xsl

```
<script language="javascript">
 function openwindow()
 {
  window.open('show.xml','abc','width=300,height=300');
 }
 function SubmitPage()
 {
 if (Validate())
  {
```

```
    document.frm.txtAction.value = "SendMail";
    document.frm.submit();
  }
}


function Validate()
{
 if(document.frm.txtFrom.value.length == 0)
 {
  document.frm.txtFrom.focus();
  document.frm.txtFrom.select();
  alert("You must enter a sender's EMail address.");
  return false;
 }
 if(document.frm.txtTo.value.length == 0)
 {
  document.frm.txtTo.focus();
  document.frm.txtTo.select();
  alert("You must enter the recepient's EMail address.");
  return false;
 }

 if(document.frm.txtBody.value.length == 0)
 {
  document.frm.txtBody.focus();
  document.frm.txtBody.select();
  alert("You must enter a message body.");
  return false;
 }

 if(document.frm.txtSubject.value.length == 0)
 {
  document.frm.txtSubject.focus();
  document.frm.txtSubject.select();
  alert("You must enter a subject.");
  return false;
 }
 return true;
}
</script>
```

You add the preceding script to the top of the page, so that when the processor throws HTML-formatted output to the browser, it executes the script. In this code, you use three functions for different tasks. The first function is named `openwindow()`. You call this function when a user clicks on the highlighted text on the top bar of the portal; this function opens a new window and displays `show.xml` in it (see Figure 4-22).

For Voice Accessibility

Mobile Portal

Figure 4-22: The top bar on every page contains the openwindow() function.

Use the second and third functions to validate the form. If a user forgets to fill in any required information, the validation function shows an alert message to the user (Figure 4-23). This message shows the name of the field and asks the user to fill in the required data before submitting the form.

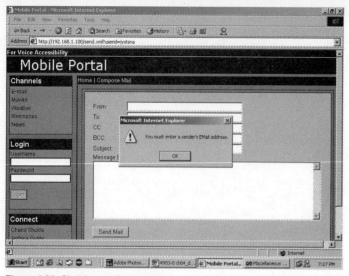

Figure 4-23: Showing an alert message

When a user presses the Send Mail button, a form is submitted to the server and a file called `send.asp` (shown in Listing 4-30) sends the e-mail message, using the I-Mail server and the J-Mail component.

Listing 4-30: send.asp

© 2001 Dreamtech Software India Inc.
All Rights Reserved

```
<%
' Get the form data

senderEmail = trim(Request.Form("txtFrom"))
recipient1= trim(Request.Form("replyto"))
recipient = trim(Request.Form("txtTo"))
textcc = trim(Request.Form("cc"))
textbcc = trim(Request.Form("bcc"))
subject = trim(Request.Form("txtSubject"))
body = trim(Request.Form("txtBody"))


' Create the JMail message Object
set msg = Server.CreateObject( "JMail.Message" )
'msg.Logging = true
'msg.silent = true
msg.From = senderEmail

if (recipient <> " ") then
msg.AddRecipient recipient
end if
if (recipient1 <> " ") then
msg.AddRecipient recipient1
end if

msg.Subject = subject
msg.body = body
if not msg.Send("192.168.1.100" ) then
```

```
    Response.write "<PRE>" & msg.log & "</PRE>"
else
    Response.write "Message sent successfully!"
end if
%>
```

In the preceding code, you first declare some variables to hold the values, which come from `send.xml`. After declaring the variables, you create an object named `jmail.message`, using `server.createobject`. You then provide all the data that has been collected from `send.xml`. Lastly, you use `msg.send` to send the e-mail message. If any problem surfaces during this operation, the code writes a log using the `msg.log`. This information helps you learn the nature of the problem.

Develop the Movie-Ticket Booking System

In this section of the Mobile Portal, a user can book movie tickets to a particular theater for a particular showtime. Most of this part of the system is written using the XML- and XSLT-based approach. In this section, you use a new approach to insert records in the database using the Updategrams. Updategrams are one of the new features introduced with Microsoft SQL Server 2000; they enable you to insert, delete, and update the records in a database using the XML-based document structure.

Display the Theater Listing

When a user clicks on the movies link on any page of the portal, it calls `movies.xml` (see Listing 4-31), which contains a query for listing all the theaters. Clicking on the link executes the query and returns the data in XML format.

Listing 4-31: movies.xml

© 2001 Dreamtech Software India Inc.
All Rights Reserved

```
<?xml version ='1.0' encoding='UTF-8'?>
<?xml:stylesheet type="text/xsl" href="movies.xsl"?>
<ROOT xmlns:sql="urn:schemas-microsoft-com:xml-sql">

  <sql:query>

SELECT  * from hall
FOR XML AUTO,ELEMENTS

  </sql:query>

</ROOT>
```

When you execute this query, it returns a listing of all the theaters, in XML format. After procuring all the data in XML format, you can transform it into HTML for displaying it in the browser by using XSLT.

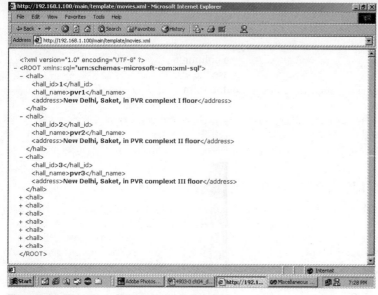

Figure 4-24: The output of the movies.xml query

When you have all the data in your hands, you write an XSL document for the transformation of this data. The code listing for `movies.xsl` is shown in Listing 4-32. Use the following line to apply the template:

```
<xsl:apply-templates select="ROOT" />
```

Listing 4-32: Code snippet from movies.xsl

© 2001 Dreamtech Software India Inc.
All Rights Reserved

```
<xsl:template match="ROOT">
<table cellspacing="0" cellpadding="0" width="80%"  align="center" border="1"
style="MARGIN: 10px 2px 0px 2px">
            <tr>
                <td valign="center"><b>Theater</b></td>
                <td valign="center"><b>Address</b></td>
            </tr>
        <xsl:for-each select="hall">
            <tr class="medium" bgcolor="#cccccc">
    <td valign="center" align="center" height="42">
     <b>
            <a class="blacklinks">
    <xsl:attribute name="href" >
        http://192.168.1.100/main/template/status.xml?tname=<xsl:value-of
select="hall_name" />
    </xsl:attribute>
    <xsl:value-of select="hall_name" />
    </a>

    </b>
    </td>
    <td class="text" valign="center">
                        <xsl:value-of select="address" />
    </td>
            </tr>
```

```
            </xsl:for-each>
        </table>
</xsl:template>
```

This is a simple code. In this code, you create a table with two columns and then go fetch the value by using the `<xsl:value-of select>` statement. You also use the `<xsl:attribute>` statement for creating links of the movie theater names. Figure 4-25 shows the formatted output of `movies.xml` after applying `movies.xsl`.

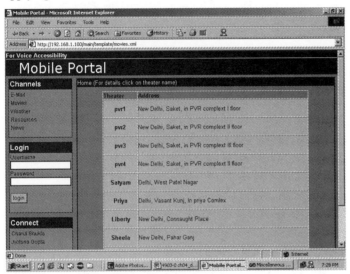

Figure 7-25: The formatted output of movies.xml

When a user clicks on the theater name, control passes to a file named `status.xml` with the parameter `tname` (which stand for the 'theater name') in the query string.

Display Detailed Status of the Theater

This section discusses how to display more details about the theater so that the user can choose the movie and book tickets for that movie. For this task, first take a look at the query in `status.xml`, shown in Listing 4-33.

Listing 4-33: status.xml

```
<?xml version ='1.0' encoding='UTF-8'?>
<?xml:stylesheet type="text/xsl" href="status.xsl"?>
<ROOT xmlns:sql="urn:schemas-microsoft-com:xml-sql">

<sql:header>
    <sql:param name='tname'>Pvr1</sql:param>

</sql:header>


<sql:query>
 select
hall.hall_name,movie.movie_name,status.movie_date,status.showtime,status.balcony
,status.rear_stall,status.upper_stall,status.middle_stall  from hall,movie,status
```

```
where status.hall_id=hall.hall_id and status.hall_id=hall.hall_id and
status.movie_id = movie.movie_id and hall_name = @tname
    FOR XML AUTO,ELEMENTS
</sql:query>

</ROOT>
```

When a user executes this query, it yields all kinds of details about the theater and the movies showing in the theater at that particular time. It also provides additional information on the number of tickets left for each category of seats in a theater.

Now put down the XSLT document used for transforming this data into HTML format. For applying the template, add the following code in `status.xsl`:

```
<xsl:apply-templates select="ROOT" />
```

Listing 4-34 contains the code for `status.xsl`.

Listing 4-34: Code snippet from status.xsl

© 2001 Dreamtech Software India Inc.
All Rights Reserved

```
<xsl:template match="ROOT">
<xsl:variable name="hallname" select="hall/hall_name" />
<b><xsl:value-of select="hall/hall_name" /></b><i><br />(tickets available)</i>
<p></p>


<xsl:for-each select="hall/movie">
<p></p>
<b><xsl:value-of select="./movie_name" /></b>
<a class="blacklinks">
<xsl:attribute name="href" >
http://192.168.1.100/main/template/booking.xml?tname=<xsl:value-of
select='$hallname' />
</xsl:attribute>
<font size="1"><i><b>(Ticket booking)</b></i></font>
</a>


<table cellspacing="0" cellpadding="0" width="80%"  align="center" border="1"
style="MARGIN: 10px 2px 0px 2px">
  <tr>
                <td valign="center"><b>Date</b></td>
                <td valign="center"><b>Showtime</b></td>
                <td valign="center"><b>Balcony</b></td>
                <td valign="center"><b>rear_stall</b></td>
                <td valign="center"><b>middle_stall</b></td>
                <td valign="center"><b>upper_stall</b></td>


        </tr>

  <xsl:for-each select="./status">
```

```
                    <tr class="medium" bgcolor="#cccccc">
        <td class="text" valign="center" height="43">
                        <xsl:value-of select="./movie_date" />

        </td>

            <td class="text" valign="center">
     <xsl:value-of select="./showtime" />
        </td>

        <td class="text" valign="center">
                        <xsl:value-of select="./balcony" />

        </td>
        <td class="text" valign="center">
                        <xsl:value-of select="./rear_stall" />

        </td>
        <td class="text" valign="center">
                        <xsl:value-of select="./middle_stall" />
        </td>
        <td class="text" valign="center">
                        <xsl:value-of select="./upper_stall" />

        </td>

                        </tr>
  </xsl:for-each>

            </table>

</xsl:for-each>
</xsl:template>
```

After applying this XSLT document to `status.xml`, you see the output shown in Figure 4-26.

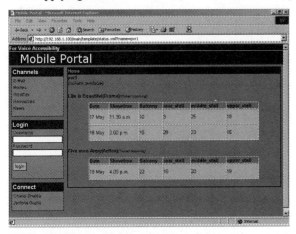

Figure 4-26: Showing the formatted output of status.xml

Booking Tickets for a Movie

After viewing the entire record, the user clicks on the ticket-booking link to book tickets for the movie of choice. Consequently, control passes to `booking.xml` (see Listing 4-35), where the user can book tickets for a particular movie.

Listing 4-35: booking.xml

```
<?xml version ='1.0' encoding='UTF-8'?>
<?xml:stylesheet type="text/xsl" href="booking.xsl"?>

<ROOT xmlns:sql="urn:schemas-microsoft-com:xml-sql">

<sql:header>
    <sql:param name='tname'>Pvr1</sql:param>

</sql:header>
  <sql:query>
 select
hall.hall_id,movie.movie_id,hall.hall_name,movie.movie_name,status.movie_date,st
atus.showtime,status.balcony,status.rear_stall,status.upper_stall,status.middle_
stall,status.movie_date from hall,movie,status

where status.hall_id=hall.hall_id and status.hall_id=hall.hall_id and
status.movie_id = movie.movie_id and hall_name = @tname
    FOR XML AUTO,ELEMENTS
</sql:query>

</ROOT>
```

Figure 4-27 shows the output after this query executes.

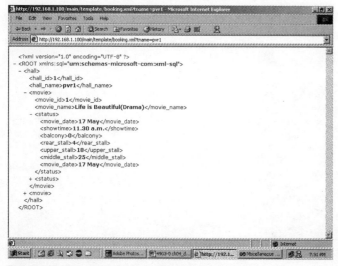

Figure 4-27: Output of the query in booking.xml

This query returns the status of the movie tickets left, as well the showtimes of the movie. You transform this data by using the XSLT document named `booking.XSL`. First, you need to make some changes in the JavaScript for this page in the `pagesubmit` and `validate` functions. The new functions are shown in Listing 4-36.

Listing 4-36: New pagesubmit function

```
function pagesubmit()
{

    if(validate())
    {
  var
fname,lname,telno,email,tname,mname,movieid,hallid,stime,showdate,stall,ticketno
;
    fname = document.frm.fname.value;
    lname = document.frm.lname.value;
    telno = document.frm.telno.value;
    email = document.frm.email.value;
    tname = document.frm.hall.options[document.frm.hall.selectedIndex].text;
    mname = document.frm.movie.options[document.frm.movie.selectedIndex].text;
    movieid = document.frm.movie.options[document.frm.movie.selectedIndex].value;
    hallid = document.frm.hall.options[document.frm.hall.selectedIndex].value;
    stime =
document.frm.showtime.options[document.frm.showtime.selectedIndex].text;
    showdate =
document.frm.showdate.options[document.frm.showdate.selectedIndex].text;
    stall = document.frm.stall.options[document.frm.stall.selectedIndex].text;
    ticketno = document.frm.ticket.value;
    document.location.href = "http://192.168.1.100/ticketbooking.asp?fname=" +
fname + "&amp;lname=" + lname + "&amp;telno=" + telno + "&amp;hall=" + tname +
"&amp;movie=" + mname + "&amp;movieid=" + movieid + "&amp;hallid=" + hallid +
"&amp;showtime=" + stime + "&amp;stall=" + stall + "&amp;ticket=" + ticketno +
"&amp;showdate=" + showdate + "&amp;email=" + email;
        }

    }
```

Some changes also occur in the `validate ()` function. The new function is as follows:

```
function validate()
{
  if(document.frm.fname.value.length == 0)
  {
   document.frm.fname.focus();
   document.frm.fname.select();
   alert("You must enter your first name.");
   return false;
  }

  if(document.frm.lname.value.length == 0)
  {
   document.frm.lname.focus();
   document.frm.lname.select();
   alert("You must enter your last name.");
   return false;
```

```
      }

      if(document.frm.telno.value.length == 0)
      {
       document.frm.telno.focus();
       document.frm.telno.select();
       alert("You must enter your telephone no.");
       return false;
      }

      if(document.frm.email.value.length == 0)
      {
       document.frm.email.focus();
       document.frm.email.select();
       alert("You must enter your e-mail address.");
       return false;
      }
      if(document.frm.ticket.value.length == 0)
      {
       document.frm.ticket.focus();
       document.frm.ticket.select();
       alert("You must enter the no. of tickets you want to book.");
       return false;
      }

      return true;


     }
```

Now look at the code for the template called "ROOT", which you use for building the right side of the screen and accepting all kinds of input from the user regarding the booking of tickets:

```
<xsl:template match="ROOT">
<xsl:value-of select='@msg' />
<b><i>Enter the following information:-</i></b>

<form name="frm">

<table cellspacing="0" cellpadding="0" width="80%"  align="center" border="0"
style="MARGIN: 10px 2px 0px 2px">
  <tr>
                <td valign="center" height="40"><b>Your First Name</b></td>
                <td valign="center"><input type="text" name="fname" /></td>
  </tr>
  <tr>
                <td valign="center" height="40"><b>Your Last Name</b></td>
                <td valign="center"><input type="text" name="lname" /></td>
  </tr>
  <tr>
                <td valign="center" height="40"><b>Telephone No.</b></td>
                <td valign="center"><input type="text" name="telno" /></td>
  </tr>
  <tr>
                <td valign="center" height="40"><b>Enter your E-mail
Address</b></td>
                <td valign="center"><input type="text" name="email" /></td>
```

```
</tr>
<tr>
                <td valign="center" height="40"><b>Theater name</b></td>
                <td valign="center"><select name="hall">
<option>
<xsl:attribute name="value" >
<xsl:value-of select="hall/hall_id" />
</xsl:attribute>
<xsl:value-of select="hall/hall_name" />
</option></select>
</td>
</tr>
<tr>
                <td valign="center" height="40"><b>Select a movie</b></td>
                <td valign="center"><select name="movie">
 <xsl:for-each select="hall/movie">
  <option>
  <xsl:attribute name="value" >
   <xsl:value-of select="movie_id" />
  </xsl:attribute>
  <xsl:value-of select="movie_name" />
        </option>
 </xsl:for-each>

 </select>
  </td>
</tr>


<tr>
                <td valign="center" height="40"><b>Select a Stall</b></td>
                <td valign="center">
  <select name="stall">
 <option>Balcony</option>
 <option>Rear_stall</option>
 <option>Middle_stall</option>
 <option>Upper_stall</option>
  </select>
  </td>
</tr>

<tr>
                <td valign="center" height="40"><b>Date</b></td>
                <td valign="center">
  <select name="showdate">
 <xsl:for-each select="hall/movie/status">
  <option>
  <xsl:attribute name="value" >
   <xsl:value-of select="movie_date" />
  </xsl:attribute>
  <xsl:value-of select="movie_date" />

 </option>
 </xsl:for-each>
  </select>
</td>
```

```
          </tr>

          <tr>
                    <td valign="center" height="40"><b>ShowTime</b></td>
                    <td valign="center">
        <select name="showtime">

        <option>9.00 a.m.</option>
        <option>11.30 a.m.</option>
        <option>12.00 p.m.</option>
        <option>2.00 p.m.</option>
        <option>3.00 p.m.</option>
        <option>4.00 p.m.</option>
        <option>6.00 p.m.</option>
        <option>9.00 p.m.</option>
         </select>
        </td>
        </tr>

          <tr>
                    <td valign="center" height="40"><b>How many tickets</b></td>
                    <td valign="center"><input type="text" name="ticket"
maxlength="2" /></td>
        </tr>
        <tr>

                    <td colspan="2" align="center"><input type="button"
value="Book" onclick="return pagesubmit()" /></td>
        </tr>

               </table>

</form>

</xsl:template>
```

In the preceding code, you can observe that you are generating some input boxes, select options, for accepting input from the user. When you apply this XSL to the `booking.xml` document, it results in the formatted output shown in Figure 4-28.

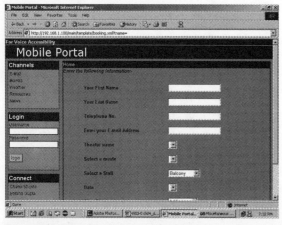

Figure 4-28: The formatted output of booking.xml

After filling in all the required data in the various fields, the user clicks the Book button and all the information passes to another XML file, which actually is an Updategram. You use the Updategram for the insertion and updating of records in the database.

Updategrams

Now for the first time, you are going to use the Updategram feature for SQL Server 2000. In this case, you use it for inserting and updating records when a user books tickets for a movie (see Listing 4-37). For more information about Updategrams, please refer to Chapter 3.

Listing 4-37: ticketbook.xml

```xml
<?xml version ='1.0' encoding='UTF-8'?>
<?xml:stylesheet type="text/xsl" href="ticketbook.xsl"?>
<ROOT xmlns:updg="urn:schemas-microsoft-com:xml-updategram">
<updg:header nullvalue="isnull">
  <updg:param name='fname' />
  <updg:param name='lname' />
  <updg:param name='telno' />
  <updg:param name='email' />
  <updg:param name='hall' />
  <updg:param name='movie' />
  <updg:param name='movieid' />
  <updg:param name='hallid' />
  <updg:param name='balcony' />
  <updg:param name='rear' />
  <updg:param name='middle' />
  <updg:param name='upper' />
  <updg:param name='showtime'/>
  <updg:param name='showdate'/>
  <updg:param name='stall'/>
  <updg:param name='ticket'/>

</updg:header>


<updg:sync>
 <updg:before>
          <status  hall_id="$hallid" movie_id="$movieid" showtime="$showtime"
/>
 </updg:before>
    <updg:after>
          <status  hall_id="$hallid" movie_id="$movieid" showtime="$showtime"
balcony="$balcony" rear_stall="$rear" middle_stall="$middle"
upper_stall="$upper" />
    </updg:after>
  </updg:sync>



<updg:sync>
 <updg:before>
 </updg:before>
 <updg:after>
```

```
    <user firstname="$fname" lastname="$lname" telno="$telno" movieid="$movieid"
hallid="$hallid" stall="$stall" movie_date="$showdate" showtime="$showtime"
totaltickets="$ticket" emailadd="$email" />
 </updg:after>

</updg:sync>

</ROOT>
```

In the preceding code, you first declare the variables for holding the values, which come from booking.xml. You then use the first <updg:sync> </updg:sync> block to update the records in the database. You use the second <updg:sync> </updg:sync> block to insert the details about the booking into the database.

You also apply XSL to this Updategram to confirm the user's booking.

Show the Order Confirmation

To display the confirmation, just add the following code in XSLT code:

```
Thanks for using Our service. Your tickets have been booked!
```

Figure 4-29 shows the confirmation screen.

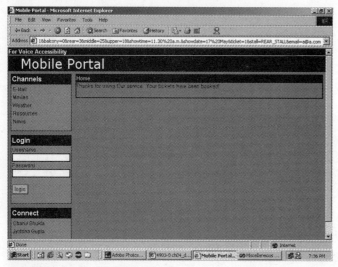

Figure 4-29: The confirmation screen

Thus ends the discussion of the movie-ticket booking section of the Mobile Portal. This entire system, as mentioned earlier, has been built on the XSL- and XML-based approach, and you easily can extend it according your needs.

Summary

This chapter provided a vivid demonstration of the XML- and XSLT-based approach to developing portals. The project undertaken in this chapter gave you a practical demonstration of the process of using interfaces on your site using the XSLT and XML data. Further, you learned how to use an XML-based query to retrieve the data from SQL Server 2000 in XML format and how to apply XSLT to the data returned from this query.

Chapter 5

Transforming the Wireless Web Application for WAP Clients Using XSLT

In this chapter, you will transform the Web application you developed in Chapter 4. In previous chapters, you gained insight on XML- and XSLT-based design of the application for PC browsers. This chapter presents changes in the application's design. You will again use the XML- and XSLT-based approach and transform the data. This time, however, you will also use Microsoft ASP (Active Server Pages) technology to generate XML data on-the-fly and then load the XSLT to display data on WAP browsers.

Limitations of Small Devices

While working with small devices such as cell phones, you face some very basic limitations. Thus, as you transform the application for WAP clients, you must keep the following limitations in mind:

♦ Devices such as cell phones have a small display screen. As a result, displaying a large amount of data will be troublesome for the user. You should, therefore, condense the data as much as possible and supply very compact and valuable information to the user.

♦ Small devices cannot perform heavy operations because this would require a large amount of memory and disk space. One of the operations we will encounter in the case of small devices is content transformation using XSLT. Unlike PC-based browsers, WAP browsers don't have a built-in XML parser. Thus, it becomes impossible to perform XSLT transformation on the client side. To solve this problem, we use Active Server Pages to generate XML data and load XSLT documents. We then perform a transformation on the server side and produce WML contents for WAP browsers.

♦ One of the major problems that you'll encounter during the development process is firing the XML query from the WAP browser to access data from SQL Server 2000 in XML format. To resolve this issue, you use Active Server Pages, which helps generate XML data and then transform this data using XSLT documents.

Design the Framework for a WAP Portal

Let's take a look at the design you'll follow while building the application. In Chapter 4, we discussed application design for a PC browser. This time, there are some design changes that are needed because of the limitations of mobile devices.

Figure 5-1 is a flow chart of the basic design of the application.

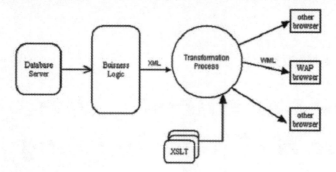

Figure 5-1: Application flow chart

As you can see in Figure 5-1, instead of using XML-based queries, we have used Active Server Pages (ASP) to generate XML data. We have then used different XSLT documents for transforming that data for WAP browsers. We have also used ASP to write the business-logic portion of the application.

Detect the Browser for Serving Content

Before generating the XML data and any kind of transformation process, you need to detect the client type. Only then will you be able to transform the content for that device in the form of WML, HDML, or cHTML, depending on the client type.

To detect the client, use checkbrows.asp (Listing 5-1). This file works as a home page document loader for the WAP client's application and plays a crucial role in the application.

Listing 5-1: checkbrows.asp

```
<%@ Language=VBScript %>

<%
'declaring variables
dim sXML
dim sXSL
dim xmlDocument,xslDocument
dim browser,bt

'creating the objects for the XMLDOM
set xmlDocument = Server.CreateObject("MICROSOFT.XMLDOM")
set xslDocument = Server.CreateObject("MICROSOFT.XMLDOM")

acc=Request.ServerVariables("HTTP_ACCEPT")

Set bt = Server.CreateObject("MSWC.BrowserType")
browser = bt.browser

uagent = Request.ServerVariables("HTTP_USER_AGENT")
browserinfo = split("/",uagent,5)
browsertype = browserinfo(0)

'check for the browser then according to that browser load the xml files
if (InStr(acc,"text/vnd.wap.wml")) then

 session("browsmode")="wml"
```

```
  sXML = "main.xml"
  sXSL = "wml.xsl"


  xmlDocument.async = false
  xslDocument.async = false


  xmlDocument.load(Server.MapPath(sXML))
  xslDocument.load(Server.MapPath(sXSL))

  Response.ContentType = "text/vnd.wap.wml"
  Response.Write "<!DOCTYPE wml PUBLIC ""-//WAPFORUM//DTD " & _
    "WML 1.1//EN"" ""http://www.wapforum.org/DTD/wml_1.1.xml"">"

    Response.Write(xmlDocument.transformNode(xslDocument))

end if

%>
```

In Listing 5-1, you first declare some variables for storing the different values for executing the code. After that, the code detects the user agent (browser) type using the `Request.servervariable(HTTP_USER_AGENT)` function. This returns a string that contains information about the user agent. The code then looks for the string `"text/vnd.wap.wml"` in that string. If it finds it there, the code will store some information in variables that you declared earlier, as follows:

```
session("browsmode")="wml"

sXML = "main.xml"
sXSL = "wml.xsl"
```

In the next few lines of code, you load both XML and XSL documents and set the content type for the WAP browser using the `Response.ContentType` method. In the next line, you send WML prolog for the WAP browser by writing the following lines of code:

```
Response.Write "<!DOCTYPE wml PUBLIC ""-//WAPFORUM//DTD " & _
  "WML 1.1//EN"" ""http://www.wapforum.org/DTD/wml_1.1.xml"">"
```

After setting the content type and sending the WML prolog, you perform the transformation process using the following line of code:

```
Response.Write(xmlDocument.transformNode(xslDocument))
```

Convert Static XML Data into WML Using XSLT

Now take a look at the XSL code (Listing 5-2) that was used for the transformation process from XML to WML. You are using the same document that you used for the PC-browser implementation of the application.

Listing 5-2: WML.XSL

```
<?xml version='1.0'?>
<xsl:stylesheet xmlns:xsl="http://www.w3.org/TR/WD-xsl">

<xsl:template match="/">
```

```
<wml>
    <card>
 <p>
  Menu
   <xsl:apply-templates select="main/linklist" />
 </p>

   </card>
</wml>
</xsl:template>

<xsl:template match="main/linklist">
<select>
<xsl:for-each select="set">
<option>
 <xsl:attribute name="onpick">
  generatexml.asp?sec=<xsl:value-of select="link" />
 </xsl:attribute>
 <xsl:value-of select="link" />
</option>
</xsl:for-each>
</select>

</xsl:template>
</xsl:stylesheet>
```

Let's see how this XSL file works in the real world. There is one noticeable change in the XSL document from the last time we used it. The change lies in the beginning lines of Listing 5-2. First you map a template in a traditional way. Then, instead of writing the HTML code, you write the code for the WML browser. These actions are clarified as follows:

```
<xsl:template match="/">

<wml>
    <card>
 <p>
  Menu
   <xsl:apply-templates select="main/linklist" />
 </p>

   </card>
</wml>
</xsl:template>
```

As you can see, this code constructs a card in which the contents for the WML browser are displayed. Under this code, you use a paragraph section and then apply a template using the `<xsl:apply-template>` function. Later in the code, you fetch the data from the same `main.xml` file that you used for the PC-browser implementation. This time, however, you do not need to fetch all the data from the file. Instead, you fetch just a list of services that you are providing for the WML browser. The process by which you fetch the records from `main.xml` is as follows:

```
<xsl:template match="main/linklist">
<select>
<xsl:for-each select="set">
<option>
 <xsl:attribute name="onpick">
  generatexml.asp?sec=<xsl:value-of select="link" />
```

```
        </xsl:attribute>
        <xsl:value-of select="link" />
    </option>
    </xsl:for-each>
    </select>

</xsl:template>
```

There is one noticeable change in this file that you can easily map in the `<xsl:attribute>` statement. While fetching the records for display, add the `generatexml.asp?sec=` string in front of their URLs. So if you choose the weather section, the URL for that becomes `generatexml.asp?sec=weather`.

Test the XSL Using the Simulator

For testing purposes, we are using an Openwave WAP simulator. In the address bar of the simulator, type the address of the `checkbrows.asp` file. When the `checkbrows.asp` loads, the transformation process is carried out and the result is displayed on the screen of the simulator. The output is shown in Figure 5-2.

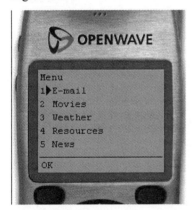

Figure 5-2: The application home page (Image of Up.SDK courtesy Openwave Systems Inc.)

The screen first displays a home page that contains four options. These options are the following:

◆ E-mail

◆ Movies

◆ Weather

◆ News

◆ Resources

Transform the Weather Section

This section discusses the transformation of the application's weather section. But before that, you must take a look at the work flow of the WML transformation. For the first time in the application, you are going to use Active Server Pages (ASP) to generate the XML data for the weather section and then transform it into WML using XSLT.

First, take a look at the `generatexml.asp` file (Listing 5-3), which you use to generate XML data based on the parameter you get from the user.

Listing 5-3: generatexml.asp

```
<%@ Language=VBScript %>

<%
dim address,section,fname,ctg,city,emode
ctg =" "
city=" "
emode =" "
section = " "
section = trim(Request.QueryString("sec"))
emode = trim(Request.QueryString("mode"))
ctg = trim(Request.QueryString("category"))
city = trim(Request.QueryString("city"))


set conn = Server.CreateObject("ADODB.Connection")
conn.ConnectionString = "Provider=SQLOLEDB.1;Persist Security Info=False;User
ID=sa;Initial Catalog=portal;Data Source=developers"
conn.Open

Set objrs = Server.CreateObject("ADODB.Recordset")
objrs.CursorType = adOpenStatic

if (section="Weather")  then
  sql = "select city from weather"
  objrs.Open sql,conn

  Set fso =Server.CreateObject("Scripting.FileSystemObject")
  fname = "tweather.xml"

  fullpath = server.MapPath("main")

  Set MyFile=fso.CreateTextFile(fullpath & "/" & fname)

  MyFile.WriteLine("<?xml version='1.0'?>")
  MyFile.WriteLine("<ROOT>")


  while not(objrs.EOF)
    city = objrs.Fields("city")
    MyFile.WriteLine("<city>" & city & "</city>")

  objrs.MoveNext
  wend

  MyFile.WriteLine("</ROOT>")
  address = "http://192.168.1.100/loadxsl.asp?name=" & fname
  Response.Redirect(address)

end if

if ((city="Sydney") OR (city="NewDelhi") OR (city="Miami") OR (city="Tokyo") OR
(city="Phoenix") OR (city="NewYork"))  then
  sql = "select * from weather where city = '" & city & "'"
```

```
objrs.Open sql,conn

Set fso =Server.CreateObject("Scripting.FileSystemObject")
fname = "tweatherdetail.xml"

fullpath = server.MapPath("main")

Set MyFile=fso.CreateTextFile(fullpath & "/" & fname)

MyFile.WriteLine("<?xml version='1.0'?>")
MyFile.WriteLine("<ROOT>")


while not(objrs.EOF)
   lowtemp = objrs.Fields("low_temp")
   hightemp = objrs.Fields("high_temp")
   srise = objrs.Fields("sunrise")
   sset = objrs.Fields("sunset")
   mrise = objrs.Fields("moonrise")
   mset = objrs.Fields("moonset")
   dhumid = objrs.Fields("dayhumidity")
   nhumid = objrs.Fields("nighthumidity")
   rain = objrs.Fields("rainfall")
   fig = objrs.Fields("figure")

   MyFile.WriteLine("<weather>")
   MyFile.WriteLine("<low_temp>" & lowtemp & "</low_temp>")
   MyFile.WriteLine("<high_temp>" & hightemp & "</high_temp>")
   MyFile.WriteLine("<sunrise>" & srise & "</sunrise>")
   MyFile.WriteLine("<sunset>" & sset & "</sunset>")
   MyFile.WriteLine("<moonrise>" & mrise & "</moonrise>")
   MyFile.WriteLine("<moonset>" & mset & "</moonset>")
   MyFile.WriteLine("<dayhumidity>" & dhumid & "</dayhumidity>")
   MyFile.WriteLine("<nighthumidity>" & nhumid & "</nighthumidity>")
   MyFile.WriteLine("<rainfall>" & rain & "</rainfall>")
   MyFile.WriteLine("<figure>" & fig & "</figure>")
   MyFile.WriteLine("</weather>")
 objrs.MoveNext
 wend

 MyFile.WriteLine("</ROOT>")
 address = "http://192.168.1.100/loadxsl.asp?name=" & fname
 Response.Redirect(address)

end if



objrs.Close
conn.Close

%>
```

This is one of the most important pieces of code in the application; in this file, you generate all kinds of XML data for the application's sections on the basis of different parameters that you receive. Let's see how it's done.

Starting at the beginning of Listing 5-3, you first declare some variables for holding the values you sent using the query string method, as shown in the following code. In the case of the weather section, some variables may not be used. You will use them later in other sections of the portal.

```
dim address,section,fname,ctg,city,emode
ctg =" "
city=" "
emode =" "
section = " "
section = trim(Request.QueryString("sec"))
emode = trim(Request.QueryString("mode"))
ctg = trim(Request.QueryString("category"))
city = trim(Request.QueryString("city"))
```

After declaring the variables, you go on to establish a connection with the database that you are using. We are using Microsoft SQL Server 2000 as a database server. If you use a different database, you have to change the connection string accordingly. The following listing shows the statements for establishing a database connection with the SQL Server 2000 database server.

```
set conn = Server.CreateObject("ADODB.Connection")
conn.ConnectionString = "Provider=SQLOLEDB.1;Persist Security Info=False;User
ID=sa;Initial Catalog=portal;Data Source=developers"
conn.Open

Set objrs = Server.CreateObject("ADODB.Recordset")
objrs.CursorType = adOpenStatic
```

After this, you check for the parameter passed to you from `checkbrows.asp` by using the query string and providing the IF condition in code. If the parameter you get is `weather`, you generate XML data for displaying the name of all the cities in the database. You then save the XML files on the local hard drive so that they are available for the transformation process that will allow you to display data on a WML browser.

First you check the condition passed to the section variable. If it is `weather`, you select the names of all the cities from the table. You then use a file system object (FSO) and create an XML file named `tweather.xml` in the main directory of the server, as seen in the following code. You use this file for saving the names of all the cities in an XML format.

```
if (section="Weather")  then
  sql = "select state from weather"
  objrs.Open sql,conn

  Set fso =Server.CreateObject("Scripting.FileSystemObject")
  fname = "tweather.xml"

  fullpath = server.MapPath("main")

  Set MyFile=fso.CreateTextFile(fullpath & "/" & fname)
```

After creating a file on the server, you now write the data into it in XML format using the following lines of code:

```
  MyFile.WriteLine("<?xml version='1.0'?>")
  MyFile.WriteLine("<ROOT>")


  while not(objrs.EOF)
    city = objrs.Fields("city")
```

```
      MyFile.WriteLine("<city>" & city & "</city>")

 objrs.MoveNext
 wend

 MyFile.WriteLine("</ROOT>")
 address = "http://192.168.1.100/loadxsl.asp?name=" & fname
 Response.Redirect(address)

end if
```

In the preceding code, you write the XML data into `tweather.xml` and then redirect control to `loadxsl.asp` (Listing 5-4). This performs the transformation of `tweather.xml` by using the `Response.redirect` method. While redirecting, you pass the name of the file as a parameter using the query string method.

Figure 5-3 is a screen shot of the code that contains the names of all the cities, in XML format, in the file `tweather.xml`.

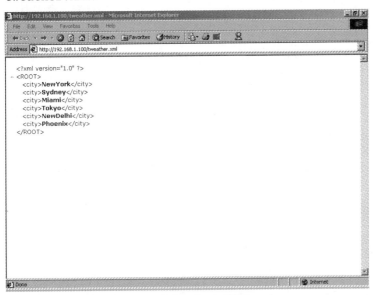

Figure 5-3: Showing the names of different cities in XML format

Listing 5-4: loadxsl.asp

```
<%@ Language=VBScript %>

<%
fname = trim(Request.QueryString("name"))


'declaring variables
dim sXML
dim sXSL
dim xmlDocument,xslDocument

'creating the objects for the XMLDOM
set xmlDocument = Server.CreateObject("MSXML2.DOMDocument")
set xslDocument = Server.CreateObject("MSXML2.DOMDocument")
```

```
sXML = fname
fname = mid(fname,1,(len(fname) - 4))


if (session.Contents(1)="wml") then
 sXSL = fname & "wml.xsl"

 xmlDocument.async = false
 xslDocument.async = false

 xmlDocument.load(Server.MapPath(sXML))
 xslDocument.load(Server.MapPath(sXSL))

 Response.ContentType = "text/vnd.wap.wml"
 Response.Write "<?xml version=""1.0"" ?>"


 Response.Write "<!DOCTYPE wml PUBLIC ""-//WAPFORUM//DTD " & _
 "WML 1.1//EN"" ""http://www.wapforum.org/DTD/wml_1.1.xml"">"

 Response.Write(xmlDocument.transformNode(xslDocument))

end if

%>
```

In `loadxsl.asp`, you first declare some variables for storing the values that you get from the query string. After this, you create two XMLDOM objects for performing the transformation process. Finally, you store the value in the sXML variable (the name of the XML document) and then use the mid function to trim off the XML document's file extension, as follows:

```
sXML = fname
fname = mid(fname,1,(len(fname) - 4))
```

Suppose you have `tweather.xml` as a value in fname. After trimming the four characters from it, you now have `tweather` as a value in fname.

In the next line, you check the mode for the transformation. In this case, the mode is WML. Having checked the mode, you set the value of the sXSL variable (the name of the XSLT document) and then load both the XML document and the XSLT document by writing the following lines of code:

```
if (session.Contents(1)="wml") then
 sXSL = fname & "wml.xsl"

 xmlDocument.async = false
 xslDocument.async = false

 xmlDocument.load(Server.MapPath(sXML))
 xslDocument.load(Server.MapPath(sXSL))
```

After checking the mode, you set the value of the sXSL variable by adding the fname in front of the wml.xsl string. So for the `tweather.xml` file, the value of the XSL file becomes `tweatherwml.xsl`, and so on. In the next few lines, you set the content type and write the WML prolog for the WML browser. In the last line of the listing, you perform a transformation process for the `tweather.xml` file and display the names of all the cities on the WML device.

Take a look at the `tweatherwml.xsl` file in Listing 5-5, which transforms the XML data into WML format.

Listing 5-5: tweatherwml.xsl

```
<?xml version='1.0'?>
<xsl:stylesheet xmlns:xsl="http://www.w3.org/1999/XSL/Transform" version="1.0">
  <xsl:output omit-xml-declaration="yes"/>
<xsl:template match="/">

<wml>
<head>
 <meta http-equiv="Cache-Control" content="max-age=time" forua="true"/>
</head>
    <card>
   <do type="accept" label="Next">
    <go href="checkbrows.asp" />
   </do>
   <p>
    <xsl:apply-templates select="ROOT" />
   </p>
    </card>
</wml>

</xsl:template>


<xsl:template match="ROOT">
Sections
<select>
<xsl:for-each select="city">
<option>
    <xsl:attribute name="onpick">
    generatexml.asp?city=<xsl:value-of select="." />
    </xsl:attribute>
    <xsl:value-of select="." />
</option>
</xsl:for-each>
</select>

</xsl:template>

</xsl:stylesheet>
```

This is a very simple XSLT document that was created for the transformation of the `tweather.xml` file. This code creates one card and then uses a `<xsl:apply-template>` function to insert all the records that you have fetched from the document.

While fetching the records from the XML file, you also set the URL for each record using the `<xsl:attribute>` function. You again call the `generatexml.asp` file, but this time with a different parameter named `city`, which holds the name of the particular city for which you will display the weather details.

Figure 5-4 shows the result of the transformation.

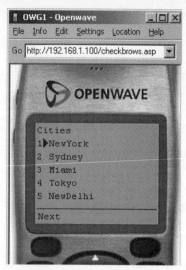

Figure 5-4: Displays the names of cities (Image of Up.SDK courtesy Openwave Systems Inc.)

The user now selects a city from one of the many names that are displayed on the browser. You again call the generatexml.asp file with the city parameter in the query string. In generatexml.asp, you check for the parameter. On the basis of the parameter, you perform the following action in generatexml.asp:

```
if ((city="Sydney") OR (city="NewDelhi") OR (city="Miami") OR (city="Tokyo") OR
(city="Phoenix") OR (city="NewYork"))  then
  sql = "select * from weather where city = '" & city & "'"
  objrs.Open sql,conn

  Set fso =Server.CreateObject("Scripting.FileSystemObject")
  fname = "tweatherdetail.xml"

  fullpath = server.MapPath("main")

  Set MyFile=fso.CreateTextFile(fullpath & "/" & fname)

  MyFile.WriteLine("<?xml version='1.0'?>")
  MyFile.WriteLine("<ROOT>")


  while not(objrs.EOF)
    lowtemp = objrs.Fields("low_temp")
    hightemp = objrs.Fields("high_temp")
    srise = objrs.Fields("sunrise")
    sset = objrs.Fields("sunset")
    mrise = objrs.Fields("moonrise")
    mset = objrs.Fields("moonset")
    dhumid = objrs.Fields("dayhumidity")
    nhumid = objrs.Fields("nighthumidity")
    rain = objrs.Fields("rainfall")
    fig = objrs.Fields("figure")

    MyFile.WriteLine("<weather>")
    MyFile.WriteLine("<low_temp>" & lowtemp & "</low_temp>")
    MyFile.WriteLine("<high_temp>" & hightemp & "</high_temp>")
```

```
        MyFile.WriteLine("<sunrise>" & srise & "</sunrise>")
        MyFile.WriteLine("<sunset>" & sset & "</sunset>")
        MyFile.WriteLine("<moonrise>" & mrise & "</moonrise>")
        MyFile.WriteLine("<moonset>" & mset & "</moonset>")
        MyFile.WriteLine("<dayhumidity>" & dhumid & "</dayhumidity>")
        MyFile.WriteLine("<nighthumidity>" & nhumid & "</nighthumidity>")
        MyFile.WriteLine("<rainfall>" & rain & "</rainfall>")
        MyFile.WriteLine("<figure>" & fig & "</figure>")
        MyFile.WriteLine("</weather>")
    objrs.MoveNext
    wend

    MyFile.WriteLine("</ROOT>")
    address = "http://192.168.1.100/loadxsl.asp?name=" & fname
    Response.Redirect(address)

end if
```

First you check for the name of the city passed to you. Then, on the basis of the city name, you collect the records from the database and save them in an XML file named `tweatherdetail.xml` using the file system object. After saving the XML file, you again redirect control to the `loadxsl.asp` file for performing the transformation process for the city's weather details.

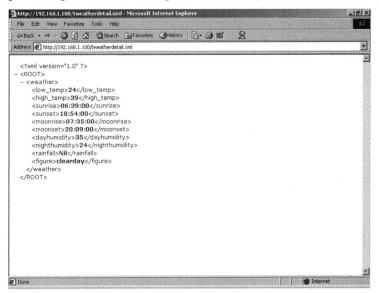

Figure 5-5: A screen shot of weather details generated by generatexml.asp

Now take a look at the code for `tweatherdetailwml.xsl` (Listing 5-6), which is used for transforming the detailed weather report for a city.

Listing 5-6: tweatherdetailwml.xsl

```
<?xml version='1.0'?>
<xsl:stylesheet xmlns:xsl="http://www.w3.org/1999/XSL/Transform" version="1.0">
  <xsl:output omit-xml-declaration="yes"/>

<xsl:template match="/">

<wml>
```

```
<head>
 <meta http-equiv="Cache-Control" content="max-age=time" forua="true"/>
</head>
    <card>
  <do type="accept" label="Next">
   <go href="checkbrows.asp" />
  </do>
  <p>
   <xsl:apply-templates select="ROOT" />
  </p>


   </card>
</wml>

</xsl:template>

<xsl:template match="ROOT">

<table columns="2">
  <xsl:for-each select="weather">

  <tr>
     <td><b>Low Temp-</b></td>
     <td><xsl:value-of select="low_temp" /></td>
   </tr>
   <tr>
     <td><b>High Temp-</b></td>
     <td><xsl:value-of select="high_temp" /></td>
  </tr>
  <tr>
     <td><b>Sunrise-</b></td>
     <td><xsl:value-of select="sunrise" /></td>
  </tr>
  <tr>
     <td> <b>Sunset-</b></td>
     <td><xsl:value-of select="sunset" /></td>
  </tr>
  <tr>
     <td> <b>Moonrise-</b></td>
     <td><xsl:value-of select="moonrise" /></td>
  </tr>
  <tr>
     <td> <b>Moonset-</b></td>
     <td><xsl:value-of select="moonset" /></td>
  </tr>
  <tr>
     <td> <b>DayHumidity-</b></td>
     <td><xsl:value-of select="dayhumidity" /></td>
  </tr>
  <tr>
      <td><b>NightHumidity-</b></td>
     <td><xsl:value-of select="nighthumidity" /></td>
   </tr>
   <tr>
     <td> <b>Rainfall-</b></td>
     <td><xsl:value-of select="rainfall" /></td>
```

```
      </tr>

</xsl:for-each>
 </table>

</xsl:template>

</xsl:stylesheet>
```

Let's now discuss the following lines of code in the XSLT documents that you have used for the transformation process. You generate a card in the same manner you did before, after generating the card for the WML document. You start fetching the records from the XML files and displaying them in a tabular format using the following lines of code in the XSLT document:

```
<xsl:template match="ROOT">

<table columns="2">
  <xsl:for-each select="weather">

  <tr>
     <td><b>Low Temp-</b></td>
     <td><xsl:value-of select="low_temp" /></td>
   </tr>
   <tr>
     <td><b>High Temp-</b></td>
     <td><xsl:value-of select="high_temp" /></td>
   </tr>
   <tr>
     <td><b>Sunrise-</b></td>
     <td><xsl:value-of select="sunrise" /></td>
   </tr>
   <tr>
     <td> <b>Sunset-</b></td>
     <td><xsl:value-of select="sunset" /></td>
   </tr>
   <tr>
     <td> <b>Moonrise-</b></td>
     <td><xsl:value-of select="moonrise" /></td>
   </tr>
   <tr>
     <td> <b>Moonset-</b></td>
     <td><xsl:value-of select="moonset" /></td>
   </tr>
   <tr>
     <td> <b>DayHumidity-</b></td>
     <td><xsl:value-of select="dayhumidity" /></td>
   </tr>
   <tr>
      <td><b>NightHumidity-</b></td>
     <td><xsl:value-of select="nighthumidity" /></td>
    </tr>
    <tr>
     <td> <b>Rainfall-</b></td>
     <td><xsl:value-of select="rainfall" /></td>
    </tr>

</xsl:for-each>
```

```
</table>
```

```
</xsl:template>
```

You construct a loop so that you can fetch all the records contained in the XML documents and display these in a tabular format. The output of the transformation is shown in Figures 5-6 and 5-7.

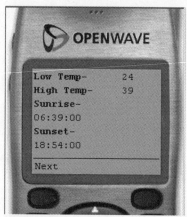

Figure 5-6: The output of the weather details (Image of Up.SDK courtesy Openwave Systems Inc.)

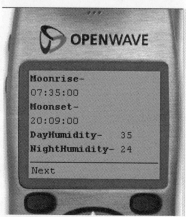

Figure 5-7: The output of the weather details (continued) (Image of Up.SDK courtesy Openwave Systems Inc.)

This completes the process of transforming the weather section for the WML client. In the next section, you will see how to transform the news section of the application for the WML client using the same methods and techniques.

Transform the News Section

The news section is divided into four categories so that the user can choose one of them and read the news for that particular category. Figure 5-8 is a diagram of the work flow of the news section.

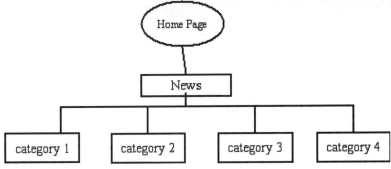

Figure 5-8: The work flow of the application's news section

Generate XML for the News Category Section

First you have to generate the XML data for the news section by using the same `generatexml.asp` file that we used before for the weather section and append code for the news section(Listing 5-7), so that you can use this data for the transformation process.

Listing 5-7: Addition to generatexml.asp

```
if (section="News")   then
  sql = "select DISTINCT category from news"
  objrs.Open sql,conn

  Set fso =Server.CreateObject("Scripting.FileSystemObject")
  fname = "tnews.xml"

  fullpath = server.MapPath("main")

  Set MyFile=fso.CreateTextFile(fullpath & "/" & fname)

  MyFile.WriteLine("<?xml version='1.0'?>")
  MyFile.WriteLine("<ROOT>")


  while not(objrs.EOF)

    category = objrs.Fields("category")

    MyFile.WriteLine("<category>" & category & "</category>")

  objrs.MoveNext
  wend

  MyFile.WriteLine("</ROOT>")
  address = "http://192.168.1.100/loadxsl.asp?name=" & fname
  Response.Redirect(address)

end if
```

You append this code in the `generatexml.asp` file after the weather section. This code also has the same functionality of generating xml data: It generates XML data for displaying the lists of categories and saves the data in XML format in a file named `tnews.xml` (Figure 5-9). After saving the data in XML format, you redirect control to `loadxsl.asp` for undertaking the transformation of the categories listing.

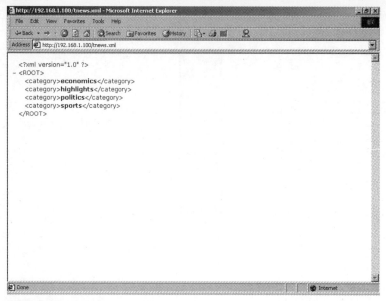

Figure 5-9: tnews.xml

Write XSLT for the News Category Section

In `loadxsl.asp`, you load the XML document, as well as the XSL document (`tnewswml.xsl`) for performing the transformation operation. Listing 5-8 is the code of the `tnewswml.xsl` file that is used for the transformation process.

Listing 5-8: tnewswml.xsl

```
<?xml version='1.0'?>
<xsl:stylesheet xmlns:xsl="http://www.w3.org/1999/XSL/Transform" version="1.0">
  <xsl:output omit-xml-declaration="yes"/>

<xsl:template match="/">

<wml>
<head>
 <meta http-equiv="Cache-Control" content="max-age=time" forua="true"/>
</head>
<card>
   <do type="accept" label="Next">
    <go href="checkbrows.asp" />
   </do>
    <p>
    <xsl:apply-templates select="ROOT" />
    </p>
</card>

</wml>
</xsl:template>


<xsl:template match="ROOT">
Sections
<select>
```

```
<xsl:for-each select="category">
<option>
   <xsl:attribute name="onpick">
   generatexml.asp?category=<xsl:value-of select="." />
   </xsl:attribute>
   <xsl:value-of select="." />
</option>
</xsl:for-each>
</select>
</xsl:template>

</xsl:stylesheet>
```

In Listing 5-8, you generate a card for the WML browser and then fetch the values from `tnews.xml` and display them on a WML browser. If the user selects any category from the ones that you have listed, you again call `generatexml.asp` to generate XML data for that particular category. When you perform the transformation process, it shows the output screen in Figure 5-10.

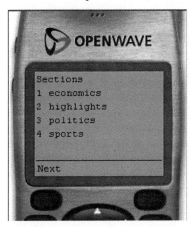

Figure 5-10: The transformation of tnews.xml (Image of Up.SDK courtesy Openwave Systems Inc.)

Generate XML for the News Details Section

When the user selects a category to browse through the news, you generate XML data for the news details for that category, as shown in the following code:

```
if ((ctg="highlight") OR (ctg="politics") OR (ctg="economics") OR
(ctg="sports"))  then
  sql = "select * from news where category = '" & ctg & "'"
  objrs.Open sql,conn

  Set fso =Server.CreateObject("Scripting.FileSystemObject")
  fname = "tnewsdetail.xml"
  fullpath = server.MapPath("main")
  Set MyFile=fso.CreateTextFile(fullpath & "/" & fname)

  MyFile.WriteLine("<?xml version='1.0'?>")
  MyFile.WriteLine("<ROOT>")

    while not(objrs.EOF)
      heading = objrs.Fields("heading")
      category = objrs.Fields("category")
      description = objrs.Fields("description")
```

```
    newsdate = objrs.Fields("news_date")
    MyFile.WriteLine("<news>")
    MyFile.WriteLine("<category>" & category & "</category>")
    MyFile.WriteLine("<heading>" & heading & "</heading>")
    MyFile.WriteLine("<description>" & description & "</description>")
    MyFile.WriteLine("<news_date>" & newsdate & "</news_date>")
    MyFile.WriteLine("</news>")
objrs.MoveNext
wend

MyFile.WriteLine("</ROOT>")
address = "http://192.168.1.100/loadxsl.asp?name=" & fname
Response.Redirect(address)

end if
```

You select all the news from the database based on the category the user selects and then use the file system object to create and save an XML file on the server. This file contains the news details in XML format. Later in the code, you again call the `loadxsl.asp` file for transforming the operation of the `tnewsdetail.xml` file, which you just created using the file system object.

Figure 5-11 is a screen shot of `tnewsdetail.xml`.

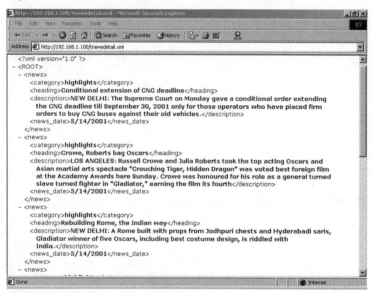

Figure 5-11: Screen shot of tnewsdetail.xml

Write XSLT for the News Details Section

Now take a look at the `tnewsdetailwml.xsl` file (Listing 5-9), which you are using for the news details section's transformation process. You load this XSLT document and the XML document from the `loadxsl.asp` file to perform the transformation operation.

Listing 5-9: tnewsdetailwml.xsl

```
<?xml version='1.0'?>
<xsl:stylesheet xmlns:xsl="http://www.w3.org/1999/XSL/Transform" version="1.0">
  <xsl:output omit-xml-declaration="yes"/>
```

```
<xsl:template match="/">

<wml>
<head>
 <meta http-equiv="Cache-Control" content="max-age=time" forua="true"/>
</head>
    <card>
  <do type="accept" label="Next">
   <go href="checkbrows.asp" />
  </do>
  <p>
   <xsl:apply-templates select="ROOT" />
  </p>


  </card>

</wml>
</xsl:template>


<xsl:template match="ROOT">

  <xsl:for-each select="news">

     <b><xsl:value-of select="heading" /></b><br />

     <xsl:value-of select="description" /><br /><br/>


  </xsl:for-each>

</xsl:template>

</xsl:stylesheet>
```

Listing 5-9 is a simple XSLT document. With it you fetch from the XML document the values of two nodes, named HEADING and DESCRIPTION, and display them in a card. The result of the transformation process is shown in Figure 5-12.

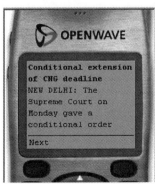

Figure 5-12: Displaying news details (Image of Up.SDK courtesy Openwave Systems Inc.)

This completes the transformation of the news area of the application for WAP clients. Now there are two more sections of the application left to transform.

Transform the E-mail Section

In this section, you will transform the e-mail section of the application. After this transformation process, the user will be able to send and receive e-mail using his WAP device. The e-mail section is divided into two parts: sending and receiving messages via the WAP device.

Design Login Forms for the E-mail Section

First, take a look at the login section for the e-mail system. There are two options on the login page, read mail and send mail. To start the transformation process, go back to the `generatexml.asp` file and append some lines of code (Listing 5-10) for generating the XML data to display the login page on the WAP device.

Listing 5-10: generatexml.asp (modified)

```
if (section="E-mail") then

  Set fso =Server.CreateObject("Scripting.FileSystemObject")
  fname = "tlogin.xml"

  fullpath = server.MapPath("main")
  Set MyFile=fso.CreateTextFile(fullpath & "/" & fname)

  MyFile.WriteLine("<?xml version='1.0'?>")
  MyFile.WriteLine("<logininfo>")
  MyFile.WriteLine("<username />")
  MyFile.WriteLine("<password />")
  MyFile.WriteLine("<login />")
  MyFile.WriteLine("</logininfo>")

  address = "http://192.168.1.100/loadxsl.asp?name=" & fname
  Response.Redirect(address)
end if
```

Here you again use a file on the server, using the file system object and saving it as `tlogin.xml`. This file is used to store some empty elements. The file's structure is exactly the same as the file that you used in the PC browser implementation of the e-mail section.

A screen shot of `tlogin.xml` is shown in Figure 5-13.

Figure 5-13: A screen shot of tlogin.xml in the PC browser

Now you again redirect control to the `loadxsl.asp` file to load both the XML and XSL documents and perform the transformation process. You use the `tloginwml.xsl` document (Listing 5-11) as the XSLT source for the transformation.

Listing 5-11: tloginwml.xsl

```
<?xml version='1.0'?>
<xsl:stylesheet xmlns:xsl="http://www.w3.org/1999/XSL/Transform" version="1.0">
<xsl:output omit-xml-declaration="yes"/>
<xsl:template match="/">
<wml>
<head>
<meta http-equiv="Cache-Control" content="max-age=time" forua="true"/>
</head>
 <card>
    <do type="accept" label="Next">
    <go href="checkbrows.asp" />
    </do>
    <p>
    <select>
     <option onpick="generatexml.asp?mode=read">Read Mails</option>
     <option onpick="generatexml.asp?mode=send">Send Mail</option>
    </select>
    </p>
 </card>
</wml>
</xsl:template>
</xsl:stylesheet>
```

In this file, you generate a simple card. This card then helps you generate two options for the user for reading and sending e-mail. When the user chooses an option, you call the `generatexml.asp` file and pass a parameter named `mode` in the query string, which contains the mode of selection. We have two options for the `mode` parameter named `read` and `send`. It can be read or sent depending on the user's choice.

When you apply this XSL to the XML document, it transforms the XML content into WML content and displays the output in Figure 5-14.

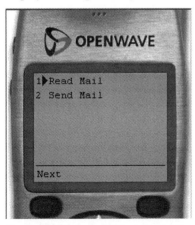

Figure 5-14: The output of tloginwml.xsl (Image of Up.SDK courtesy Openwave Systems Inc.)

Read E-mail Using WAP Clients

When the user selects the Read Mail option, you again call the `generatexml.asp` file with a parameter in the query string named `mode` that contains the selected mode. Now you again append more lines of code in the `generatexml.asp` file to check the mode and the browser type. Then finally, on the basis of the browser type, you call one more file named `loadmailwml.asp` with some parameters in the query string. The following is the code that you append in `generatexml.asp`.

```
if (emode <> " ") then

  Set fso =Server.CreateObject("Scripting.FileSystemObject")
  fname = "tlogin.xml"

  fullpath = server.MapPath("main")
  Set MyFile=fso.CreateTextFile(fullpath & "/" & fname)

  MyFile.WriteLine("<?xml version='1.0'?>")
  MyFile.WriteLine("<logininfo>")
  MyFile.WriteLine("<username />")
  MyFile.WriteLine("<password />")
  MyFile.WriteLine("<login />")
  MyFile.WriteLine("</logininfo>")

  if (session.Contents(1)="wml") then
  address = "http://192.168.1.100/loadmailwml.asp?name=" & fname & "&mode=" &
emode
  end if
  if (session.Contents(1)="imode") then
  address = "http://192.168.1.100/loadmailimode.asp?name=" & fname & "&mode=" &
emode
  end if
  if (session.Contents(1)="hdml") then
  address = "http://192.168.1.100/loadmailhdml.asp?name=" & fname & "&mode=" &
emode
  end if
  Response.Redirect(address)
end if
```

Listing 5-12 shows what you are doing in the `loadmailwml.asp` file.

Listing 5-12: loadmailwml.asp

```
<%@ Language=VBScript %>

<%
fname = trim(Request.QueryString("name"))
emode = trim(Request.QueryString("mode"))


'declaring variables
dim client
dim sXML
dim sXSL
dim xmlDocument,xslDocument

'creating the objects for the XMLDOM
set xmlDocument = Server.CreateObject("MSXML2.DOMDocument")
```

```
set xslDocument = Server.CreateObject("MSXML2.DOMDocument")

sXML = fname
fname = mid(fname,1,(len(fname) - 4))

if (emode="read") then
 sXSL = "readwml.xsl"
end if

xmlDocument.async = false
xslDocument.async = false

xmlDocument.load(Server.MapPath(sXML))
xslDocument.load(Server.MapPath(sXSL))

Response.ContentType = "text/vnd.wap.wml"
Response.Write "<?xml version=""1.0"" ?>"


Response.Write "<!DOCTYPE wml PUBLIC ""-//WAPFORUM//DTD " & _
"WML 1.1//EN"" ""http://www.wapforum.org/DTD/wml_1.1.xml"">"

Response.Write(xmlDocument.transformNode(xslDocument))
%>
```

Take a closer look at Listing 5-12. In the beginning of the file, you declared some variables for storing the values that come to you from the query string. After declaring the variables, you create two objects for the XMLDOM; these objects perform a transformation process later.

In the next part of the code, you check for the mode the user selected — read mode or send mode. You then assign a value to the sXSL variable on the basis of the mode the user selected. In this case, you assign readwml.xsl to the sXSL variable.

After assigning the values, you load both the XML and XSL documents and set the content type for the browser. You also write some WML prolog for the WML browser. Then you start the transformation process in the last line of code.

The XSL document in Listing 5-13 contains code for asking the user for his username and password. These are used to verify the account information with the server.

Listing 5-13: readwml.xsl

```
<?xml version='1.0'?>
<xsl:stylesheet xmlns:xsl="http://www.w3.org/1999/XSL/Transform" version="1.0">
<xsl:output omit-xml-declaration="yes"/>

<xsl:template match="/">
<wml>
<head>
<meta http-equiv="Cache-Control" content="max-age=time" forua="true"/>
</head>
<card  newcontext="true">

<onevent type="onenterforward">
 <refresh>
     <setvar name="username" value="" />
     <setvar name="password" value="" />
  </refresh>
```

```
</onevent>

    <do type="accept" label="Next">
      <go href="receivewml.asp?uname=$username&amp;pwd=$password" />
    </do>

    <p>
     <xsl:apply-templates select="logininfo" />
    </p>

</card>

</wml>
</xsl:template>


<xsl:template match="logininfo">

  <xsl:for-each select="username">
     Enter the User Name
     <input name="username" />
  </xsl:for-each>


  <xsl:for-each select="password">
     Enter the Password
     <input type="password" name="password" />
  </xsl:for-each>

</xsl:template>

</xsl:stylesheet>
```

In this document, you first generate a card and then declare two variables for passing the values to the next file in which you are going to verify the username and password. The following lines of code can easily do this:

```
<onevent type="onenterforward">
 <refresh>
     <setvar name="username" value="" />
     <setvar name="password" value="" />
  </refresh>
</onevent>
```

When the user fills in the form with his username and password, you then generate a <do> condition on-the-fly and specify the target file name, as in the following code:

```
    <do type="accept" label="Next">
      <go href="receivewml.asp?uname=$username&amp;pwd=$password" />
    </do>

    <p>
     <xsl:apply-templates select="logininfo" />
    </p>
```

In the preceding code, you specify receivewml.asp as an action file when the user fills in the form and use the <xsl:apply> template statement to construct the form for the user. In the following code, you are constructing a form for the user to fill in the information. When this style sheet is executed, it first asks for the username; then it asks for the password (Figures 5-15 and 5-16).

```
<xsl:template match="logininfo">

  <xsl:for-each select="username">
      Enter the User Name
      <input name="username" />
  </xsl:for-each>


  <xsl:for-each select="password">
      Enter the Password
      <input type="password" name="password" />
  </xsl:for-each>

</xsl:template>
```

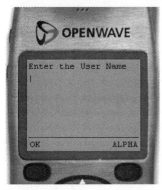

Figure 5-15: Username request (Image of Up.SDK courtesy Openwave Systems Inc.)

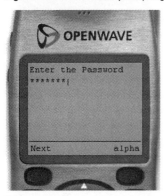

Figure 5-16: Password entry (Image of Up.SDK courtesy Openwave Systems Inc.)

When the user selects Next after filling in the password, it calls the `receivewml.asp` file. The username and password are passed on to it as a parameter in the query string.

Now take a look at the `receivewml.asp` code in Listing 5-14.

Listing 5-14: receivewml.asp

```
<%@LANGUAGE=VBSCRIPT %>

<%
```

```
uname = trim(Request.QueryString("uname"))
pwd = trim(Request.QueryString("pwd"))


Response.ContentType = "text/vnd.wap.wml"

Response.Write "<!DOCTYPE wml PUBLIC ""-//WAPFORUM//DTD " & _
"WML 1.1//EN"" ""http://www.wapforum.org/DTD/wml_1.1.xml"">"
%>

<wml>
<head>
 <meta http-equiv="Cache-Control" content="max-age=time" forua="true"/>
</head>
<card>
<do type="accept" label="Next">
 <go href="checkbrows.asp" />
</do>

<p>
<b><i>Inbox</i></b><br />--------

<%
Set pop3 = Server.CreateObject( "Jmail.pop3" )
pop3.Connect uname, pwd, "http://192.168.1.100"

c = 1
if (pop3.Count>0) then
while (c<=pop3.Count)
 Set msg = pop3.Messages.item(c)

 ReTo = ""
 ReCC = ""

 Set Recipients = msg.Recipients
 delimiter = ", "



 For i = 0 To Recipients.Count - 1
  If i = Recipients.Count - 1 Then
   delimiter = ""
  End If

  Set re = Recipients.item(i)
  If re.ReType = 0 Then
   ReTo = ReTo & re.Name & " (" & re.EMail & ")" & delimiter
  else
   ReCC = ReTo & re.Name & " (" & re.EMail & ")" & delimiter
  End If
 Next


%>
```

```
<table columns="2">

 <tr>
  <td><b>Subject</b><br /></td>
  <td><%= msg.Subject %></td>
 </tr>

 <tr>
  <td><b>From</b><br /></td>
  <td><%= msg.FromName %></td>
 </tr>


 <tr>
  <td><b>Body</b><br /></td>
  <td><%= msg.Body %></td>
 </tr>
</table>

<%
c = c + 1
wend
end if
pop3.Disconnect

%>
</p>
</card>
</wml>
```

In Listing 5-14, you are creating an object to connect with the mail server and then verify the user's username, password, and so on. After this, you collect all the messages from the inbox.

After fetching the messages from the inbox, you display the messages one by one in a tabular format on the WAP browser using the following lines of code:

```
<table columns="2">

 <tr>
  <td><b>Subject</b><br /></td>
  <td><%= msg.Subject %></td>
 </tr>

 <tr>
  <td><b>From</b><br /></td>
  <td><%= msg.FromName %></td>
 </tr>


 <tr>
  <td><b>Body</b><br /></td>
  <td><%= msg.Body %></td>
 </tr>
</table>
```

This code gives the output in Figure 5-17.

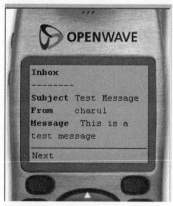

Figure 5-17: Displaying mail on the WAP device (Image of Up.SDK courtesy Openwave Systems Inc.)

Transform the Compose Mail Section for WAP Clients

Now you move on to the next part of the mail program, which is the send mail section. This code enables the user to send e-mail using his WAP device. Let's see how the process actually occurs. If the user selects the Send Mail option, you again call loadmailwml.asp (Listing 5-15).

Listing 5-15: loadmailwml.asp (modified version)

```
<%@ Language=VBScript %>

<%
fname = trim(Request.QueryString("name"))
emode = trim(Request.QueryString("mode"))


'declaring variables
dim client
dim sXML
dim sXSL
dim xmlDocument,xslDocument

'creating the objects for the XMLDOM
set xmlDocument = Server.CreateObject("MSXML2.DOMDocument")
set xslDocument = Server.CreateObject("MSXML2.DOMDocument")
sXML = fname
fname = mid(fname,1,(len(fname) - 4))

if (emode="read") then
 sXSL = "readwml.xsl"
end if

if (emode="send") then
 sXSL = "sendwml.xsl"
end if


xmlDocument.async = false
xslDocument.async = false

xmlDocument.load(Server.MapPath(sXML))
xslDocument.load(Server.MapPath(sXSL))
```

```
Response.ContentType = "text/vnd.wap.wml"
Response.Write "<?xml version=""1.0"" ?>"


Response.Write "<!DOCTYPE wml PUBLIC ""-//WAPFORUM//DTD " & _
"WML 1.1//EN"" ""http://www.wapforum.org/DTD/wml_1.1.xml"">"

Response.Write(xmlDocument.transformNode(xslDocument))
%>
```

If the user selects the Send Mail option, you set a new value to the sXSL variable, as shown in the following lines of code:

```
if (emode="send") then
 sXSL = "sendwml.xsl"
end if
```

This loads the XML document with a different style sheet document named sendwml.xsl (Listing 5-16).

Listing 5-16: sendwml.xsl

```
<?xml version='1.0'?>
<xsl:stylesheet xmlns:xsl="http://www.w3.org/1999/XSL/Transform" version="1.0">
<xsl:output omit-xml-declaration="yes"/>

<xsl:template match="/">

<wml>

<head>
 <meta http-equiv="Cache-Control" content="max-age=time" forua="true"/>
</head>
 <card  newcontext="true">
   <onevent type="onenterforward">
    <refresh>
      <setvar name="username" value="" />
      <setvar name="password" value="" />
    </refresh>
   </onevent>

  <do type="accept" label="Next">
   <go href="sendwml.asp?uname=$username&amp;pwd=$password" />
  </do>

  <p>
   <xsl:apply-templates select="logininfo" />
  </p>

 </card>

</wml>
</xsl:template>

<xsl:template match="logininfo">

 <xsl:for-each select="username">
```

```
  Enter the User Name
  <input name="username" />
 </xsl:for-each>

 <xsl:for-each select="password">

  Enter the Password
  <input type="password" name="password" />
 </xsl:for-each>

</xsl:template>

</xsl:stylesheet>
```

This file contains almost the same code as `receivewml.asp`; but this code has a slight change. When the user presses the Next button after filling in his username and the password, the code calls the `sendwml.asp` file with the username and password as a parameter in the query string, as seen in the following code:

```
<do type="accept" label="Next">
  <go href="sendwml.asp?uname=$username&amp;pwd=$password" />
</do>
```

Now take a look at the file `sendwml.asp` (Listing 5-17), which you use to collect the data from the user, such as message, subject, and the recipient's e-mail address.

Listing 5-17: sendwml.asp

```
<%@LANGUAGE=VBSCRIPT %>

<%
uname = trim(Request.QueryString("uname"))
pwd = trim(Request.QueryString("pwd"))

Response.ContentType = "text/vnd.wap.wml"

Response.Write "<!DOCTYPE wml PUBLIC ""-//WAPFORUM//DTD " & _
"WML 1.1//EN"" ""http://www.wapforum.org/DTD/wml_1.1.xml"">"
%>
<wml>
<head>
 <meta http-equiv="Cache-Control" content="max-age=time" forua="true"/>
</head>
    <card  newcontext="true">
  <onevent type="onenterforward">
  <refresh>
   <setvar name="semail" value="<%=uname%>" />
   <setvar name="remail" value="" />
   <setvar name="subject" value="" />
   <setvar name="body" value="" />
  </refresh>
    </onevent>

 <do type="accept" label="Next">
```

```
   <go
href="sendmail.asp?sender=$semail&amp;recipient=$remail&amp;subject=$subject&amp
;body=$body" />

 </do>

 <p>
 To
 <input name="remail" />

 Subject
 <input name="subject" />

 Message
 <input name="body" />
 </p>
</card>
</wml>
```

When this code executes, it collects the data from the user to send a mail message and then ports all the data to `sendmail.asp` for sending the mail. Figures 5-18 to 5-20 show the output of `sendwml.asp`.

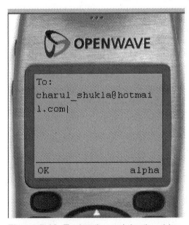

Figure 5-18: Typing the recipient's address (Image of Up.SDK courtesy Openwave Systems Inc.)

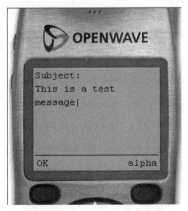

Figure 5-19: Entering the subject of the e-mail message (Image of Up.SDK courtesy Openwave Systems Inc.)

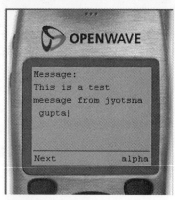

Figure 5-20: Filling up the message field (Image of Up.SDK courtesy Openwave Systems Inc.)

After the user fills in all the fields and presses the Next button, you use the `sendmail.asp` file (Listing 5-18) to send the mail and display the confirmation message.

Listing 5-18: sendmail.asp

```
<%@LANGUAGE = VBSCRIPT%>
<%


senderEmail = trim(Request.QueryString("sender"))
subject     = trim(Request.QueryString("subject"))
recipient   = trim(Request.QueryString("recipient"))
body = trim(Request.QueryString("body"))

Response.ContentType = "text/vnd.wap.wml"

Response.Write "<!DOCTYPE wml PUBLIC ""-//WAPFORUM//DTD " & _
"WML 1.1//EN"" ""http://www.wapforum.org/DTD/wml_1.1.xml"">" %>
<wml>
<head>
 <meta http-equiv="Cache-Control" content="max-age=time" forua="true"/>
</head>
<card>
<do type="accept" label="Next">
 <go href="checkbrows.asp" />
</do>
<p>
<%
' Create the JMail message Object
set msg = Server.CreateOBject( "JMail.Message" )


msg.From = senderEmail
msg.AddRecipient recipient
msg.Subject = subject
msg.body = body


if not msg.Send("192.168.1.100") then
    Response.write msg.log
else
    Response.write "Message sent succesfully!"
```

```
end if

%>
</p>
</card>
</wml>
```

When this file executes, it sends an e-mail message to the desired address and then displays the confirmation message to the sender on his WAP device.

This code uses the `Dimac Jmail` component for sending the e-mail message. First you create an object on the server and then provide all the necessary information for sending the e-mail. If the server is unable to send the e-mail, you display the error log message; otherwise, you display a confirmation to the user on the WAP device. The following lines of code can easily do this:

```
<%
' Create the JMail message Object
set msg = Server.CreateOBject( "JMail.Message" )


msg.From = senderEmail
msg.AddRecipient recipient
msg.Subject = subject
msg.body = body


if not msg.Send("192.168.1.100") then
    Response.write msg.log
else
    Response.write "Message sent successfully!"
end if

%>
```

Figure 5-21 shows the output the user will see if the task is successfully completed.

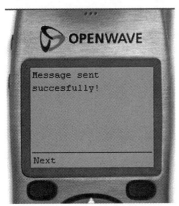

Figure 5-21: Displaying the confirmation message (Image of Up.SDK courtesy Openwave Systems Inc.)

When the user selects the Next option, it takes him back to the application's home page.

Transform the Movie Reservation System

This is the last section that you need to transform to make it compliant for WAP devices. By using this system, the user can browse the movie listings for the nearest theaters as well as book the tickets online using his WAP-enabled device.

Displaying the Theater Listings

When the user uses this program, it first displays a list of all the theaters available in the database. Before writing any other code, you first append some code in the generatexml.asp file, which belongs to the movies section of our application. Examine the code in Listing 5-19.

Listing 5-19: generatexml.asp (modified)

```
if (section="Movies") then

  sql = "select * from hall"
  objrs.Open sql,conn

  Set fso =Server.CreateObject("Scripting.FileSystemObject")
  fname = "tmovies.xml"

  fullpath = server.MapPath("main")

  Set MyFile=fso.CreateTextFile(fullpath & "/" & fname)

  MyFile.WriteLine("<?xml version='1.0'?>")
  MyFile.WriteLine("<ROOT>")


  while not(objrs.EOF)
    hallid = objrs.Fields("hall_id")
    hallname = objrs.Fields("hall_name")
    add = objrs.Fields("address")
    MyFile.WriteLine("<hall>")
    MyFile.WriteLine("<hall_id>" & hallid & "</hall_id>")
    MyFile.WriteLine("<hall_name>" & hallname & "</hall_name>")
    MyFile.WriteLine("<address>" & add & "</address>")
    MyFile.WriteLine("</hall>")
  objrs.MoveNext
  wend

  MyFile.WriteLine("</ROOT>")
  address = "http://192.168.1.100/loadxsl.asp?name=" & fname
  Response.Redirect(address)

end if
```

First, you check for the section and follow it by selecting all the records from the hall table. Then by using the file system object, you save all the data in XML format in a file named tmovies.xml. After storing all the data in XML format, you redirect control to loadxsl.asp, which loads the tmovieswml.xsl file for the transformation process. Figure 5-22 is a screen shot of tmovies.xml.

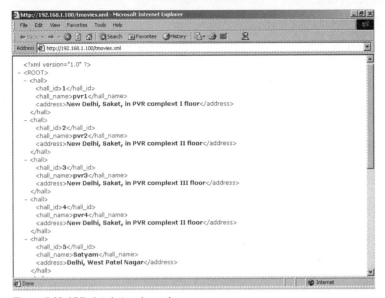

Figure 5-22: XML data in tmovies.xml

Now look at the code for `tmovieswml.xsl` in Listing 5-20. You use this XSLT code to fetch the hall names from the XML document and display them on the WAP device.

Listing 5-20: tmovieswml.xsl

```
<?xml version='1.0'?>
<xsl:stylesheet xmlns:xsl="http://www.w3.org/1999/XSL/Transform" version="1.0">
<xsl:output omit-xml-declaration="yes"/>

<xsl:template match="/">

<wml>
    <card>
  <p>
   <xsl:apply-templates select="ROOT" />
  </p>
   </card>

</wml>
</xsl:template>


<xsl:template match="ROOT">
<b>select a theatre name</b>
 <select>
<xsl:for-each select="hall">

  <option>
    <xsl:attribute name="onpick" >
     loadmoviewml.asp?hall=<xsl:value-of select="hall_name" />
    </xsl:attribute>
    <xsl:value-of select="hall_name" />
  </option>
```

```
</xsl:for-each>
</select>
</xsl:template>

</xsl:stylesheet>
```

In the beginning, you generated a card for the WML browser; under that card you used an `<xsl:apply>` statement to apply a template. In the second half of the code, you construct a loop condition to fetch all the records from the XML documents, as follows:

```
<select>
<xsl:for-each select="hall">

  <option>
    <xsl:attribute name="onpick" >
     loadmoviewml.asp?hall=<xsl:value-of select="hall_name" />
    </xsl:attribute>
    <xsl:value-of select="hall_name" />
  </option>


</xsl:for-each>
</select>
```

Under this loop, you use `<xsl:attribute>` to create the URL for all the theater names that you have fetched from the document. During the transformation process, the output in Figure 5-23 appears on the device screen.

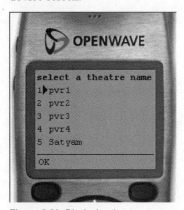

Figure 5-23: Displaying theater names on the device (Image of Up.SDK courtesy Openwave Systems Inc.)

The user now selects the theaters of his choice from the available list. When the user selects a theater, you call `loadmoviewml.asp` (Listing 5-21) and pass the name of the theater as a parameter using the query string. When `loadmoviewml.asp` is executed on the server, it generates an XML file that contains data on all the movies that currently are being shown in that particular theater.

Listing 5-21: loadmoviewml.asp

```
<%@ Language=VBScript %>

<%
dim address,section,fname

hallname = trim(Request.QueryString("hall"))
```

```
set conn = Server.CreateObject("ADODB.Connection")
conn.ConnectionString = "Provider=SQLOLEDB.1;Persist Security Info=False;User
ID=sa;Initial Catalog=portal;Data Source=developers"
conn.Open

Set objrs = Server.CreateObject("ADODB.Recordset")
objrs.CursorType = adOpenStatic

Set rsmovie = Server.CreateObject("ADODB.Recordset")
rsmovie.CursorType = adOpenStatic

Set rsstatus = Server.CreateObject("ADODB.Recordset")
rsstatus.CursorType = adOpenStatic
Set rshall = Server.CreateObject("ADODB.Recordset")
rshall.CursorType = adOpenStatic

  sqlhall = "select hall_id from hall where hall_name='" & hallname & "'"
  rshall.Open sqlhall,conn

  while not (rshall.EOF)
   id = rshall.Fields("hall_id")

  rshall.MoveNext
  wend
  sqlmovies = "select DISTINCT(status.movie_id),movie.movie_name from
movie,status where status.movie_id=movie.movie_id and status.hall_id=" & id
  rsmovie.Open sqlmovies,conn

  Set fso =Server.CreateObject("Scripting.FileSystemObject")
  fname = "tmovielist.xml"

  fullpath = server.MapPath("main")

  Set MyFile=fso.CreateTextFile(fullpath & "/" & fname)

  MyFile.WriteLine("<?xml version='1.0'?>")
  MyFile.WriteLine("<ROOT>")
  MyFile.WriteLine("<hall>")
  MyFile.WriteLine("<hall_name>" & hallname & "</hall_name>")
  MyFile.WriteLine("<hall_id>" & id & "</hall_id>")

  while not (rsmovie.EOF)
   MyFile.WriteLine("<movie>")
   moviename = rsmovie.Fields("movie_name")
   MyFile.WriteLine("<movie_name>" & moviename & "</movie_name>")
   movieid = rsmovie.Fields("movie_id")
   MyFile.WriteLine("<movie_id>" & movieid & "</movie_id>")
   sqlstatus = "select
movie_date,showtime,balcony,rear_stall,middle_stall,upper_stall from status
where movie_id=" & movieid & "and hall_id=" & id
   rsstatus.Open sqlstatus,conn
    while not (rsstatus.EOF)
     MyFile.WriteLine("<status>")
     moviedate = rsstatus.Fields("movie_date")
     showtime = rsstatus.Fields("showtime")
     balcony = rsstatus.Fields("balcony")
```

```
        rearstall = rsstatus.Fields("rear_stall")
        middlestall = rsstatus.Fields("middle_stall")
        upperstall = rsstatus.Fields("upper_stall")
        MyFile.WriteLine("<movie_date>" & moviedate & "</movie_date>")
        MyFile.WriteLine("<showtime>" & showtime & "</showtime>")
        MyFile.WriteLine("<balcony>" & balcony & "</balcony>")
        MyFile.WriteLine("<rear_stall>" & rearstall & "</rear_stall>")
        MyFile.WriteLine("<middle_stall>" & middlestall & "</middle_stall>")
        MyFile.WriteLine("<upper_stall>" & upperstall & "</upper_stall>")
        MyFile.WriteLine("</status>")
        rsstatus.MoveNext
      wend
   MyFile.WriteLine("</movie>")
   rsstatus.Close
   rsmovie.MoveNext
   wend
   MyFile.WriteLine("</hall>")
   MyFile.WriteLine("</ROOT>")


   conn.Close

   address = "http://192.168.1.100/loadxsl.asp?name=" & fname
   Response.Redirect(address)
%>
```

We will discuss the code in parts to enable you to understand better how it works. First consider the following portion:

```
dim address,section,fname

hallname = trim(Request.QueryString("hall"))

set conn = Server.CreateObject("ADODB.Connection")
conn.ConnectionString = "Provider=SQLOLEDB.1;Persist Security Info=False;User
ID=sa;Initial Catalog=portal;Data Source=developers"
conn.Open

Set objrs = Server.CreateObject("ADODB.Recordset")
objrs.CursorType = adOpenStatic

Set rsmovie = Server.CreateObject("ADODB.Recordset")
rsmovie.CursorType = adOpenStatic

Set rsstatus = Server.CreateObject("ADODB.Recordset")
rsstatus.CursorType = adOpenStatic
Set rshall = Server.CreateObject("ADODB.Recordset")
rshall.CursorType = adOpenStatic

  sqlhall = "select hall_id from hall where hall_name='" & hallname & "'"
  rshall.Open sqlhall,conn
```

In the first line of the code, you have stored the value of the theater in a variable named `hallname`. You then establish a connection with the database and create some objects and set certain variables. In the last two lines, you select the `hall_id` from the hall table on the basis of the value you used in the `hallname` variable. You then select the `movie_name` and `movie_id` of the movies currently on the screen in that theater by providing the following SQL statement:

```
while not (rshall.EOF)
  id = rshall.Fields("hall_id")

rshall.MoveNext
wend
sqlmovies = "select DISTINCT(status.movie_id),movie.movie_name from
movie,status where status.movie_id=movie.movie_id and status.hall_id=" & id
rsmovie.Open sqlmovies,conn
```

After collecting all the records from the different tables, you write this data in XML format so that you can use it for the transformation process later. You are following a quite complex structure for storing the XML data in the file. However, this gives you better control of the XML document while transforming it for the WAP browser later.

```
Set fso =Server.CreateObject("Scripting.FileSystemObject")
fname = "tmovielist.xml"

fullpath = server.MapPath("main")

Set MyFile=fso.CreateTextFile(fullpath & "/" & fname)

MyFile.WriteLine("<?xml version='1.0'?>")
MyFile.WriteLine("<ROOT>")
MyFile.WriteLine("<hall>")
MyFile.WriteLine("<hall_name>" & hallname & "</hall_name>")
MyFile.WriteLine("<hall_id>" & id & "</hall_id>")

while not (rsmovie.EOF)
  MyFile.WriteLine("<movie>")
  moviename = rsmovie.Fields("movie_name")
  MyFile.WriteLine("<movie_name>" & moviename & "</movie_name>")
  movieid = rsmovie.Fields("movie_id")
  MyFile.WriteLine("<movie_id>" & movieid & "</movie_id>")
  sqlstatus = "select
movie_date,showtime,balcony,rear_stall,middle_stall,upper_stall from status
where movie_id=" & movieid & "and hall_id=" & id
  rsstatus.Open sqlstatus,conn
    while not (rsstatus.EOF)
      MyFile.WriteLine("<status>")
      moviedate = rsstatus.Fields("movie_date")
      showtime = rsstatus.Fields("showtime")
      balcony = rsstatus.Fields("balcony")
      rearstall = rsstatus.Fields("rear_stall")
      middlestall = rsstatus.Fields("middle_stall")
      upperstall = rsstatus.Fields("upper_stall")
      MyFile.WriteLine("<movie_date>" & moviedate & "</movie_date>")
      MyFile.WriteLine("<showtime>" & showtime & "</showtime>")
      MyFile.WriteLine("<balcony>" & balcony & "</balcony>")
      MyFile.WriteLine("<rear_stall>" & rearstall & "</rear_stall>")
      MyFile.WriteLine("<middle_stall>" & middlestall & "</middle_stall>")
      MyFile.WriteLine("<upper_stall>" & upperstall & "</upper_stall>")
      MyFile.WriteLine("</status>")
      rsstatus.MoveNext
    wend
  MyFile.WriteLine("</movie>")
  rsstatus.Close
  rsmovie.MoveNext
```

```
wend
MyFile.WriteLine("</hall>")
MyFile.WriteLine("</ROOT>")
```

Figure 5-24 is a screen shot of `tmovielist.xml`.

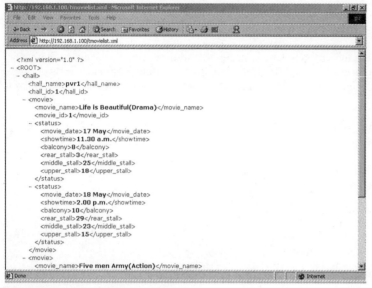

Figure 5-24: A screen shot of tmovielist.xml

You save all the data in the file named `tmovielist.xml` and finally redirect control to `loadxsl.asp` to begin the transformation process.

Generate WML Cards on-the-Fly

Now take a look at the code for `tmovielistwml.xsl` in Listing 5-22, which you used for the transformation process with `tmovielistwml.xsl`.

Listing 5-22: tmovielistwml.xsl

```
<?xml version='1.0'?>
<xsl:stylesheet xmlns:xsl="http://www.w3.org/1999/XSL/Transform" version="1.0">
<xsl:output omit-xml-declaration="yes"/>

<xsl:template match="/">

<wml>

<head>
 <meta http-equiv="Cache-Control" content="max-age=time" forua="true"/>
</head>


   <card>
     <p>
       <xsl:apply-templates select="ROOT" />
     </p>
   </card>
```

```xml
<xsl:apply-templates select="ROOT/hall" />

</wml>
</xsl:template>

<xsl:template match="ROOT">
<xsl:variable name="hallname" select="hall/hall_name" />
<b>List of movies on <xsl:value-of select="hall/hall_name" /></b><br />


<select>
<xsl:for-each select="hall/movie">

<option>
<xsl:attribute name="onpick">#m<xsl:value-of select="./movie_id"
/></xsl:attribute>
<xsl:value-of select="./movie_name" />
</option>

</xsl:for-each>
</select>


</xsl:template>


<xsl:template match="ROOT/hall">
<xsl:variable name="hname" select="hall_name" />
<xsl:variable name="hid" select="hall_id" />
<xsl:for-each select="movie">

<card>

<xsl:attribute name="id" >m<xsl:value-of select="./movie_id" /></xsl:attribute>
<xsl:attribute name="newcontext" >true</xsl:attribute>
<onevent type="onenterforward">

  <refresh>
  <setvar>
  <xsl:attribute name="name">hallid</xsl:attribute>
  <xsl:attribute name="value"><xsl:value-of select="$hid" /></xsl:attribute>
  </setvar>
  <setvar>
  <xsl:attribute name="name">hallname</xsl:attribute>
  <xsl:attribute name="value"><xsl:value-of select="$hname" /></xsl:attribute>
  </setvar>
  <setvar>
  <xsl:attribute name="name">movieid</xsl:attribute>
  <xsl:attribute name="value"><xsl:value-of select="./movie_id"
/></xsl:attribute>
  </setvar>
  <setvar>
  <xsl:attribute name="name">moviename</xsl:attribute>
  <xsl:attribute name="value"><xsl:value-of select="./movie_name"/>
  </xsl:attribute>
  </setvar>
```

```
<setvar>
<xsl:attribute name="name">username</xsl:attribute>
<xsl:attribute name="value"></xsl:attribute>
</setvar>
<setvar>
<xsl:attribute name="name">email</xsl:attribute>
<xsl:attribute name="value"></xsl:attribute>
</setvar>
<setvar>
<xsl:attribute name="name">stime</xsl:attribute>
<xsl:attribute name="value"></xsl:attribute>
</setvar>
<setvar>
<xsl:attribute name="name">sdate</xsl:attribute>
<xsl:attribute name="value"></xsl:attribute>
</setvar>
<setvar>
<xsl:attribute name="name">stallname</xsl:attribute>
<xsl:attribute name="value"></xsl:attribute>
</setvar>
<setvar>
<xsl:attribute name="name">ticket</xsl:attribute>
<xsl:attribute name="value"></xsl:attribute>
</setvar>
</refresh>


</onevent>

<do type="accept" label="Next">
<go
href="ticketbookingwml.asp?uname=$username&amp;mail=$email&amp;hallid=$hallid&am
p;hallname=$hallname&amp;movieid=$movieid&amp;moviename=$moviename&amp;showdate=
$sdate&amp;showtime=$stime&amp;stall=$stallname&amp;no=$ticket"  />
</do>

<p>
<b>Status:--</b>
<br /><br />
<b><xsl:value-of select="./movie_name" /></b><br />
------------


<table columns="2">
<xsl:for-each select="status">
 <tr>
  <td><b>Date</b></td>
  <td><xsl:value-of select="movie_date" /></td>
 </tr>
 <tr>
  <td><b>Showtime</b></td>
  <td><xsl:value-of select="showtime" /></td>
 </tr>
 <tr>
  <td><b>Balcony</b></td>
  <td><xsl:value-of select="balcony" /></td>
```

```
  </tr>
  <tr>
   <td><b>Rear Stall</b></td>
   <td><xsl:value-of select="rear_stall" /></td>
  </tr>
  <tr>
   <td><b>Middle Stall</b></td>
   <td><xsl:value-of select="middle_stall" /></td>
  </tr>
  <tr>
   <td><b>Upper Stall</b></td>
   <td><xsl:value-of select="upper_stall" /></td>
    </tr>
</xsl:for-each>

</table>

<br />
<b>Enter the information for ticket booking</b><br />

Enter the your Name
<input name="username" />
Enter the E-mail Address
<input name="email" />
Enter the ShowDate
<input name="sdate" />
Enter the ShowTime
<input name="stime" />
Enter the Stall Name
<input name="stallname" />
Enter the No. of Tickets
<input name="ticket" />
</p>

</card>
</xsl:for-each>

</xsl:template>
</xsl:stylesheet>
```

This is one of the longest style sheets you will use in this chapter. In this XSL document, you are generating WML cards on-the-fly on the basis of receiving different parameters. Now we'll discuss this code in parts.

First, you generate a simple card and apply a template called ROOT to it. After closing the card, you apply another template named ROOT/hall and then close all the tags, including the WML and <xsl:template> tag, as follows:

```
<xsl:template match="/">

<wml>

<head>
 <meta http-equiv="Cache-Control" content="max-age=time" forua="true"/>
</head>
```

```
  <card>
    <p>
     <xsl:apply-templates select="ROOT" />
    </p>
  </card>

<xsl:apply-templates select="ROOT/hall" />

</wml>
</xsl:template>
```

You then fetch all the movie titles and theater names from the XML document and display them using the following lines of code:

```
<xsl:template match="ROOT">
<xsl:variable name="hallname" select="hall/hall_name" />
<b>List of movies on <xsl:value-of select="hall/hall_name" /></b><br />


<select>
<xsl:for-each select="hall/movie">

<option>
<xsl:attribute name="onpick">#m<xsl:value-of select="./movie_id"
/></xsl:attribute>
<xsl:value-of select="./movie_name" />
</option>

</xsl:for-each>
</select>


</xsl:template>
```

This displays the titles of all the movies for a particular theater on the WAP browser. If the user wants to view the details of a particular movie, you call the second card in the deck; the name of this card is generated on-the-fly by using the `movie id`. For starting the `movie id`, you add `m`. So if the `movie id` is `1234`, the card name for that movie will be `m1234`. Every time the user selects a movie, you generate a card named for that movie on-the-fly. The output of the code is shown in Figure 5-25.

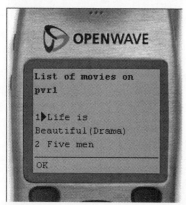

Figure 5-25: Displaying movie titles for a theater (Image of Up.SDK courtesy Openwave Systems Inc.)

When the user selects one of the movies from the available options, you call the card by the name that you just generated using the `movie id`.

In the next step of the process, you generate a card on-the-fly, which holds the ID that you called in the first template. See how we actually did it by examining the following code:

```
<xsl:template match="ROOT/hall">
<xsl:variable name="hname" select="hall_name" />
<xsl:variable name="hid" select="hall_id" />
<xsl:for-each select="movie">

<card>

<xsl:attribute name="id" >m<xsl:value-of select="./movie_id" /></xsl:attribute>
<xsl:attribute name="newcontext" >true</xsl:attribute>
<onevent type="onenterforward">

  <refresh>
  <setvar>
  <xsl:attribute name="name">hallid</xsl:attribute>
  <xsl:attribute name="value"><xsl:value-of select="$hid" /></xsl:attribute>
  </setvar>
  <setvar>
  <xsl:attribute name="name">hallname</xsl:attribute>
  <xsl:attribute name="value"><xsl:value-of select="$hname" /></xsl:attribute>
  </setvar>
  <setvar>
  <xsl:attribute name="name">movieid</xsl:attribute>
  <xsl:attribute name="value"><xsl:value-of select="./movie_id"
/></xsl:attribute>
  </setvar>
  <setvar>
  <xsl:attribute name="name">moviename</xsl:attribute>
  <xsl:attribute name="value"><xsl:value-of select="./movie_name"/>
  </xsl:attribute>
  </setvar>
  <setvar>
  <xsl:attribute name="name">username</xsl:attribute>
  <xsl:attribute name="value"></xsl:attribute>
  </setvar>
  <setvar>
  <xsl:attribute name="name">email</xsl:attribute>
  <xsl:attribute name="value"></xsl:attribute>
  </setvar>
  <setvar>
  <xsl:attribute name="name">stime</xsl:attribute>
  <xsl:attribute name="value"></xsl:attribute>
  </setvar>
  <setvar>
  <xsl:attribute name="name">sdate</xsl:attribute>
  <xsl:attribute name="value"></xsl:attribute>
  </setvar>
  <setvar>
  <xsl:attribute name="name">stallname</xsl:attribute>
  <xsl:attribute name="value"></xsl:attribute>
  </setvar>
  <setvar>
```

```
<xsl:attribute name="name">ticket</xsl:attribute>
<xsl:attribute name="value"></xsl:attribute>
</setvar>
</refresh>


</onevent>

<do type="accept" label="Next">
<go
href="ticketbookingwml.asp?uname=$username&amp;mail=$email&amp;hallid=$hallid&am
p;hallname=$hallname&amp;movieid=$movieid&amp;moviename=$moviename&amp;showdate=
$sdate&amp;showtime=$stime&amp;stall=$stallname&amp;no=$ticket" />
</do>
```

In this code, you first set some variables using the `<xsl:variable>` element. You then start a `<card>` tag and set its different attributes, such as ID and newcontext, using the `<xsl:attribute>` element. By doing this, the card gets the movie id as the ID of the card, with m as a prefix; for example, m1234.

After assigning the card ID, you now set a series of variables using the `<setvar>` element in conjunction with the `<xsl:attribute>` element. You use these variables for storing the values when the user books the ticket.

After declaring all the variables, you declare a `<do>` element and set the ticketbookingwml.asp file as an HREF of the `<go>` element, which comes under the `<do>` element. You also pass all the values you stored in the variables using the query string method. The following lines of code are used to display the details about the movie the user selects:

```
<p>
<b>Status:--</b>
<br /><br />
<b><xsl:value-of select="./movie_name" /></b><br />
-----------


<table columns="2">
<xsl:for-each select="status">
 <tr>
  <td><b>Date</b></td>
  <td><xsl:value-of select="movie_date" /></td>
 </tr>
 <tr>
  <td><b>Showtime</b></td>
  <td><xsl:value-of select="showtime" /></td>
 </tr>
 <tr>
  <td><b>Balcony</b></td>
  <td><xsl:value-of select="balcony" /></td>
 </tr>
 <tr>
  <td><b>Rear Stall</b></td>
  <td><xsl:value-of select="rear_stall" /></td>
 </tr>
 <tr>
  <td><b>Middle Stall</b></td>
  <td><xsl:value-of select="middle_stall" /></td>
 </tr>
```

```
 <tr>
  <td><b>Upper Stall</b></td>
  <td><xsl:value-of select="upper_stall" /></td>
   </tr>
</xsl:for-each>

</table>
```

In this code, you simply construct a loop for displaying all the available status nodes in the XML document for that particular movie. So when this code executes, it will display the status for all the dates for that movie in the selected theater, as shown in Figures 5-26 and 5-27.

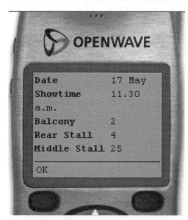

Figure 5-26: Displaying the status for one day (Image of Up.SDK courtesy Openwave Systems Inc.)

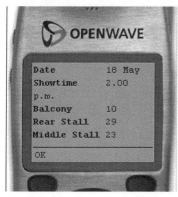

Figure 5-27: Displaying the status for a second day (Image of Up.SDK courtesy Openwave Systems Inc.)

In the next process, you design a simple data-entry form for the user so that he or she can book the tickets online. You begin the process by collecting various details and then placing all the data in `ticketbookingwml.asp`, as follows:

```
<b>Enter the information for ticket booking</b><br />

Enter your name
<input name="username" />
Enter your e-mail address
<input name="email" />
Enter the show date
<input name="sdate" />
Enter the show time
```

```
<input name="stime" />
Enter the stall name
<input name="stallname" />
Enter the no. of tickets
<input name="ticket" />
```

This is the simple code that is used for the data-entry form. When this code executes, it shows the output in Figure 5-28 on the WAP browser:

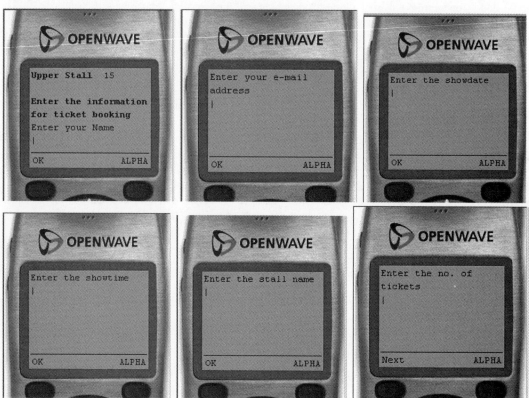

Figure 5-28: Data-entry form screens (All images of Up.SDK courtesy Openwave Systems Inc.)

After the user fills in all the required information and presses Enter, you post all the data to `ticketbookingwml.asp` to perform a check on the number of tickets available, as well as the number of tickets the user entered.

Write Interfaces for the Booking Section

In this section, you verify the number of tickets available and the number that the user has requested. Listing 5-23 can easily do this.

Listing 5-23: ticketbookingwml.asp

```
<%@ Language=VBScript %>
<HTML>
<HEAD>
<META NAME="GENERATOR" Content="Microsoft Visual Studio 6.0">
</HEAD>
<BODY>
```

```
<%
uname = trim(Request.QueryString("uname"))
email = trim(Request.QueryString("mail"))
hall = trim(Request.QueryString("hallname"))
hallid = trim(Request.QueryString("hallid"))
movieid = trim(Request.QueryString("movieid"))
movie = trim(Request.QueryString("moviename"))
showdate = trim(Request.QueryString("showdate"))
showtime = trim(Request.QueryString("showtime"))
stall = UCase(trim(Request.QueryString("stall")))
ticket = trim(Request.QueryString("no"))



dim conn,objrs,tsql,balcony,rear,middle,upper,total

set conn = Server.CreateObject("ADODB.Connection")
conn.ConnectionString = "Provider=SQLOLEDB.1;Persist Security Info=False;User
ID=sa;Initial Catalog=portal;Data Source=developers"
conn.Open

Set objrs = Server.CreateObject("ADODB.Recordset")
tsql="select balcony,rear_stall,middle_stall,upper_stall from status where
hall_id=" & hallid & "and movie_id=" & movieid & "and showtime='" & showtime &
"'and movie_date='" & showdate &"'"
objrs.CursorType = adOpenStatic
objrs.Open tsql,conn

while not (objrs.EOF )
 if (stall = "BALCONY") then
  total = UCase(trim(objrs.Fields("Balcony")))
 end if

 if (stall = "REAR_STALL") then
  total = UCase(trim(objrs.Fields("rear_stall")))
 end if

 if (stall = "MIDDLE_STALL") then
  total = UCase(trim(objrs.Fields("middle_stall")))
 end if

 if (stall = "UPPER_STALL") then
  total = UCase(trim(objrs.Fields("upper_stall")))
 end if

 balcony = objrs.Fields("Balcony")
 rear = objrs.Fields("rear_stall")
 middle = objrs.Fields("middle_stall")
 upper = objrs.Fields("upper_stall")

objrs.MoveNext
wend
```

```
if (CInt(total) < CInt(ticket) ) then
 Response.Redirect "http://192.168.1.100/bookwml.asp?name=fail"
else
 if (stall = "BALCONY") then
  balcony = CInt(balcony) - CInt(ticket)
 end if

 if (stall = "MIDDLE_STALL") then
  middle = CInt(middle) - CInt(ticket)
 end if

 if (stall = "REAR_STALL") then
  rear = CInt(rear) - CInt(ticket)
 end if

 if (stall = "UPPER_STALL") then
  upper = CInt(upper) - CInt(ticket)
 end if

 conn.Execute("update status set balcony=" & balcony & ",rear_stall=" & rear &
",middle_stall=" & middle & ",upper_stall=" & upper & "where hall_id=" & hallid
& "and movie_id=" & movieid & "and showtime='" & showtime & "'")
 'conn.Execute("Insert user
(firstname,movieid,hallid,stall,movie_date,showtime,totaltickets,emailadd)
values('" & uname & "'," & movieid  & "," & hallid & ",'" & stall & "','" &
showdate & "','" & showtime & "'," & tickets & ",'" & email & "')")
 Response.Redirect "http://192.168.1.100/bookwml.asp?name=pass"

end if
conn.Close
%>

</BODY>
</HTML>
```

The ASP code in Listing 5-22 is fairly straightforward. If the user requests more tickets than the actual number that is available, he is redirected to `bookwml.asp` with the parameter `name="fail"`. If the operation is successful, you change the parameter name with the value `name="pass"` and redirect the user to `bookwml.asp` (Listing 5-24).

Listing 5-24: bookwml.asp

```
<%@ Language=VBScript %>

<%
name = trim(Request.QueryString("name"))


'declaring variables
dim sXML
dim sXSL
dim xmlDocument,xslDocument

'creating the objects for the XMLDOM
set xmlDocument = Server.CreateObject("MICROSOFT.XMLDOM")
set xslDocument = Server.CreateObject("MICROSOFT.XMLDOM")
```

```
sXML = "book.xml"

if(name = "fail") then
 sXSL = "bookfail.xsl"
else
 sXSL = "bookpass.xsl"
end if

xmlDocument.async = false
xslDocument.async = false

xmlDocument.load(Server.MapPath(sXML))
xslDocument.load(Server.MapPath(sXSL))

Response.ContentType = "text/vnd.wap.wml"
Response.Write "<?xml version=""1.0"" ?>"


Response.Write "<!DOCTYPE wml PUBLIC ""-//WAPFORUM//DTD " & _
"WML 1.1//EN"" ""http://www.wapforum.org/DTD/wml_1.1.xml"">"

Response.Write(xmlDocument.transformNode(xslDocument))

%>
```

With the help of this file, you display the confirmation message based on the parameter you get from
`ticketbookingwml.asp`. In case of failure, you load `book.xml` (Listing 5-25) with `bookfail.xsl`
(Listing 5-26). In case of successful completion, you load `book.xml` with `bookpass.xsl` (Listing 5-27).

Listing 5-25: book.xml

```
<?xml version="1.0" encoding="utf-8"?>
<ROOT>
</ROOT>
```

Listing 5-26: bookpasswml.xsl

```
<?xml version='1.0'?>
<xsl:stylesheet xmlns:xsl="http://www.w3.org/TR/WD-xsl">

<xsl:template match="/">

<wml>
    <card newcontext="true">
 <do type="accept" label="Next">
  <go href="checkbrows.asp" />
 </do>
  <p>
   <xsl:apply-templates select="ROOT" />
  </p>
    </card>

</wml>
</xsl:template>
```

```
<xsl:template match="ROOT">
Try again!
</xsl:template>
</xsl:stylesheet>
```

Listing 5-27: bookpasswml.xsl

```
<?xml version='1.0'?>
<xsl:stylesheet xmlns:xsl="http://www.w3.org/TR/WD-xsl">
<xsl:template match="/">

<wml>
    <card>

  <do type="accept" label="Next">
   <go href="checkbrows.asp" />
  </do>
  <p>
   <xsl:apply-templates select="ROOT" />
  </p>
   </card>

</wml>
</xsl:template>

<xsl:template match="ROOT">
Congratulations! Tickets have been booked.
</xsl:template>
</xsl:stylesheet>
```

Figures 5-29 and 5-30 are the screen shots of book.xml with both style sheets.

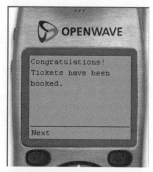

Figure 5-29: Congratulations screen (Image of Up.SDK courtesy Openwave Systems Inc.)

Figure 5-30: Try again screen (Image of Up.SDK courtesy Openwave Systems Inc.)

Summary

In this chapter, you have learned the techniques to develop applications for the WAP market using the XML- and XSLT-based approach. You have used Active Server Pages as a business-logic layer of your application to transform all the data using XSLT. Consequently, the data becomes neutral and can be used for any kind of transformation process that you intend to carry out.

Chapter 6

Learning HDML: A Case Study

Handheld Device Markup Language (HDML) is the markup language that enables users to access the Internet efficiently via small display screens on mobile handheld devices such as mobile phones, pagers, and wireless personal digital assistants (PDAs). HDML is specifically targeted for devices with tiny display screens, limited memory, less processing power, and low network bandwidth for communication.

HDML supports the infrastructure and Internet protocols, thereby offering a viable solution for wireless applications and other Web-enabled handheld devices. The two buttons (PREV and NEXT) and the pull-down history menu are the result of the navigation model HDML provides. Mobile devices retrieve HDML as a container — called a *deck* — which requires a transaction-based, explicit navigation model. Each card in the deck is regarded as a separate Web page in the mobile browser such devices use.

HDML basically focuses on retrieving day-to-day, routine information such as e-mail alerts; stock updates; trading, hotel, and restaurant information; the latest news and weather forecasts; sports information; train schedules; flight information; and so on.

Introduction to HDML

The concept of HDML was formulated and developed by Unwired Planet in May 1997. (Unwired Planet evolved into Phone.com and most recently Openwave Systems Inc.) Because Unwired Planet is a World Wide Web Consortium (W3C) member, it submitted a complete list of HDML specifications to the consortium for approval. You can refer to these HDML specifications at the following Web site:
`http://www.w3.org/pub/WWW/TR/NOTE-Submission-HDML-spec.html`

Even though the W3C has duly acknowledged these HDML specifications, they are still considered a "working draft." As of now, no further developments have been implemented in this regard. In June 1999, the WAP Forum launched the Wireless Markup Language (WML) for WAP-enabled devices. You can get additional information on the forum at `www.wapforum.org`.

Phone.com later developed a mobile browser that was an operating environment for Web-enabled devices. This browser, popularly known as the UP.Browser, completely supports HDML.

Figure 6-1 demonstrates the way Web-enabled mobile phones access HDML sites.

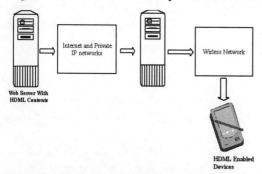

Figure 6-1: A working model of HDML

When the user enters the URL in the mobile handset, a request is sent to the UP.Link Gateway (server). This server then communicates an HTTP request to the Web server with the help of an HDML service through the Internet and private IP networks. The page then returns through HTTP from the Web server to the UP.Link Gateway. The UP.Link Gateway transmits the data further via the Handheld Device Transport Protocol (HDTP) to the wireless network, and finally to the Web-enabled mobile phone.

WML versus HDML

HDML and WML share the same programming model for Web-enabled mobile devices that access the Internet. However, some pertinent technical differences remain. These differences are shown in the following table.

HDML	WML
Not a subset of XML.	A subset of XML.
Does not have a Document Type Definition (DTD).	Has a Document Type Definition (DTD).
Supports bookmarks.	Does not support bookmarks.
Supports nested activities.	Does not support nested activities.
Supports key accelerators for links.	Does not support key accelerators for links.
Supports mobile originated pre-fetch.	Does not support mobile originated pre-fetch.
Does not support script language.	Supports WMLScript script language.
Does not support timers.	Supports timers.
Does not support multiple-choice lists.	Supports multiple-choice lists.
Does not support images in labels and choices.	Supports images in labels and choices.

A Brief Look at the HDML Clients on the Market

HDML is targeted for Web-enabled handheld devices with limited resources, such as mobile phones and PDAs.

Mobile Phones

The mobile phone has emerged as a handheld device with great promise. It uses a short-wave analog or digital transmission, in which a subscriber has a wireless connection from a mobile telephone to a nearby transmitter.

In the past several years, the market for mobile phones has grown considerably; consequently, the prices of mobile phones and mobile phone service have dropped considerably. Due to these factors, mobile phones are coming to be seen as necessities rather than luxury items.

Personal Digital Assistants

PDAs are mobile devices that run many useful applications, which are capable of computing, storing, and retrieving information. PDAs are generally used for maintaining schedule calendars and address-book information. Brands of PDAs include Palmtops, Handsprings, and so on. They are usually equipped with a small built-in keyboard and require an operating system, usually Windows CE or their own operating environment. Most PDAs have advanced features such as touchscreens that permit handwriting on the screens.

Write an HDML Document

The complete Software Development Kit required for executing HDML documents (UP.SDK) is available for free download at `http://developer.openwave.com/download/index.html`. Listing 6-1, `ex1.hdml`, demonstrates the first example in HDML:

Listing 6-1: ex1.hdml

© 2001 Dreamtech Software India Inc.
All Rights Reserved

```
<hdml version="3.0">
   <display>
      <!-- First Basic Example -->
      Hello World! <br>
      <wrap>Welcome to the world of HDML
   </display>
</hdml>
```

In the preceding code

- ◆ The line `<hdml version="3.0">` defines the initial header of the HDML deck. In fact, all HDML decks begin with an `<hdml>` tag and end with an `</hdml>` tag.
- ◆ `<!-- First Basic Example -->` is a comment, which HDML ignores.
- ◆ `<display>` displays formatted text.

The syntax of `<display>` is described in the following code and table.

```
<display
name="string"
title="string"
markable="boolean"
 bookmark="url">
actions (The actions to execute when the user presses a function key)
text (The formatted text the card displays.)
</display>
```

Attribute Name	Value	Description
Name	String	A unique name for the card.
title	String	The bookmark name that appears, by default, when the user marks the card.
markable	Boolean	This specifies whether the card can be marked. TRUE allows the card to be marked.
bookmark	url	Stands for the URL that the phone adds to the bookmark list if the user marks the card. The phone adds the URL of the current card to the bookmark list if the bookmark option is not specified.

`<wrap>` starts a new line. If the text is too long to fit the phone display, the phone word-wraps the text to the next line. All the lines that follow the current line are word-wrapped until a `<line>` statement is encountered.

Figure 6-2 shows the output of Listing 6-1 as rendered by the UP.Simulator.

Figure 6-2: The output of ex1.hdml (Image of Up.SDK courtesy Openwave Systems Inc.)

Accepting Data Input Using HDML

The user can enter data in an entry display card using the `entry` statement. This statement can use the various format specifiers (discussed later in this chapter) for the data the user enters. It specifies a variable that stores the field's value as well as a default value.

The syntax of the `entry` statement is as follows:

```
<entry
name="string"
title="string"
markable="boolean"
  bookmark="url"
format="fmt"
fill="val"
default="default"
key="var"
noecho="boolean"
emptyok="boolean">
actions (The actions to execute when the user presses a function key)
text (The formatted text displayed above the entry line)
</entry>
```

The preceding syntax is described in the following table.

Attribute Name	Value	Description
name	String	A unique card name.
title	String	The bookmark name that appears when the user marks the card.
markable	Boolean	Specifies whether the card can be marked. TRUE allows the card to be marked.
Bookmark	url	The URL that the phone adds to the bookmark list when the user marks the card. If the `bookmark` option is not specified, the phone adds the current card's URL to the bookmark list.
format	Fmt	Format specifier for the entered data. *M is the default format (can be any number of mixed-case alphabetic and numeric characters).
Fill	Val	Specifies the text direction in the entry field. LEFT, the default, fills the entry field from left to right. RIGHT fills the entry field from right to left.

Attribute Name	Value	Description
Default	Default	An editable string that appears in the entry field when the phone initially displays the card.
Key	Var	The variable name to which the phone stores the entered data. If the specified variable already has a value, the value appears as the default in the entry field.
Noecho	Boolean	TRUE hides the text the user enters in the phone. An asterisk (*) is displayed after the user enters a character. FALSE is the default.
Emptyok	Boolean	TRUE accepts the empty input in the phone even if a format is specified with the format option. FALSE is the default.

The following example accepts the input from the user and then displays the accepted data.

Listing 6-2: example.hdml

© 2001 Dreamtech Software India Inc.
All Rights Reserved

```
<hdml version=3.0 public=true>
 <entry name="rno" key="rollno">
  <action type="accept" label="next"task="go" dest="#nm">
  Roll No.:
 </entry>

 <entry name="nm" key="name">
  <action type="accept" label="next"task="go" dest="#add">
  Name:
 </entry>

 <entry name="add" key="address">
  <action type="accept" label="next"task="go" dest="#d_o_b">
  City:
 </entry>

 <entry name="d_o_b" key="dob">
  <action type="accept" label="next"task="go" dest="#detail">
   Birthdate:
 </entry>

 <display name="detail">
  <action type="accept" label="ok" task="go" dest="#rno">
  <center>DETAILS -<br>
  $rollno<br>
  $name<br>
  $address<br>
  $dob
 </display>
</hdml>
```

The output of the preceding code, as rendered by the UP.Simulator, is shown in Figure 6-3.

Figure 6-3: Output of Listing 6-2 (Images of Up.SDK courtesy Openwave Systems Inc.)

Submit Forms and Use Dynamic Data in HDML

The following example demonstrates how the HDML data is submitted to the server script using the method and postdata attributes of the action statement. If the dest attribute in the action statement specifies a URL, the method used to request the URL can be either GET or POST. The default is GET.

Listing 6-3: example1.hdml
© 2001 Dreamtech Software India Inc.
All Rights Reserved

```
<hdml version=3.0 public=true>
   <entry name="rno" key="rollno">
      <action type="accept" label="next"task="go"   dest="#nm">
      Roll No.:
   </entry>

   <entry name="nm" key="name">
      <action type="accept" label="next"task="go"  dest="#add">
      Name:
```

```
      </entry>

      <entry name="add" key="address">
         <action type="accept" label="next"task="go" dest="#d_o_b">
         Address:
      </entry>

      <entry name="d_o_b" key="dob">
         <action type="accept" label="next"task="go" dest="#detail">
         Date of Birth:
      </entry>

      <display name="detail">
         <action type="accept" label="ok" task="go" method=post
postdata=$rollno&$name
                    dest="http://localhost/demoname.asp">
 <center>DETAILS -<br>
 $rollno<br>
 $name<br>
 $address<br>
 $dob
      </display>
</hdml>
```

Navigation between Different Cards

Navigation is possible between different cards. Listing 6-4 defines a deck with two display cards. The `action` statement assigns the CANCEL task and SOFT1 assigns the key label. Note that this action applies to both of the cards in the deck. An action is defined by the first card, which states that the phone will request the second card when the user presses the ACCEPT key, and vice versa. (Note that the `action` statement is discussed later in this chapter.)

Listing 6-4: example2.hdml

```
<hdml version=3.0>
   <action type=SOFT1 task=CANCEL label=Cancel>
   <display name=card1>
      <action type=ACCEPT task=GO dest=#card2>
      Inside Card 1
   </display>

   <display name=card2>
      <action type=ACCEPT task=GO dest=#card1>
    Inside Card 2
   </display>
</hdml>
```

Figure 6-4 shows the output of the preceding code on the UP.Simulator.

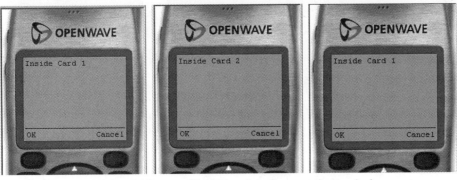

Figure 6-4: Output of Listing 6-4 (Images of Up.SDK courtesy Openwave Systems Inc.)

Navigation Control in HDML

Listing 6-5 demonstrates navigation control. The action in the second card instructs the phone to request the destination URL when the user presses the ACCEPT key.

Listing 6-5: example3.hdml

© 2001 Dreamtech Software India Inc.
All Rights Reserved

```
<hdml version=3.0>
   <display name=card1>
      <action type=ACCEPT task=GO dest=#card2>
      Inside Card 1
   </display>

   <display name=card2>
      <action type=ACCEPT task=GOSUB dest="file://c:/hdml/ex2.hdml">
      Inside Card 2
   </display>
</hdml>
```

Figure 6-5 shows the output of the preceding code as rendered by the UP.Simulator.

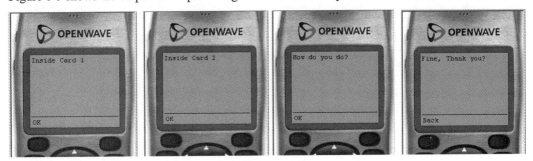

Figure 6-5: Output of Listing 6-5 (Images of Up.SDK courtesy Openwave systems Inc.)

Some More HDML Examples

The `<line>` element starts a new line. If the text is too long to fit the phone display, the user can view the entire line as the phone automatically scrolls it horizontally. All the lines that follow the current line are scrolled until a `<wrap>` statement is encountered.

Listing 6-6: example4.hdml
© 2001 Dreamtech Software India Inc.
All Rights Reserved

```
<hdml version=3.0>
    <display name=deck1>
        <line>Demonstrating the usage of line
    </display>
</hdml>
```

Figure 6-6 shows the output of the preceding code as rendered by the UP.Simulator.

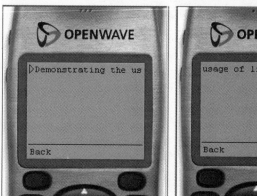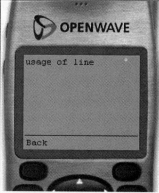

Figure 6-6: Output of Listing 6-6 (Images of Up.SDK courtesy Openwave Systems Inc.)

Formatted Text

Numerous statements can be applied to the text for wrapping and scrolling, and also for how the text is finally placed on the phone's display screen. Consider Listing 6-7.

Listing 6-7: example5.hdml
© 2001 Dreamtech Software India Inc.
All Rights Reserved

```
<hdml version="3.0">
 <display>
  <center>A B C
  <line>D&amp;F<tab>2&lt;5<tab>2&gt;1<tab>&dol;5
  <line><right>&quot;JKL&quot;
 </display>
</hdml>
```

In this example

- ♦ <center> center-aligns the formatted text.
- ♦ <right> right-aligns the formatted text.
- ♦ < represents < (less than).
- ♦ > represents > (greater than).
- ♦ " represents " (a quotation mark).
- ♦ & represents & (an ampersand).
- ♦ &dol; represents $ (a dollar sign).
- ♦   represents a regular space.

- &#*n*; represents any ASCII character (*n* is the ASCII code).

Note that the characters <, >, ", $, and & are reserved. The output of the preceding code as rendered by the UP.Simulator is shown in Figure 6-7.

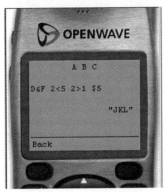

Figure 6-7: Output of Listing 6-7 (Image of Up.SDK courtesy Openwave Systems Inc.)

Format Specifiers in HDML

HDML supports the following format specifiers. Note that if the first character the user enters is preceded by a number or a * specifier, it is uppercase by default. (The user can change it to lowercase.)

- A — represents any uppercase alphabetic character or punctuation (no numbers).
- a — represents any lowercase alphabetic character or punctuation (no numbers).
- N — represents any numeric character (no symbols).
- X — represents any numeric, symbolic, or uppercase alphabetic character or punctuation.
- x —represents any numeric, symbolic, or lowercase alphabetic character or punctuation.
- M — represents any alphabetic character in any case, and any numeric or symbolic character.
- m — represents any alphabetic character in any case, and any numeric or symbolic character.

Listing 6-8 demonstrates how format specifiers are used in HDML:

Listing 6-8: example6.hdml
© 2001 Dreamtech Software India Inc.
All Rights Reserved

```
<hdml version=3.0 public=true>
 <entry name="rno" key="rollno" format="NNNNN">
  <action type="accept" label="next"task="go" dest="#nm">
  Roll No.(5 digits):
 </entry>

 <entry name="nm" key="name" format="AAAAAAAAAA">
  <action type="accept" label="next"task="go" dest="#add">
  Name(10 alphabets):
 </entry>

<entry name="add" key="address" format="XXXXXXXXXX">
  <action type="accept" label="next"task="go" dest="#d_o_b">
  City(10 characters):
 </entry>
```

```
<entry name="d_o_b" key="dob" format="NN/NN/NN">
 <action type="accept" label="next"task="go" dest="#detail">
  Birthdate(mm/dd/yy):
</entry>

<display name="detail">
 <action type="accept" label="ok" task="go" dest="#rno">
 <center>Description<br>
 $rollno<br>
 $name<br>
 $address<br>
 $dob
 </display>
</hdml>
```

The output of the preceding code as rendered by the UP.Simulator is shown in Figure 6-8.

Figure 6-8: Output of Listing 6-8 (Images of Up.SDK courtesy Openwave Systems Inc.)

Cards and Decks

In HDML a deck contains one or more than one card. Each card specifies a single interaction between the user and the phone. At one time only one card can be loaded on the phone. The following types of cards are supported by HDML:

Card Name	Description
Display cards	Displays information.
Entry cards	Displays a message and permits the user to enter a text string.
Choice cards	Displays multiple options from which the user can select only one.
No-display cards	Executes an action that does not appear on the phone display.

A phone always goes to the first card in the deck and displays the information the card provides. The user can interact by choosing an option or entering text and then pressing a function key, depending on the card type.

Consider Listing 6-9, ex2.hdml, which contains an HDML deck that contains two display cards. The phone always loads the first card in the deck. When the user presses the SOFT key, the next card is displayed.

Listing 6-9: ex2.hdml

© 2001 Dreamtech Software India Inc.
All Rights Reserved

```
<hdml version="3.0">
   <display>
       <action type=ACCEPT task="GO" dest="#card2">
       How do you do?
   </display>

   <display name="card2">
       Fine, Thank you…
   </display>
</hdml>
```

<action> associates a task with the function keys (ACCEPT, PREV, HELP, SEND, DELETE, SOFT1 or SOFT2) on a mobile phone. When the user presses the function key, the associated task is executed. The syntax of the <action> element is shown in the following code and table.

```
<action
label="string"
type="key"
task="task"
 dest="url"
rel="val"
method="type"
postdata="data"
accept-charset="ac"
vars="varlist"
receive="varlist"
retvals="varlist"
next="url"
cancel="url"
friend="boolean"
```

```
sendreferer="boolean"
clear="boolean"
number="no"
src="url"
icon="string">
```

Attribute Name	Value	Description
Label	string	Defines the key's label, which should be five characters or shorter. The default label for the ACCEPT key is OK.
Type	key	The key to associate to the task. For the ACCEPT key, the label is optional. For the HELP, SOFT1, SOFT2, SEND, DELETE keys, the label is required. For the PREV key, the label is ignored.
task	task	The task to execute. **Task:** GO **Description:** Requests the URL specified by the DEST option. If you use the GO task, you can specify only a relative URL for the DEST option. **Options:** DEST, VARS, SENDREFERER, REL, METHOD, POSTDATA, ACCEPT-CHARSET Task: GOSUB **Description:** Pushes a new activity onto the activity stack and requests the URL specified by the DEST option. When the nested activity returns, the phone puts the nested activity's return values into the variables specified by the RECEIVE option. If the nested activity cancels, the phone requests the URL specified by the CANCEL option. **Options:** DEST, VARS, SENDREFERER, FRIEND, RECEIVE, NEXT, CANCEL, REL, METHOD, POSTDATA, ACCEPT-CHARSET **Task:** RETURN **Description:** Returns from a nested activity to the previous activity with the return values specified by the RETVALS option.

		Options: RETVALS, DEST, CLEAR, REL, METHOD, POSTDATA, ACCEPT-CHARSET **Task:** CANCEL **Description:** Cancels the current activity, requesting the URL specified by the previous activity's CANCEL option. If no CANCEL option is specified, the phone requests the current card in the previous activity. **Options:** DEST, CLEAR, REL, METHOD, POSTDATA, ACCEPT-CHARSET **Task:** PREV **Description:** Displays the previous card in the activity history. If the current card is the first card in the current activity, PREV has the same effect as CANCEL. **Task:** CALL **Description:** Switches the phone to voice mode and dials the number specified by the NUMBER option. **Options:** NUMBER **Task:** NOOP **Description:** Do nothing. This is useful for disabling the default behavior of the specified action.
dest	url	The URL to request in GO and GOSUB tasks.
rel	val	Instructs the phone to pre-fetch the URL specified by the DEST option. The phone loads and caches the URL while the user is viewing the current card.
method	type	If the DEST option specifies a URL, the method used to request the URL: GET or POST. The default is GET.
postdata	data	The data to post if the method option specifies POST. The arguments with ampersands (&) are delimited if the data contains multiple arguments.
accept-charset	ac	Specifies the character set that the HDML application expects data returned from the phone to use.
vars	varlist	A list of variables to set for the current activity (if the task is GO) or the nested activity (if the task is GOSUB).

Attribute Name	Value	Description
receive	varlist	A list of variables delimited by semicolons, in which the phone stores the return values from a GOSUB task.
retvals	varlist	A list of values delimited by semicolons that an activity invoked with GOSUB returns to the invoking activity. The RETVALS option is allowed only with the RETURN task.
next	url	The URL to request after a nested activity returns.
cancel	url	The URL to request after a nested activity invoked by a GOSUB task cancels.
friend	boolean	A Boolean value specifying whether the nested activity specified in a GOSUB task is "friendly." TRUE indicates that the nested activity is friendly. The default is FALSE.
sendreferer	boolean	A Boolean value specifying whether the UP.Browser should provide the URL of the current deck when requesting the URL specified by the DEST or NEXT options. If TRUE, the UP.Browser specifies the deck's URL in the Referer header of the request. The default is FALSE.
clear	boolean	A Boolean value specifying whether a RETURN or CANCEL task from a nested activity unsets all the calling activity's variables. TRUE unsets the calling activity's variables. The default is FALSE.
number	no	The phone number to call for a CALL task.
Src	url	The URL of an image to display in place of the softkey label.
Icon	string	The name of a local image to display in place of the softkey label.

The output as rendered by the UP.Simulator is shown in Figure 6-9.

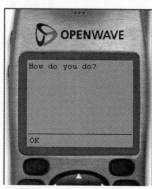 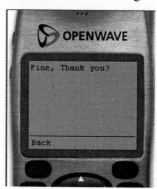

Figure 6-9: Output of Listing 6-9 (Images of Up.SDK courtesy Openwave Systems Inc.)

Listing 6-10, ex4.hdml, demonstrates how HDML assigns tasks to individual choices in a choice card.

Listing 6-10: ex4.hdml

```
<hdml version=3.0>
    <action type=ACCEPT task=GO dest=#other_info>
    <choice>
        <ce task=GO dest=#movies>Movie Information
        <ce task=GO dest=#stock>Stock Market Update
        <ce task=GO dest=#weather>Weather Forecast
    </choice>
```

```
    <display name=movies>
        <center>Gladiator - Plaza
        <wrap>Titanic - Odeon
        <wrap>Pearl Harbor - Eros
    </display>

    <display name=stock>
 <center>
 Wipro:Up<br>2%(Rs.2000)
 <wrap>Infosys:Up 5%(Rs.5000)
 <wrap>Satyam:Down 2.5%(Rs.1300)
    </display>

    <display name=weather>
        <center>New Delhi
        <wrap>Temperature<br>Max:33.6C Min:24.7c
        <wrap>
        <wrap>Relative Humidity Max:75% Min:30%
    </display>
</hdml>
```

In this listing, <choice> displays multiple options from which the user can select. <choice> must include one or more <CE> statements for defining the option (or options). The following code and table show the syntax of the <choice> element.

```
<choice
name="string"
title="string"
markable="boolean"
 bookmark="url"
key="var"
ikey="var"
method="meth"
default="default"
idefault="index">
actions (The actions to execute when the user presses a function)
text (The formatted text the card displays.)
<ce>text
</choice>
```

Attribute Name	Value	Description
Name	string	A unique name for the card.
Title	string	The default bookmark name that appears when the user marks the card.
Markable	boolean	Specifies whether the card can be marked. TRUE allows the card to be marked.
Bookmark	url	The URL that the phone adds to the bookmark list if the user marks the card. If the bookmark option is not specified, the phone adds the URL of the current card to the bookmark list.
Key	var	The variable name that gets the value (if any) of the selected option.
Ikey	var	The variable name containing an index number that indicates the default option. The number 1 specifies the first option, 2 specifies the second option, and so on.
Method	meth	Indicates the choice method. NUMBER is specified to make the list numbered. If the list is numbered, the user can select an option by pressing the option number or

		by scrolling to it and pressing a key. ALPHA is specified to make the list unnumbered. If the list is unnumbered, the user can select an option only by scrolling to it and pressing the ACCEPT key.
Default	default	The variable's default value, indicated by the key option.
Idefault	index	The index of the default entry. This is used if the ikey option is not specified or it specifies a variable with no value to it. If the ikey option is specified, idefault is ignored.

<ce> means choice entry that defines options that a user can select in a choice card. The syntax of the <ce> element is as follows:

```
<ce
value="val"
label="string"
task="task"
  dest="url"
rel="val"
method="type"
postdata="data"
accept-charset="ac"
vars="varlist"
receive="varlist"
retvals="varlist"
next="url"
cancel="url"
friend="boolean"
sendreferer="boolean"
clear="boolean"
number="no">
text (The formatted text the card displays.)
```

Attribute Name	Value	Description
Value	val	Stores information to the variable specified by the KEY option (specified in the <CHOICE> statement that contains the <CE> statement) when the user selects the option.
Label	string	Defines the key's label, which should be five characters or shorter. The default label for the ACCEPT key is OK.
Type	key	The key to associate to the task. For the ACCEPT key, the label is optional. For the HELP, SOFT1, SOFT2, SEND, and DELETE keys, the label is required. For the PREV key, the label is ignored.
Task	task	The task to execute. This task takes precedence over any actions specified at the card or deck level. Possible values are the same as specified by <action>.
Dest	url	The URL to request in GO and GOSUB tasks.
Rel	val	Instructs the phone to pre-fetch the URL specified by the DEST option. The phone loads and caches the URL while the user is viewing the current card.
Method	type	If the DEST option specifies a URL, the method used to request the URL: GET or POST. The default is GET.
Postdata	data	The data to post if the method option specifies POST. The arguments with ampersands (&) are delimited if the data contains multiple arguments.

`Accept-charset`	`ac`	Specifies the character set that the HDML application expects data returned from the phone to use.
`Vars`	`varlist`	A list of variables to set for the current activity (if the task is `GO`) or nested activity (if the task is `GOSUB`).
`Receive`	`varlist`	A list of variables delimited by semicolons, in which the phone stores the return values from a `GOSUB` task.
`Retvals`	`varlist`	A list of values delimited by semicolons that an activity invoked with `GOSUB` returns to the invoking activity. The `RETVALS` option is allowed only with the `RETURN` task.
`Next`	`url`	The URL to request after a nested activity returns.
`Cancel`	`url`	The URL to request after a nested activity invoked by a `GOSUB` task cancels.
`Friend`	`Boolean`	A Boolean value specifying whether the nested activity specified in a `GOSUB` task is "friendly." `TRUE` indicates that the nested activity is friendly. The default is `FALSE`.
`Sendreferer`	`Boolean`	A Boolean value specifying whether the UP.Browser should provide the URL of the current deck when requesting the URL specified by the `DEST` or `NEXT` options. If `TRUE`, the UP.Browser specifies the deck's URL in the `Referer` header of the request. The default is `FALSE`.
`Clear`	`Boolean`	A Boolean value specifying whether a `RETURN` or `CANCEL` task from a nested activity unsets all the calling activity's variables. `TRUE` unsets the calling activity's variables. The default is `FALSE`.
`Number`	`no`	The phone number to call for a `CALL` task.

The output as rendered by the UP.Simulator is shown in Figure 6-10.

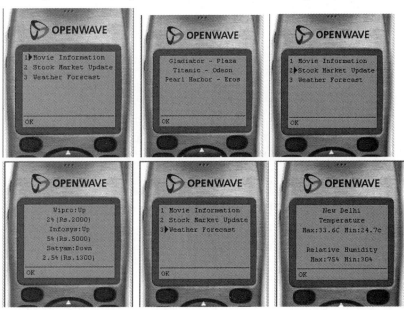

Figure 6-10: Output of Listing 6-10 (Images of Up.SDK courtesy Openwave Systems Inc.)

Activities

In HDML, an activity consists of one or more steps that represent the tasks the user performs. Typically, each step in an activity corresponds to an HDML card. The HDML for an activity uses GO tasks for navigating between cards. The following HDML documents, `act1.hdml` and `act2.hdml`, demonstrate activities.

Listing 6-11: example7.hdml

© 2001 Dreamtech Software India Inc.
All Rights Reserved

```
<hdml version=3.0>
 <display name=deck1>
  <action type=ACCEPT task=GO dest=act2.hdml>
  <center>Activities<br>
          Deck 1
 </display>
</hdml>
```

Listing 6-12: example8.hdml

© 2001 Dreamtech Software India Inc.
All Rights Reserved

```
<hdml version=3.0>
 <display name=deck2>
  <action type=ACCEPT task=GO dest=act1.hdml>
  <center>Activities<br>
          Deck 2
 </display>
</hdml>
```

Figure 6-11 shows the output as rendered by the UP.Simulator.

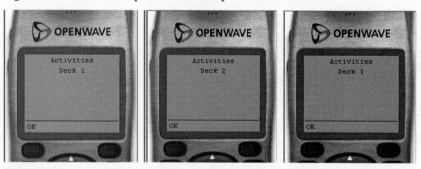

Figure 6-11: Output of Listings 6-11 and 6-12 (Images of Up.SDK courtesy Openwave systems Inc.)

Listing 6-13 shows that HDML cards can navigate backward when the user selects PREV. The phone pushes the card onto the history stack each time the user navigates forward to a card; it pops the current card off the stack each time the user selects PREV to navigate backward, thereby leaving the previous card at the top of the history stack.

Listing 6-13: example9.hdml

© 2001 Dreamtech Software India Inc.
All Rights Reserved

```
<hdml version=3.0>
  <entry name=email_address key=mailid>
 <action type=ACCEPT task=GO dest=#email_subject>
```

```
Email ID:
  </entry>

  <entry name=email_subject key=subject>
<action type=ACCEPT task=GO dest=#email_body>
Subject:
  </entry>

  <entry name=email_body key=body>
<action type=ACCEPT label="Back" task=GO dest=#email_address>
Message:
  </entry>
</hdml>
```

Figure 6-12 shows the output as rendered by the UP.Simulator.

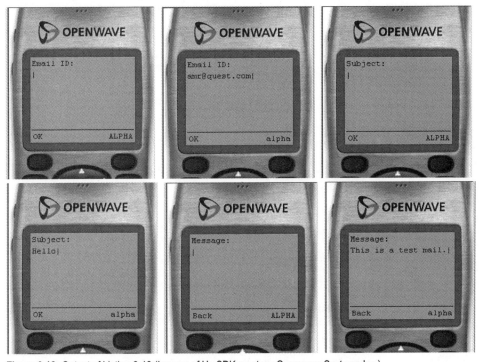

Figure 6-12: Output of Listing 6-13 (Images of Up.SDK courtesy Openwave Systems Inc.)

Variables in HDML

Variables in HDML represent information that the phone substitutes into HDML at runtime, thereby changing the display content and navigation dynamically based on user input. Variables can be set explicitly, or the user can set them with choice or entry cards. Variable values can be substituted into display text and certain HDML options; for example, the destination option (DEST).

The naming conventions followed by variables in HDML are as follows:

- ◆ Nonalphanumeric characters ($, <, >, =, /, \, &, *, and #) are not allowed.
- ◆ Case is important in the sense that sample and Sample are two different variables.
- ◆ Variable names should be relatively short because they are stored in the device memory.

A list of variables can be set using the VARS task option when navigating from one card to another.

Consider the following example, which demonstrates how the initial display card sets the value of three variables when the user selects ACCEPT. The variable values are displayed in the second display card.

Listing 6-14: example10.hdml

```
<hdml version=3.0>
   <display>
      <action type=ACCEPT task=GO dest=#VALUES
              vars=V1=Tea&V2=Coffee&V3=Milk>
      Click OK for a complete list of variables.
   </display>

   <display name=values>
      <wrap>VARIABLE1 = $V1
      <wrap>VARIABLE2 = $V2
      <wrap>VARIABLE3 = $V3
   </display>
</hdml>
```

The output as rendered by the UP.Simulator is shown in Figure 6-13.

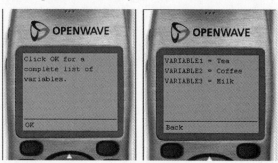

Figure 6-13: Output of Listing 6-14 (Images of Up.SDK courtesy Openwave Systems Inc.)

The variables can be referenced into formatted text, option values, choices, and entry-card defaults. In the following example, the user sets the value of a variable in an entry card. The variable's value is then substituted in the display text of another card. Variables can be substituted in any of the following ways:

♦ $var1 —Substitutes the value of var1, escaping nonalphunumeric characters $(var1) according to URL conventions.

♦ $(var1:esc)—Substitutes the value of var1, escaping nonalphunumeric characters according to URL conventions.

♦ $(var1:noesc)—Substitutes the value of var1, without escaping characters.

Listing 6-15: example11.hdml

```
<hdml version=3.0 markable=false>
   <entry name=name_entry key=user>
      <action type=ACCEPT task=GO dest="#welcome_message">
      Enter your name:
   </entry>
```

```
<display name=welcome_message>
    <center>Hi! $(user)
</display>
</hdml>
```

The output as rendered by the UP.Simulator is shown in Figure 6-14.

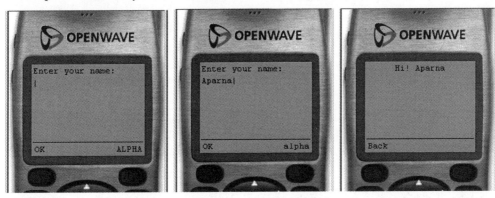

Figure 6-14: Output of Listing 6-15 (Images of Up.SDK courtesy Openwave Systems Inc.)

Listing 6-16 demonstrates that the `noesc` option is used when a variable is used for specifying a complete destination URL. After the user selects ACCEPT, the device requests the entered URL.

Listing 6-16: example12.hdml
© 2001 Dreamtech Software India Inc.
All Rights Reserved

```
<hdml version=3.0>
    <entry key=url>
        <action type=ACCEPT task=GOSUB dest="$(url:noesc)">
        &quot;Enter HDML URL Address:&quot;
    </entry>
</hdml>
```

The output as rendered by the UP.Simulator is shown in Figure 6-15.

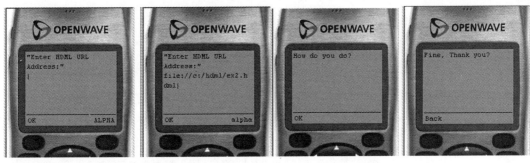

Figure 6-15: Output of Listing 6-16 (Images of Up.SDK courtesy Openwave Systems Inc.)

Listing 6-17 demonstrates how an anchored link associates a task with a string of formatted text.

Listing 6-17: example13.hdml
© 2001 Dreamtech Software India Inc.
All Rights Reserved

```
<hdml1 version=3.0>
   <display>
      Demonstrating Links -
      <br>1st Link
      <a task=GO dest=ex1.hdml label=Link1 accesskey=1>Example1</a>
      <br>2nd Link
      <a task=GO dest=ex2.hdml label=Link2 accesskey=2>Example2</a>
   </display>
</hdml>
```

The anchored link <a> element is specified in the formatted text. The > marker is displayed on the mobile phone when the user scrolls to a link. If a label is specified by the link, the ACCEPT label changes to that label. If the user presses ACCEPT, the task specified by the link is executed.

The syntax of the <a> element is as follows:

```
<a
label="string"
accesskey="key"
task="boolean"
 dest="url"
rel="val"
method="type"
postdata="data"
accept-charset="ac"
vars="varlist"
receive="varlist"
retvals="varlist"
next="url"
cancel="url"
friend="boolean"
sendreferer="boolean"
clear="boolean"
number="no">
text (The link's text which is displayed in square brackets [ ] by the phone)
</a>
```

Attribute Name	Value	Description
Label	string	An optional part that appears on the ACCEPT key when the link is selected. It should be five characters or shorter. The default label for the ACCEPT key is OK.
Accesskey	key	Represents a number (0 to 9) appearing on the left side of the screen next to the link. If the corresponding key is pressed on the keypad, the phone executes the task defined by the link.
Task	task	The task to be executed. This task takes precedence over any actions specified at the card or deck level. Possible values are the same as specified by <action>.
Dest	url	The URL to request in GO and GOSUB tasks.
rel	val	Instructs the phone to pre-fetch the URL specified by the DEST option. The phone loads and caches the URL while the user is viewing the current card.
Method	type	If the DEST option specifies a URL, the method used to request the URL: GET or POST. The default is GET.
Postdata	data	The data to post if the method option specifies POST. The arguments with ampersands (&) are delimited if the data contains multiple arguments.

Attribute Name	Value	Description
`accept-charset`	`ac`	Specifies the character set that the HDML application expects data returned from the phone to use.
`Vars`	`varlist`	A list of variables to set for the current activity (if the task is `GO`) or nested activity (if the task is `GOSUB`).
`Receive`	`varlist`	A list of variables, delimited by semicolons, in which the phone stores the return values from a `GOSUB` task.
`Retvals`	`varlist`	A list of values delimited by semicolons that an activity invoked with `GOSUB` returns to the invoking activity. The `RETVALS` option is allowed only with the `RETURN` task.
`Next`	`url`	The URL to request after a nested activity returns.
`Cancel`	`url`	The URL to request after a nested activity invoked by a `GOSUB` task cancels.
`Friend`	`boolean`	A Boolean value specifying whether the nested activity specified in a `GOSUB` task is "friendly." `TRUE` indicates that the nested activity is friendly. The default is `FALSE`.
`Sendreferer`	`boolean`	A Boolean value specifying whether the UP.Browser should provide the URL of the current deck when requesting the URL specified by the `DEST` or `NEXT` options. If `TRUE`, the UP.Browser specifies the deck's URL in the `Referer` header of the request. The default is `FALSE`.
`Clear`	`boolean`	A Boolean value specifying whether a `RETURN` or `CANCEL` task from a nested activity unsets all the calling activity's variables. `TRUE` unsets the calling activity's variables. The default is `FALSE`.
`Number`	`no`	The phone number to call for a `CALL` task.

The output as rendered by the UP.Simulator is shown in Figure 6-16.

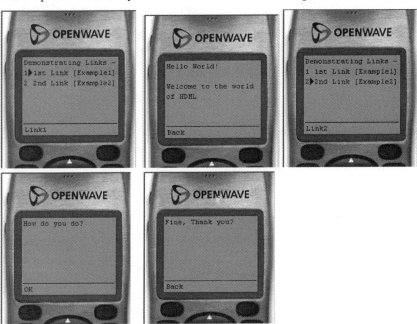

Figure 6-16: Output of Listing 6-17 (Images of Up.SDK courtesy Openwave Systems Inc.)

The src or icon option within an <action> statement can be specified for displaying images as softkey labels. Listing 6-18 illustrates how to use icons for the ACCEPT and SOFT1 function keys. The label option must be specified so that some label can be displayed on the device in case an image cannot be found.

Listing 6-18: example14.hdml

© 2001 Dreamtech Software India Inc.
All Rights Reserved

```
<hdml version=3.0>
   <choice key=choice>
      <action type=ACCEPT icon=rightarrow2 task=GO dest=#$choice>
      <action type=SOFT1 label=Call icon=phone1 task=CALL number=$choice>
      <ce  value="7777777">Ashish
      <ce  value="8888888">Madan
      <ce  value="9999999">Ravinder
   </choice>

   <display name=7777777>
        <action type=ACCEPT icon=rightarrow2>
<img icon="smileyface" alt=":-)" >
Mr. Ashish Baveja
777 Golf Links<br>
New Delhi-110014
India
   </display>

   <display name=8888888>
        <action type=ACCEPT icon=rightarrow2>
<img icon="smileyface" alt=":-)" >
Mr. M.K. Dhyani
888 South Ex II<br>
New Delhi-110024
India
   </display>

   <display name=9999999>
        <action type=ACCEPT icon=rightarrow2>
<img icon="smileyface" alt=":-)" >
Mr. R. Srivastav
999 Green Park<br>
New Delhi-110015
India
   </display>
</hdml>
```

The output as rendered by the UP.Simulator is shown in Figure 6-17.

Figure 6-17: Output of Listing 6-18 (Images of Up.SDK courtesy Openwave Systems Inc.)

The image statement supports the following icon names:

```
exclamation1, exclamation2, question1, question2, lefttri1, righttri1, lefttri2,
righttri2, littlesqaure1, littlesquare2, isymbol, wineglass, speaker,
dollarsign, moon1, bolt, medsquare1, medsquare2, littlediamond1, littlediamond2,
bigsquare1, bigsquare2, littlecircel1. littlecircle2, wristwatch, plus, minus,
star1, uparrow1, downarrow1, circleslash, downtri1, uptri1, downtri2, uptri2,
bigdiamond1, bigdiamond2, biggestsquare1, biggestsquare2, bigcircle1,
bigcircle2, uparrow2, downarrow2, sun, baseball, clock, moon2, bell, pushpin,
smallface, heart, martini, bud, trademark, multiply, document1, hourglass1,
hourglass2, floppy1, snowflake, cross1, cross2, rightarrow1, leftarrow1, mug,
divide, calendar, smileyface, star2, rightarrow2, leftarrow2, gem, checkmark1,
dog, star3, sparkle, lightbulb, bird, folder1, head1, copyright, registered,
briefcase, folder2, phone1, voiceballoon, creditcard, uptri3, downtri3, usa,
note3, clipboard, cup, camera1, rain, football, book1, stopsign, trafficlight,
book2, book3, book4, document2, scissors, day, ticket, cloud, envelope1, check,
videocam, camcorder, house, flower, knife, vidtape, glasses, roundarrow1,
roundarrow2, magnifyglass, key, note1, note2, boltnut, shoe, car, floppy2,
chart, graph1, mailbox, flashlight, rolocard, check2, leaf, hound, battery,
scroll, thumbtack, lockkey, dollar, lefthand, righthand, tablet, paperclip,
present, tag, meal1, books, truck, pencil, uplogo, envelope2, wrench, outbox,
inbox, phone2, factory, ruler1, ruler2, graph2, meal2, phone3, plug, family,
```

link, package, fax, partcloudy, plane, boat, dice, newspaper, train, blankfull, blankhalf, blankquarter

The following example displays various image icons.

Listing 6-19: example15.hdml
© 2001 Dreamtech Software India Inc.
All Rights Reserved

```
<HDML VERSION="3.0">
    <DISPLAY>
        <wrap><center>Displaying various image icons<br>
        <line>
        <img icon="smileyface" alt=":-)">
        <img icon="clock">
        <img icon="book1">
        <img icon="righthand">
        <img icon="uparrow1">
        <img icon="downarrow1">
        <img icon="floppy1"><br>
        <img icon="bell">
        <img icon="heart">
        <img icon="pushpin">
        <img icon="calendar">
        <img icon="boat">
        <img icon="plane">
    </DISPLAY>
</HDML>
```

The output as rendered by the UP.Simulator is shown in Figure 6-18.

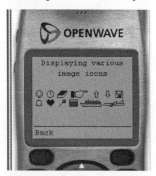

Figure 6-18: Output of Listing 6-19 (Image of Up.SDK courtesy Openwave Systems Inc.)

Case Study

The following case study is based on an imaginary online fast-food processing outlet that distributes through mobile services using HDML. There are four HDML documents: cs1.hdml, cs2.hdml, cs3.hdml, and cs4.hdml. cs1.hdml offers the user three options, namely pizza, burger, and sandwich. After selecting any of the items, the user is prompted for the category of the item (veg or non veg) and then for the type of the item, such as regular, medium, or large. Now the user is sequentially prompted for personal details such as name, address, ZIP code, credit card number, and credit card expiration date. Finally, the user is prompted for the confirmation or cancellation of the order. If the user selects the option buy, the order is confirmed. If the user selects the second option, cancel, the transaction is cancelled and the user returns to the main menu.

Listing 6-20 provides the user with the initial menu for selection and invokes cs2.hdml.

Listing 6-20: cs1.hdml

```
<hdml version=3.0 public=true>
   <choice key=opt>
      <action type=accept label="select an item" task=go
dest="file://c:/hdml/cs2.hdml"
        vars="ch=$opt" >
      <ce value=pizza>pizza
      <ce value=burger>burger
      <ce value=sandwich>sandwich
   </choice>
</hdml>
```

Listing 6-21 provides the user with various categories for the selected item and invokes cs3.hdml.

Listing 6-21: cs2.hdml

```
<hdml version=3.0 public=true>
  <choice key=ty>
     <action type=accept label="select a type" task=go
dest="file://c:/hdml/cs3.hdml"
        vars="val=$ty-$ch">
     <ce value=veg>veg $ch
     <ce value=non-veg>non veg $ch
  </choice>
</hdml>
```

Listing 6-22 prompts the user for the appropriate type of the selected item and invokes cs4.hdml.

Listing 6-22: cs3.hdml

```
<hdml version=3.0 public=true>
   <choice key=sz>
      <action type=accept label="select a size" task=go dest=#$sz-$ty-$ch>
      <ce value=regular>regular $val
      <ce value=medium>medium $val
      <ce value=large>large $val
   </choice>

   <display name=regular-non-veg-pizza>
       <action type=accept task="go" dest="file://c:/hdml/cs4.hdml">
Enter your
       <img icon="smileyface" alt=":-)">
details:
   </display>

   <display name=medium-non-veg-pizza>
       <action type=accept task="go" dest="file://c:/hdml/cs4.hdml">
Enter your
<img icon="smileyface" alt=":-)">
details:
   </display>
```

```
    <display name=large-non-veg-pizza>
        <action type=accept task="go" dest="file://c:/hdml/cs4.hdml">
Enter your
<img icon="smileyface" alt=":-)">
details:
    </display>


    <display name=regular-veg-pizza>
        <action type=accept task="go" dest="file://c:/hdml/cs4.hdml">
Enter your
<img icon="smileyface" alt=":-)">
details:
    </display>


    <display name=medium-veg-pizza>
        <action type=accept task="go" dest="file://c:/hdml/cs4.hdml">
Enter your
<img icon="smileyface" alt=":-)">
details:
    </display>


    <display name=large-veg-pizza>
        <action type=accept task="go" dest="file://c:/hdml/cs4.hdml">
Enter your
<img icon="smileyface" alt=":-)">
details:
    </display>
    <display name=regular-non-veg-burger>
        <action type=accept task="go" dest="file://c:/hdml/cs4.hdml">
Enter your
<img icon="smileyface" alt=":-)">
details:
    </display>
    <display name=medium-non-veg-burger>
        <action type=accept task="go" dest="file://c:/hdml/cs4.hdml">
Enter your
<img icon="smileyface" alt=":-)">
details:
    </display>
    <display name=large-non-veg-burger>
        <action type=accept task="go" dest="file://c:/hdml/cs4.hdml">
Enter your
<img icon="smileyface" alt=":-)">
details:
    </display>


    <display name=regular-veg-burger>
        <action type=accept task="go" dest="file://c:/hdml/cs4.hdml">
Enter your
<img icon="smileyface" alt=":-)">
details:
    </display>


    <display name=medium-veg-burger>
        <action type=accept task="go" dest="file://c:/hdml/cs4.hdml">
```

```
Enter your
<img icon="smileyface" alt=":-)">
details:
  </display>

  <display name=large-veg-burger>
      <action type=accept task="go" dest="file://c:/hdml/cs4.hdml">
Enter your
<img icon="smileyface" alt=":-)">
details:
  </display>

  <display name=regular-non-veg-sandwich>
      <action type=accept task="go" dest="file://c:/hdml/cs4.hdml">
Enter your
<img icon="smileyface" alt=":-)">
details:
  </display>

  <display name=medium-non-veg-sandwich>
      <action type=accept task="go" dest="file://c:/hdml/cs4.hdml">
Enter your
<img icon="smileyface" alt=":-)">
details:
  </display>

  <display name=large-non-veg-sandwich>
      <action type=accept task="go" dest="file://c:/hdml/cs4.hdml">
Enter your
<img icon="smileyface" alt=":-)">
details:
  </display>

  <display name=regular-veg-sandwich>
      <action type=accept task="go" dest="file://c:/hdml/cs4.hdml">
Enter your
<img icon="smileyface" alt=":-)">
details:
  </display>

  <display name=medium-veg-sandwich>
      <action type=accept task="go" dest="file://c:/hdml/cs4.hdml">
Enter your
<img icon="smileyface" alt=":-)">
details:
  </display>

  <display name=large-veg-sandwich>
      <action type=accept task="go" dest="file://c:/hdml/cs4.hdml">
Enter your
<img icon="smileyface" alt=":-)">
details:
  </display>
</hdml>
```

Listing 6-23 accepts the user's personal details for transactional purposes. It brings control to the menu for the final confirmation or for cancellation of the order, thereby returning to the main menu, cs1.hdml.

Listing 6-23: cs4.hdml

```
<hdml version=3.0 public=true>
 <entry name="first" key="fnam">
  <action type="prev" task="go" dest="#shield" >
  <action type="accept" task="go" label="ok" dest="#add" >
  Name:
 </entry>

 <entry name="add" key="add">
  <action type="accept" task="go" dest="#z" >
  Address:
 </entry>

 <entry name="z" key="zip" format="nnnnnn">
  <action type="accept" task="go" dest="#cc" >
  Zip Code:
 </entry>

 <entry name="cc" key="cc" format="nnnnnnnnnnnnnnn*n">
  <action type="accept" task="go" dest="#exp" >
  Credit Card:
 </entry>

 <entry name="exp" format="nn\/nnnn" key="exp">
  <action type="accept" task="go" dest="#confirm" >
  Expiry (mm/yyyy):
 </entry>

 <display name="confirm">
  <action type="accept" task="return" label="buy" >
  <action type="soft1" task="go" dest="#first" label="edit">
  Item: $sz-$ty-$ch<br>
  Name: $fnam<br>
  Address: $add<br>
  Zip Code: $zip<br>
  <a task="go" dest="#ok" label="buy">buy</a>
  <a task="go" dest="#shield" label="cancel">cancel</a>
 </display>

 <display name="shield">
  <action type="accept" label="yes" task="cancel" dest=#abort>
  <action type="soft1" label="no" task="go" dest="#confirm">
  Are you sure you want to cancel the $val order?
 </display>

 <display name="ok">
              <action type="accept" label="home" task="go"
dest="file://c:/hdml/cs1.hdml" >
  Thank you for your purchase.
```

```
</display>

<display name="abort">
 <action type="soft1" label="home" task="go" dest="file://c:/hdml/cs1.hdml">
 Sorry! your order was cancelled.
</display>
</hdml>
```

The output as rendered by the UP.Simulator is shown in Figure 6-19.

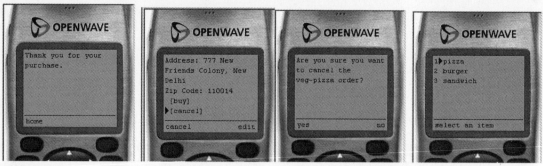

Figure 6-19: Output of Listings 6-20 to 6-23 (Images of Up.SDK courtesy Openwave Systems Inc.)

Summary

HDML furnishes a highly capable markup language for wireless and other Web-enabled handheld devices with limited resources. Moreover, HDML serves environments with limited visual context. Even though HDML is a bit older then WML, a majority of the handheld, Web-enabled mobile devices presently available are only HDML-capable.

Chapter 7

Transforming the Wireless Web Application for HDML Clients

In this chapter, you'll take a look at the process of transforming the Web application for HDML clients. Phone.com in the United States introduced HDML, and soon it became popular because of if its simple and easy coding style, as well as its wide user base in America. Chapter 6 provided introductory information on HDML and substantiated it with a case study. In this chapter, you will see the transformation of the current application for HDML clients, so that i-mode users can access the application from their devices.

Framework of the Application

Before you actually begin writing the code, you should take a look at the design framework for the HDML transformation of the application. You are following the same design that you followed in Chapter 5 for transforming the application for WAP clients.

Figure 7-1 shows the basic design of the framework for the application's HDML transformation.

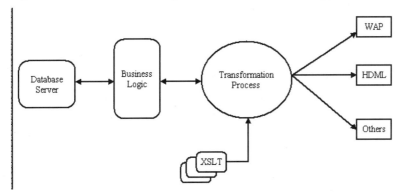

Figure 7-1: The design of the framework for an HDML transformation

This time you are not writing everything from scratch because you already have a complete set of working files. Here you will add some code in some of the Active Server Pages, which you can then use to write the XSL documents for HDML clients completely from scratch. So by the end of this chapter, you will have completed two transformations of the application.

Detect the Browser for Serving Contents

As in the preceding chapter, you first detect the HDML clients from the entire range of devices accessing the application. After you have determined the client type, you can choose different XSL documents on the basis of that client type.

To detect HDML clients, you add some lines of code in `checkbrows.asp` (Listing 7-1). As you know, this document works as a home page for the application. Hence, by including the code for detecting the HDML client, the file is enabled for both HDML and WML clients.

Listing 7-1: checkbrows.asp (modified for HDML)

```
<%@ Language=VBScript %>

<%
'declaring variables
dim sXML
dim sXSL
dim xmlDocument,xslDocument
dim browser,bt

'creating the objects for the XMLDOM
set xmlDocument = Server.CreateObject("MICROSOFT.XMLDOM")
set xslDocument = Server.CreateObject("MICROSOFT.XMLDOM")

acc=Request.ServerVariables("HTTP_ACCEPT")

Set bt = Server.CreateObject("MSWC.BrowserType")
browser = bt.browser

uagent = Request.ServerVariables("HTTP_USER_AGENT")
browserinfo = split("/",uagent,5)
browsertype = browserinfo(0)

'check for the browser then according to that browser load the xml files
if (InStr(acc,"text/vnd.wap.wml")) then

  session("browsmode")="wml"

  sXML = "main.xml"
  sXSL = "wml.xsl"

  xmlDocument.async = false
  xslDocument.async = false

  xmlDocument.load(Server.MapPath(sXML))
  xslDocument.load(Server.MapPath(sXSL))

  Response.ContentType = "text/vnd.wap.wml"
  Response.Write "<!DOCTYPE wml PUBLIC ""-//WAPFORUM//DTD " & _
    "WML 1.1//EN"" ""http://www.wapforum.org/DTD/wml_1.1.xml"">"

    Response.Write(xmlDocument.transformNode(xslDocument))

else
if (InStr(acc,"hdml")) then

  session("browsmode")="hdml"
  Response.ContentType = "text/x-hdml"

  sXML = "main.xml"
  sXSL = "hdml.xsl"
```

```
   xmlDocument.async = false
   xslDocument.async = false

   xmlDocument.load(Server.MapPath(sXML))
   xslDocument.load(Server.MapPath(sXSL))

   Response.Write(xmlDocument.transformNode(xslDocument))
  end if
 end if

%>
```

In Listing 7-1, you have added an `if` condition to detect HDML clients. You search for the HDML in the string returned by the browser. If you find HDML in that string, you store it as a value in the session variable named `browsmode`. The following line sets the content type for HDML browsers:

```
  session("browsmode")="hdml"
  Response.ContentType = "text/x-hdml"
```

You then load the XML and XSL documents, as shown in the following section. You are using the same document that you have used before in the application, but this time you load a different XSL document, named `hdml.xsl`, for HDML clients.

Transform the Home Page

Now take a look at the XSL code that you used for the transformation of the application's home page. The XML document is still the same `main.xml` file, shown in Listing 7-2.

Listing 7-2: main.xml

```
<?xml version='1.0'?>
<xsl:stylesheet xmlns:xsl="http://www.w3.org/1999/XSL/Transform" version="1.0">

<xsl:template match="/">

<HDML VERSION='3.0'>
<display name="main">
Main Menu
<a type="accept" task="go" dest="generatexml.asp?sec=Weather"
lable="view">Weather</a>
<a type="accept" task="go" dest="generatexml.asp?sec=News" lable="view">News
Section</a>
<a type="accept" task="go" dest="generatexml.asp?sec=E-mail" lable="view">E-
Mail</a>
<a type="accept" task="go" dest="generatexml.asp?sec=movies" lable="view">Movie
Ticket Booking</a>
</display>

</HDML>
</xsl:template>
</xsl:stylesheet>
```

In Listing 7-2, you first generate an HDML deck and follow it by generating all the HDML code used within the XSL file. You then start a `display` element. Within that `display` element, you generate all the links for the HDML browser to browse the application, as shown in the following code:

```
<display name="main">
Main Menu
```

```
<a type="accept" task="go" dest="generatexml.asp?sec=Weather"
lable="view">Weather</a>
<a type="accept" task="go" dest="generatexml.asp?sec=News" lable="view">News
Section</a>
<a type="accept" task="go" dest="generatexml.asp?sec=E-mail" lable="view">E-
Mail</a>
<a type="accept" task="go" dest="generatexml.asp?sec=movies" lable="view">Movie
Ticket Booking</a>
</display>
```

In these links, you are displaying four options for the user. When the user clicks any of the listed sections, he calls the `generatexml.asp` file for the first time in the transformation. This file generates the XML data on the basis of the parameter passed to it.

When the transformation takes place, the HDML browser shows the output in Figure 7-2.

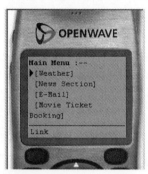

Figure 7-2: Displaying the home page for HDML clients (Image of Up.SDK courtesy Openwave Systems Inc.)

This home page displays all four sections of the application after the transformation. These sections are the following:

- Weather
- News section
- E-mail
- Movie ticket booking

In the next section, you'll see the transformation of the application's weather section. By doing this, you can provide weather information to HDML clients.

Transform the Weather Section for HDML Clients

In this section, you first see how to generate the XML data for the weather section of the application. You again use the `generatexml.asp` file (Listing 7-3) to generate the XML data. You then transform this data into HDML.

Listing 7-3: generatexml.asp

```
<%@ Language=VBScript %>

<%
dim address,section,fname,ctg,city,emode
ctg =" "
city=" "
emode =" "
section = " "
```

```
section = trim(Request.QueryString("sec"))
emode = trim(Request.QueryString("mode"))
ctg = trim(Request.QueryString("category"))
city = trim(Request.QueryString("city"))


set conn = Server.CreateObject("ADODB.Connection")
conn.ConnectionString = "Provider=SQLOLEDB.1;Persist Security Info=False;User
ID=sa;Initial Catalog=portal;Data Source=developers"
conn.Open

Set objrs = Server.CreateObject("ADODB.Recordset")
objrs.CursorType = adOpenStatic

if (section="Weather")  then
  sql = "select city from weather"
  objrs.Open sql,conn

  Set fso =Server.CreateObject("Scripting.FileSystemObject")
  fname = "tweather.xml"

  fullpath = server.MapPath("main")

  Set MyFile=fso.CreateTextFile(fullpath & "/" & fname)

  MyFile.WriteLine("<?xml version='1.0'?>")
  MyFile.WriteLine("<ROOT>")


  while not(objrs.EOF)
    city = objrs.Fields("city")
    MyFile.WriteLine("<city>" & city & "</city>")

  objrs.MoveNext
  wend

  MyFile.WriteLine("</ROOT>")
  address = "http://192.168.1.100/loadxsl.asp?name=" & fname
  Response.Redirect(address)

end if


if ((city="Sydney") OR (city="NewDelhi") OR (city="Miami") OR (city="Tokyo") OR
(city="Phoenix") OR (city="NewYork"))  then
  sql = "select * from weather where city = '" & city & "'"
  objrs.Open sql,conn

  Set fso =Server.CreateObject("Scripting.FileSystemObject")
  fname = "tweatherdetail.xml"

  fullpath = server.MapPath("main")

  Set MyFile=fso.CreateTextFile(fullpath & "/" & fname)

  MyFile.WriteLine("<?xml version='1.0'?>")
```

```
MyFile.WriteLine("<ROOT>")


while not(objrs.EOF)
   lowtemp = objrs.Fields("low_temp")
   hightemp = objrs.Fields("high_temp")
   srise = objrs.Fields("sunrise")
   sset = objrs.Fields("sunset")
   mrise = objrs.Fields("moonrise")
   mset = objrs.Fields("moonset")
   dhumid = objrs.Fields("dayhumidity")
   nhumid = objrs.Fields("nighthumidity")
   rain = objrs.Fields("rainfall")
   fig = objrs.Fields("figure")

   MyFile.WriteLine("<weather>")
   MyFile.WriteLine("<low_temp>" & lowtemp & "</low_temp>")
   MyFile.WriteLine("<high_temp>" & hightemp & "</high_temp>")
   MyFile.WriteLine("<sunrise>" & srise & "</sunrise>")
   MyFile.WriteLine("<sunset>" & sset & "</sunset>")
   MyFile.WriteLine("<moonrise>" & mrise & "</moonrise>")
   MyFile.WriteLine("<moonset>" & mset & "</moonset>")
   MyFile.WriteLine("<dayhumidity>" & dhumid & "</dayhumidity>")
   MyFile.WriteLine("<nighthumidity>" & nhumid & "</nighthumidity>")
   MyFile.WriteLine("<rainfall>" & rain & "</rainfall>")
   MyFile.WriteLine("<figure>" & fig & "</figure>")
   MyFile.WriteLine("</weather>")
 objrs.MoveNext
 wend

 MyFile.WriteLine("</ROOT>")
 address = "http://192.168.1.100/loadxsl.asp?name=" & fname
 Response.Redirect(address)

end if


objrs.Close
conn.Close

%>
```

In Listing 7-3, you generate XML data for the weather section of the application and then save all this data in an XML file, just as you did in Chapter 5, using the file system object (FSO). The corresponding code is as follows:

```
Set fso =Server.CreateObject("Scripting.FileSystemObject")
fname = "tweather.xml"

fullpath = server.MapPath("main")

Set MyFile=fso.CreateTextFile(fullpath & "/" & fname)
```

The preceding code creates a file named `tweather.xml` on the server; this file contains all the XML data for displaying the first page of the weather section of the application.

Figure 7-3 shows a screen shot of `tweather.xml` in Internet Explorer.

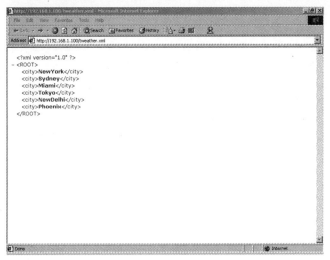

Figure 7-3: Displaying the names of different cities in XML format

You now have an XML file on the server. In the next few lines of code, you call `loadxsl.asp` to perform the transformation of this file. In this code, you pass a parameter in a query string called `name`. In this parameter you send the name of the XML file as a value.

```
address = "http://192.168.1.100/loadxsl.asp?name=" & fname
Response.Redirect(address)
```

Listing 7-4 provides the code for `loadxsl.asp`.

Listing 7-4: loadxsl.asp (modified for HDML)

```
<%@ Language=VBScript %>

<%
fname = trim(Request.QueryString("name"))


'declaring variables
dim sXML
dim sXSL
dim xmlDocument,xslDocument

'creating the objects for the XMLDOM
set xmlDocument = Server.CreateObject("MSXML2.DOMDocument")
set xslDocument = Server.CreateObject("MSXML2.DOMDocument")

sXML = fname
fname = mid(fname,1,(len(fname) - 4))


if (session.Contents(1)="wml") then
 sXSL = fname & "wml.xsl"

 xmlDocument.async = false
 xslDocument.async = false

 xmlDocument.load(Server.MapPath(sXML))
```

```
xslDocument.load(Server.MapPath(sXSL))

Response.ContentType = "text/vnd.wap.wml"
Response.Write "<?xml version=""1.0"" ?>"


Response.Write "<!DOCTYPE wml PUBLIC ""-//WAPFORUM//DTD " & _
"WML 1.1//EN"" ""http://www.wapforum.org/DTD/wml_1.1.xml"">"

Response.Write(xmlDocument.transformNode(xslDocument))

end if

if (session.Contents(1)="hdml") then
 sXSL = fname & "hdml.xsl"

 xmlDocument.async = false
 xslDocument.async = false

 Response.ContentType = "text/x-hdml"

 xmlDocument.load(Server.MapPath(sXML))
 xslDocument.load(Server.MapPath(sXSL))


 Response.Write(xmlDocument.transformNode(xslDocument))
end if

%>
```

In Listing 7-4, you have added a new `if` condition for HDML clients, along with the WML clients you are already dealing with. If the client is HDML based, you set the value of both the `sXML` and `sXSL` variables.

While setting the value of the `sXSL` variable, which will load the XSL file that corresponds with the XML file you are loading, you change the code as follows, to make it compatible with HDML clients:

```
if (session.Contents(1)="hdml") then
 sXSL = fname & "hdml.xsl"

 xmlDocument.async = false
 xslDocument.async = false

 Response.ContentType = "text/x-hdml"

 xmlDocument.load(Server.MapPath(sXML))
 xslDocument.load(Server.MapPath(sXSL))
```

First you check for the browser type in the code. After this, you set the value of the `sXSL` variable by adding the value `fname` as a prefix to the `hdml.xsl` file. So if you have `tweather` as a value in the `fname` variable, the value of the `sXSL` variable becomes `tweatherhdml.xsl`, and so on for the application's other files.

In the next line, you set the content type for HDML browsers by using `Response.contentType`.

```
Response.ContentType = "text/x-hdml"
```

Having set the content type, you now load both the XML and XSL documents. In the last line, you carry out the transformation.

Listing 7-5 shows the `tweatherhdml.xsl` file that you use in the transformation.

Listing 7-5: tweather.xsl

```
<?xml version='1.0'?>
<xsl:stylesheet xmlns:xsl="http://www.w3.org/1999/XSL/Transform" version="1.0">


<xsl:output method="text"/>



<xsl:template match="/">
<![CDATA[<HDML VERSION='3.0'>]]>

<![CDATA[<DISPLAY>]]>
  <![CDATA[<ACTION TYPE="ACCEPT" Task="GO" DEST="checkbrows.asp">]]>
 <xsl:apply-templates select="ROOT" />
<![CDATA[</DISPLAY>]]>
<![CDATA[</HDML>]]>

</xsl:template>

<xsl:template match="ROOT">
Cities
<![CDATA[<BR>]]>
<xsl:for-each select="state">
<![CDATA[<a>]]>
    <xsl:attribute name="href">
    generatexml.asp?city=<xsl:value-of select="." />
    </xsl:attribute>
    <xsl:value-of select="." />
<![CDATA[</a>]]>
<![CDATA[<BR>]]>

</xsl:for-each>
</xsl:template>

</xsl:stylesheet>
```

It's very important that you understand the `tweather.xsl` file in Listing 7-5, so let's take it step by step. In the first few lines of code, after declaring the XSLT document header, you use the `<XSL:ouput>` method to control the output this file will generate.

```
<xsl:output method="text"/>
```

Let's begin our discussion on the process of generating cards for HDML clients using XSLT. Take a look at the following lines of code:

```
<![CDATA[<HDML VERSION='3.0'>]]>

<![CDATA[<DISPLAY>]]>
  <![CDATA[<ACTION TYPE="ACCEPT" Task="GO" DEST="checkbrows.asp">]]>
 <xsl:apply-templates select="ROOT" />
<![CDATA[</DISPLAY>]]>
<![CDATA[</HDML>]]>
```

As you can see, you use the CDATA section whenever you need to generate HDML elements using XSLT. The main reason for using the CDATA section is the problem of displaying some HDML tags using XSLT. Some of the HDML tags, such as the <action> tag, do not have an end tag. So if you generate these kinds of tags using an XSLT processor, the XSLT processor can't parse the document and starts giving errors.

After generating a card for display in the HDML client browser, you now go on to fetch the records from the XML file and display them in the HDML browser by using the following lines of code:

```
<xsl:template match="ROOT">
Cities
<![CDATA[<BR>]]>
<xsl:for-each select="city">
<![CDATA[<a>]]>
    <xsl:attribute name="href">
    generatexml.asp?city=<xsl:value-of select="." />
    </xsl:attribute>
    <xsl:value-of select="." />
<![CDATA[</a>]]>
<![CDATA[<BR>]]>

</xsl:for-each>
</xsl:template>
```

In the preceding code, you have again used a CDATA section to display the records and construct a loop to get the names of all the cities.

When this transformation takes place, it shows output on the HDML browser, as shown in Figure 7-4.

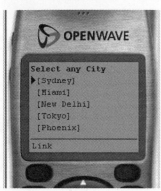

Figure 7-4: Displaying the names of cities (Image of Up.SDK courtesy Openwave Systems Inc.)

The user now selects one of cities listed on the browser. When the user selects a city, the program again calls the generatexml.asp file to generate the detailed weather report for that city in XML format. The following code shows how to generate a detailed weather report on the basis of the parameter the user enters.

```
if ((city="Sydney") OR (city="NewDelhi") OR (city="Miami") OR (city="Tokyo") OR
(city="Phoenix") OR (city="NewYork"))   then
  sql = "select * from weather where state = '" & city & "'"
  objrs.Open sql,conn

  Set fso =Server.CreateObject("Scripting.FileSystemObject")
  fname = "tweatherdetail.xml"

  fullpath = server.MapPath("main")
```

```
Set MyFile=fso.CreateTextFile(fullpath & "/" & fname)

MyFile.WriteLine("<?xml version='1.0'?>")
MyFile.WriteLine("<ROOT>")


while not(objrs.EOF)
   lowtemp = objrs.Fields("low_temp")
   hightemp = objrs.Fields("high_temp")
   srise = objrs.Fields("sunrise")
   sset = objrs.Fields("sunset")
   mrise = objrs.Fields("moonrise")
   mset = objrs.Fields("moonset")
   dhumid = objrs.Fields("dayhumidity")
   nhumid = objrs.Fields("nighthumidity")
   rain = objrs.Fields("rainfall")
   fig = objrs.Fields("figure")

   MyFile.WriteLine("<weather>")
   MyFile.WriteLine("<low_temp>" & lowtemp & "</low_temp>")
   MyFile.WriteLine("<high_temp>" & hightemp & "</high_temp>")
   MyFile.WriteLine("<sunrise>" & srise & "</sunrise>")
   MyFile.WriteLine("<sunset>" & sset & "</sunset>")
   MyFile.WriteLine("<moonrise>" & mrise & "</moonrise>")
   MyFile.WriteLine("<moonset>" & mset & "</moonset>")
   MyFile.WriteLine("<dayhumidity>" & dhumid & "</dayhumidity>")
   MyFile.WriteLine("<nighthumidity>" & nhumid & "</nighthumidity>")
   MyFile.WriteLine("<rainfall>" & rain & "</rainfall>")
   MyFile.WriteLine("<figure>" & fig & "</figure>")
   MyFile.WriteLine("</weather>")
objrs.MoveNext
wend

MyFile.WriteLine("</ROOT>")
address = "http://192.168.1.100/loadxsl.asp?name=" & fname
Response.Redirect(address)

end if
```

In the beginning of the preceding code, you check the value of the parameter for the city the user selected. On the basis of that selection, you collect records from the database and save them in XML format in a file named `tweather.xml` (Figure 7-5) using the file system object (FSO). After saving all the data, you redirect control to the `loadxml.asp` file for performing the transformation.

Listing 7-6, `tweatherdetailhdml.xsl`, shows the style sheet you use to transform the `tweatherdetail.xml` file into an HDML-compatible format.

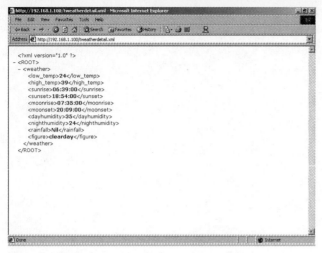

Figure 7-5: Displaying tweatherdetail.xml in Internet Explorer

Listing 7-6: tweatherdetailhdml.xsl

```
<?xml version='1.0'?>
<xsl:stylesheet xmlns:xsl="http://www.w3.org/1999/XSL/Transform" version="1.0">

<xsl:output method="text"/>

<xsl:template match="/">
<![CDATA[<HDML VERSION='3.0'>]]>

<![CDATA[<DISPLAY>]]>
  <![CDATA[<ACTION TYPE="ACCEPT" Task="GO" DEST="checkbrows.asp">]]>
 <xsl:apply-templates select="ROOT" />
<![CDATA[</DISPLAY>]]>
<![CDATA[</HDML>]]>

</xsl:template>

<xsl:template match="ROOT">

  <xsl:for-each select="weather">

<![CDATA[<B>]]>Low Temp-<![CDATA[</B>]]>
<xsl:value-of select="low_temp" /><![CDATA[<BR>]]>
<![CDATA[<B>]]>High Temp-<![CDATA[</B>]]>
<xsl:value-of select="high_temp" /><![CDATA[<BR>]]>
<![CDATA[<B>]]>Sunrise-<![CDATA[</B>]]>
<xsl:value-of select="sunrise" /><![CDATA[<BR>]]>
  <![CDATA[<B>]]>Sunset-<![CDATA[</B>]]>
<xsl:value-of select="sunset" /><![CDATA[<BR>]]>
<![CDATA[<B>]]>Moonrise-<![CDATA[</B>]]>
<xsl:value-of select="moonrise" /><![CDATA[<BR>]]>
  <![CDATA[<B>]]>Moonset-<![CDATA[</B>]]>
<xsl:value-of select="moonset" /><![CDATA[<BR>]]>
<![CDATA[<B>]]>DayHumidity-<![CDATA[</B>]]>
<xsl:value-of select="dayhumidity" /><![CDATA[<BR>]]>
```

```
    <![CDATA[<B>]]>NightHumidity-<![CDATA[</B>]]>
  <xsl:value-of select="nighthumidity" /><![CDATA[<BR>]]>
    <![CDATA[<B>]]>Rainfall-<![CDATA[</B>]]>
  <xsl:value-of select="rainfall" /><![CDATA[<BR>]]>

</xsl:for-each>
</xsl:template>
</xsl:stylesheet>
```

In the first few lines of Listing 7-6, you generate an HDML card using a CDATA section and then apply a template within that card. This template is used for fetching the records from the XML document and displaying them in the HDML browser. This is shown in the following code:

```
<xsl:template match="/">
<![CDATA[<HDML VERSION='3.0'>]]>

<![CDATA[<DISPLAY>]]>
  <![CDATA[<ACTION TYPE="ACCEPT" Task="GO" DEST="checkbrows.asp">]]>
  <xsl:apply-templates select="ROOT" />
<![CDATA[</DISPLAY>]]>
<![CDATA[</HDML>]]>

</xsl:template>
```

You now fetch the records from the XML document and display them in the browser. The following code performs this task:

```
<xsl:template match="ROOT">
  <xsl:for-each select="weather">

  <![CDATA[<B>]]>Low Temp-<![CDATA[</B>]]>
  <xsl:value-of select="low_temp" /><![CDATA[<BR>]]>
  <![CDATA[<B>]]>High Temp-<![CDATA[</B>]]>
  <xsl:value-of select="high_temp" /><![CDATA[<BR>]]>
  <![CDATA[<B>]]>Sunrise-<![CDATA[</B>]]>
  <xsl:value-of select="sunrise" /><![CDATA[<BR>]]>
   <![CDATA[<B>]]>Sunset-<![CDATA[</B>]]>
  <xsl:value-of select="sunset" /><![CDATA[<BR>]]>
  <![CDATA[<B>]]>Moonrise-<![CDATA[</B>]]>
  <xsl:value-of select="moonrise" /><![CDATA[<BR>]]>
   <![CDATA[<B>]]>Moonset-<![CDATA[</B>]]>
  <xsl:value-of select="moonset" /><![CDATA[<BR>]]>
  <![CDATA[<B>]]>DayHumidity-<![CDATA[</B>]]>
  <xsl:value-of select="dayhumidity" /><![CDATA[<BR>]]>
   <![CDATA[<B>]]>NightHumidity-<![CDATA[</B>]]>
  <xsl:value-of select="nighthumidity" /><![CDATA[<BR>]]>
    <![CDATA[<B>]]>Rainfall-<![CDATA[</B>]]>
  <xsl:value-of select="rainfall" /><![CDATA[<BR>]]>

</xsl:for-each>
</xsl:template>
```

In this template, named "ROOT", you constructed a loop for fetching all the records from the XML document; you then displayed these in an HDML browser using more than one CDATA section under the loop. The output of this code is shown in Figure 7-6.

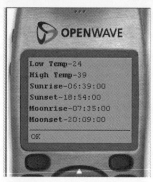

Figure 7-6: Detailed weather information (Image of Up.SDK courtesy Openwave Systems Inc.)

Transform the News Section for HDML Clients

In this section, you will transform the news section of the application for HDML clients. There are four subsections in the news section, called *categories*, which allow the user to read the news in the section of his choice. Figure 7-7 shows the navigation flow of the news section.

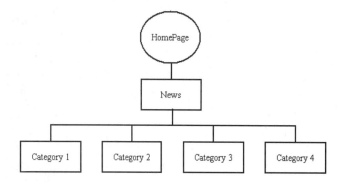

Figure 7-7: The navigation flow of the news section

Generate XML Data for the News Section

In the news section, you again use the generatexml.asp file to generate XML data, as shown in Listing 7-7.

Listing 7-7: generatexml.asp

```
if (section="News")  then
  sql = "select DISTINCT category from news"
  objrs.Open sql,conn

  Set fso =Server.CreateObject("Scripting.FileSystemObject")
  fname = "tnews.xml"

  fullpath = server.MapPath("main")

  Set MyFile=fso.CreateTextFile(fullpath & "/" & fname)

  MyFile.WriteLine("<?xml version='1.0'?>")
```

```
MyFile.WriteLine("<ROOT>")


while not(objrs.EOF)

   category = objrs.Fields("category")

   MyFile.WriteLine("<category>" & category & "</category>")

objrs.MoveNext
wend

MyFile.WriteLine("</ROOT>")
address = "http://192.168.1.100/loadxsl.asp?name=" & fname
Response.Redirect(address)

end if
```

This code collects all the records from the database and saves them in a file named `tnews.xml` (Figure 7-8). Saving the records in XML format allows you to transform all the records to an HDML-compliant format.

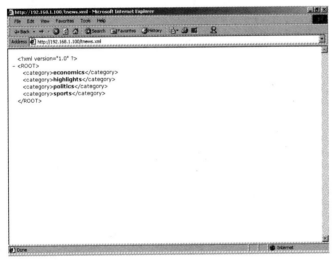

Figure 7-8: tnews.xml

After saving all the data in XML format, the program now calls the `loadxsl.asp` file to perform the transformation operations for the HDML client. You use the `tnewshdml.xsl` file in Listing 7-8 for the transformation.

Listing 7-8: tnewshdml.xsl

```
<?xml version='1.0'?>
<xsl:stylesheet xmlns:xsl="http://www.w3.org/1999/XSL/Transform" version="1.0">

<xsl:output method="text"/>

<xsl:template match="/">
<![CDATA[<HDML VERSION='3.0'>]]>

<![CDATA[<DISPLAY>]]>
```

```
        <![CDATA[<ACTION TYPE="ACCEPT" Task="GO" DEST="checkbrows.asp">]]>
     <xsl:apply-templates select="ROOT" />
 <![CDATA[</DISPLAY>]]>
 <![CDATA[</HDML>]]>


</xsl:template>

<xsl:template match="ROOT">

   <xsl:for-each select="category">

<![CDATA[<a>]]>
     <xsl:attribute name="href">
     generatexml.asp?category=<xsl:value-of select="." />
     </xsl:attribute>
     <xsl:value-of select="." />
<![CDATA[</a>]]>

   </xsl:for-each>

</xsl:template>
</xsl:stylesheet>
```

In Listing 7-8, you first generate an HDML card using the CDATA section. In this card, you use the
<XSL:apply> template command to include a template in the card. The following lines of code will
clarify the process:

```
<![CDATA[<HDML VERSION='3.0'>]]>

<![CDATA[<DISPLAY>]]>
   <![CDATA[<ACTION TYPE="ACCEPT" Task="GO" DEST="checkbrows.asp">]]>
 <xsl:apply-templates select="ROOT" />
<![CDATA[</DISPLAY>]]>
<![CDATA[</HDML>]]>
```

After generating the HDML card, you go on to fetch all the records from the XML document and display
these on the HDML browser screen by using the following lines of code:

```
<xsl:template match="ROOT">

   <xsl:for-each select="category">

<![CDATA[<a>]]>
     <xsl:attribute name="href">
     generatexml.asp?category=<xsl:value-of select="." />
     </xsl:attribute>
     <xsl:value-of select="." />
<![CDATA[</a>]]>

   </xsl:for-each>

</xsl:template>
```

As you can see in the preceding code, you first construct a loop statement to get all the available
categories from the XML document. When the user selects any category from the available choices, you
again call the generatexml.asp file to generate news for the selected category. The output of the
preceding code is shown in Figure 7-9.

Figure 7-9: The output of tnews.xml on the HDML browser (Image of Up.SDK courtesy Openwave Systems Inc.)

Generate XML for the News Section

After the user selects a category of news to read, you generate XML data using the `generatexml.asp` file. The code for carrying out this task is as follows:

```
if ((ctg="highlight") OR (ctg="politics") OR (ctg="economics") OR
(ctg="sports"))   then
  sql = "select * from news where category = '" & ctg & "'"
  objrs.Open sql,conn

  Set fso =Server.CreateObject("Scripting.FileSystemObject")
  fname = "tnewsdetail.xml"
  fullpath = server.MapPath("main")
  Set MyFile=fso.CreateTextFile(fullpath & "/" & fname)

  MyFile.WriteLine("<?xml version='1.0'?>")
  MyFile.WriteLine("<ROOT>")

    while not(objrs.EOF)
      heading = objrs.Fields("heading")
      category = objrs.Fields("category")
      description = objrs.Fields("description")
      newsdate = objrs.Fields("news_date")
      MyFile.WriteLine("<news>")
      MyFile.WriteLine("<category>" & category & "</category>")
      MyFile.WriteLine("<heading>" & heading & "</heading>")
      MyFile.WriteLine("<description>" & description & "</description>")
      MyFile.WriteLine("<news_date>" & newsdate & "</news_date>")
      MyFile.WriteLine("</news>")
    objrs.MoveNext
  wend

  MyFile.WriteLine("</ROOT>")
  address = "http://192.168.1.100/loadxsl.asp?name=" & fname
  Response.Redirect(address)

end if
```

In the preceding code, you select all the news items from the database and save them in XML format using the file system object (FSO) on the server. You will save the news in a file named `tnewsdetail.xml`. After saving, you call the `loadxsl.asp` file to perform the transformation of the `tnewsdetail.xml` file. Figure 7-10 shows the `tnewsdetail.xml` file.

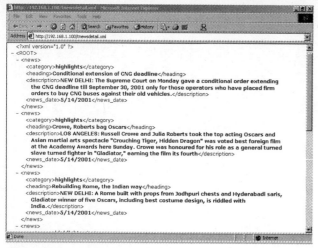

Figure 7-10: tnewsdetail.xml

Prepare XSLT for the News Details Section

In this section, let's discuss the XSLT code we are using for displaying the news details when a user clicks on any of the available categories in news section main page. Here we are using `tnewsdetailhdml.xsl` for displaying the news details. The code listing of `tnewsdetails.xsl` is given in Listing 7-9.

Listing 7-9: tnewsdetailhdml.xsl

```xml
<?xml version='1.0'?>
<xsl:stylesheet xmlns:xsl="http://www.w3.org/1999/XSL/Transform" version="1.0">

<xsl:output method="text"/>

<xsl:template match="/">
<![CDATA[<HDML VERSION='3.0'>]]>


<![CDATA[<DISPLAY>]]>
  <![CDATA[<ACTION TYPE="ACCEPT" Task="GO" DEST="checkbrows.asp">]]>
 <xsl:apply-templates select="ROOT" />
<![CDATA[</DISPLAY>]]>
<![CDATA[</HDML>]]>

</xsl:template>

<xsl:template match="ROOT">

  <xsl:for-each select="news">

      <![CDATA[<B>]]><xsl:value-of select="heading"
/><![CDATA[</B>]]><![CDATA[<BR>]]>
        <xsl:value-of select="description" /><![CDATA[<BR>]]>


  </xsl:for-each>
```

```
</xsl:template>


</xsl:stylesheet>
```

In Listing 7-9, you first generate a simple HDML card and apply a template named ROOT to that card. In this template, you then construct a loop for getting all the records from the XML document one by one and then displaying all these records on the HDML browser, as shown in Figure 7-11.

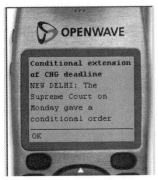

Figure 7-11: Displaying detailed news on an HDML browser (Image of Up.SDK courtesy Openwave Systems Inc.)

Transform the E-mail Section for HDML Clients

In this section, you will transform the e-mail section of the application so that users with HDML-compliant devices can read and send e-mail.

Preparing the Login Section for E-mail

Now take a look at the login form for the e-mail section, which is used in both the mail-reading and mail-sending parts of the e-mail section. You again use the generatexml.asp file to generate the XML data for the login form, shown in the following code:

```
if (section="E-mail") then

   Set fso =Server.CreateObject("Scripting.FileSystemObject")
   fname = "tlogin.xml"

   fullpath = server.MapPath("main")
   Set MyFile=fso.CreateTextFile(fullpath & "/" & fname)

   MyFile.WriteLine("<?xml version='1.0'?>")
   MyFile.WriteLine("<logininfo>")
   MyFile.WriteLine("<username />")
   MyFile.WriteLine("<password />")
   MyFile.WriteLine("<login />")
   MyFile.WriteLine("</logininfo>")

   address = "http://192.168.1.100/loadxsl.asp?name=" & fname
   Response.Redirect(address)
end if
```

After generating the XML data, you save all the data in the file named tlogin.xml (Figure 7-12) using the file system object.

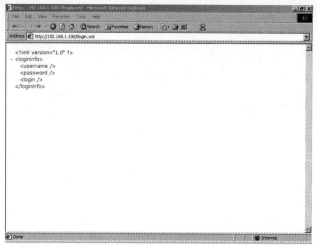

Figure 7-12: tlogin.xml

After generating the XML data, we redirect control to the `loadxsl.asp` file to perform the transformation. The file you are using as an XSLT document is `tloginhdml.xsl`, shown in Listing 7-10.

Listing 7-10: tloginhdml.xsl

```
<?xml version='1.0'?>
<xsl:stylesheet xmlns:xsl="http://www.w3.org/1999/XSL/Transform" version="1.0">

<xsl:output method="text"/>

<xsl:template match="/">
<![CDATA[<HDML VERSION='3.0'>]]>

<![CDATA[<DISPLAY>]]>
  <![CDATA[<ACTION TYPE="ACCEPT" Task="GO" DEST="checkbrows.asp">]]>
  <![CDATA[<a  href="generatexml.asp?mode=read">]]>
  Read Mail<![CDATA[</a>]]>
  <![CDATA[</BR>]]>
  <![CDATA[<a  href="generatexml.asp?mode=send">]]>
  send Mails<![CDATA[</a>]]>
<![CDATA[</DISPLAY>]]>
<![CDATA[</HDML>]]>

</xsl:template>

</xsl:stylesheet>
```

In Listing 7-10, you are again using CDATA to display information on the HDML browser. You are simply generating two links for reading and sending the messages from the HDML device by using the following lines of code:

```
<![CDATA[<a  href="generatexml.asp?mode=read">]]>
Read Mail<![CDATA[</a>]]>
<![CDATA[</BR>]]>
<![CDATA[<a  href="generatexml.asp?mode=send">]]>
```

```
send Mails<![CDATA[</a>]]>
```

If the user selects one of the available options, you pass the information to the generatexml.asp file to generate the data for the selected mode.

When this piece of code executes on the server, it actually performs the transformation. Figure 7-13 shows the output on the browser.

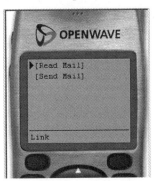

Figure 7-13: The output of tloginhdml.xsl (Image of Up.SDK courtesy Openwave Systems Inc.)

Read Mail on an HDML Device

Now take a look at the read mail section of the code. When the user selects this mode, the program calls the generatexml.asp file with a parameter in the query string named mode with the value read in it. Now the generatexml.asp file generates XML data as follows:

```
if (emode <> " ") then

  Set fso =Server.CreateObject("Scripting.FileSystemObject")
  fname = "tlogin.xml"

  fullpath = server.MapPath("main")
  Set MyFile=fso.CreateTextFile(fullpath & "/" & fname)

  MyFile.WriteLine("<?xml version='1.0'?>")
  MyFile.WriteLine("<logininfo>")
  MyFile.WriteLine("<username />")
  MyFile.WriteLine("<password />")
  MyFile.WriteLine("<login />")
  MyFile.WriteLine("</logininfo>")

  if (session.Contents(1)="wml") then
  address = "http://192.168.1.100/loadmailwml.asp?name=" & fname & "&mode=" &
emode
  end if
  if (session.Contents(1)="imode") then
  address = "http://192.168.1.100/loadmailimode.asp?name=" & fname & "&mode=" &
emode
  end if
  if (session.Contents(1)="hdml") then
  address = "http://192.168.1.100/loadmailhdml.asp?name=" & fname & "&mode=" &
emode
  end if
  Response.Redirect(address)
end if
```

In the preceding code, you first check the selected mode and then the browser type. After checking these files, you call one more file named `loadmailhdml.asp` for performing some tasks before displaying the messages to the user. You also send some parameters such as `name`, `fname`, and `emode` with the URL. To see what actually happens here, take a look at the `loadmailhdml.asp` file shown in Listing 7-11.

Listing 7-11: loadmailhdml.asp

```
<%@ Language=VBScript %>

<%
fname = trim(Request.QueryString("name"))
emode = trim(Request.QueryString("mode"))



'declaring variables
dim client
dim sXML
dim sXSL
dim xmlDocument,xslDocument


'creating the objects for the XMLDOM
set xmlDocument = Server.CreateObject("MSXML2.DOMDocument")
set xslDocument = Server.CreateObject("MSXML2.DOMDocument")

Response.ContentType = "text/x-hdml"
sXML = fname


if (emode="read") then
  sXSL = "readhdml.xsl"
end if

xmlDocument.async = false
xslDocument.async = false

xmlDocument.load(Server.MapPath(sXML))
xslDocument.load(Server.MapPath(sXSL))

Response.Write(xmlDocument.transformNode(xslDocument))

%>
```

In the beginning of Listing 7-11, you use variables to hold the data coming to you in the query string from the previous file. You now create an `XMLDOM` object to perform the transformation for HDML clients. In the next part of the code, you check for the mode the user selected and then assign values to the different variables. In this case, you have set the value of the `sXSL` variable to `readhdml.xsl`, as shown in Listing 7-12. We use this file to get the username and password from the user. After setting the values in variables, you set the content type for the HDML browser and then load both the XSL and XML documents, using `MSXML DOM` to perform the transformation.

Listing 7-12: readhdml.xsl

```
<?xml version="1.0" ?>
<xsl:stylesheet xmlns:xsl="http://www.w3.org/1999/XSL/Transform" version="1.0">
<xsl:output method="text"/>
```

```
<xsl:template match="/">
<![CDATA[<HDML VERSION='3.0'>]]>

<xsl:apply-templates select="logininfo/username" />

<xsl:apply-templates select="logininfo/password" />

<![CDATA[</HDML>]]>

</xsl:template>

<xsl:template match="logininfo/username">

<xsl:for-each select=".">
<![CDATA[<ENTRY KEY="username" NAME="user">]]>

  <![CDATA[<ACTION TYPE="ACCEPT" Task="GO" DEST="#pass" DEFAULT="">]]>

  Enter your login name



<![CDATA[</ENTRY>]]>
</xsl:for-each>
</xsl:template>


<xsl:template match="logininfo/password">

<xsl:for-each select=".">
<![CDATA[<ENTRY KEY="password" NAME="pass" DEFAULT="">]]>


  <![CDATA[<ACTION TYPE="ACCEPT" TASK="GO"
DEST="receivehdml.asp?uname=$username&amp;pwd=$password">]]>

  Enter your password

<![CDATA[</ENTRY>]]>
</xsl:for-each>

</xsl:template>
</xsl:stylesheet>
```

Take a closer look at Listing 7-12. In the beginning of the file, you generate a simple HDML card. Then within the card, you apply the two templates for displaying the data. The following lines of code do this:

```
<xsl:template match="/">
<![CDATA[<HDML VERSION='3.0'>]]>

<xsl:apply-templates select="logininfo/username" />

<xsl:apply-templates select="logininfo/password" />

<![CDATA[</HDML>]]>
```

```
</xsl:template>
```

Having generated the card and applied the two templates inside it, you now write the code for the first template that you use to display the input box for the username on the HDML device. You generate an input control for entering the username in the CDATA section in the template.

```
<xsl:template match="logininfo/username">
<xsl:for-each select=".">
<![CDATA[<ENTRY KEY="username" NAME="user">]]>
  <![CDATA[<ACTION TYPE="ACCEPT" Task="GO" DEST="#pass" DEFAULT="">]]>
  Enter your login name
<![CDATA[</ENTRY>]]>
</xsl:for-each>
</xsl:template>
```

The next template contains code that again generates an input box, this time for the user's password. Once again, this input control is generated in the CDATA section.

```
<xsl:template match="logininfo/password">

<xsl:for-each select=".">
<![CDATA[<ENTRY KEY="password" NAME="pass" DEFAULT="">]]>


  <![CDATA[<ACTION TYPE="ACCEPT" TASK="GO"
DEST="receivehdml.asp?uname=$username&amp;pwd=$password">]]>

  Enter your password

<![CDATA[</ENTRY>]]>
</xsl:for-each>

</xsl:template>
```

When this style sheet transforms the XML file, it shows the output in Figures 7-14 and 7-15 on the HDML browser.

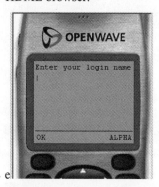

Figure 7-14: The username input screen (Image of Up.SDK courtesy Openwave Systems Inc.)

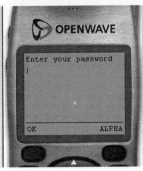

Figure 7-15: The password input screen (Image of Up.SDK courtesy Openwave Systems Inc.)

After the user enters this information, he or she selects Next on the HDML device. The `receivehdml.asp` file (Listing 7-13) is then called with the username and password as parameters in the query string.

Listing 7-13: receivehdml.asp

```
<%@LANGUAGE=VBSCRIPT %>
<%
uname = trim(Request.QueryString("uname"))
pwd = trim(Request.QueryString("pwd"))

Response.ContentType = "text/x-hdml"

%>
<HDML VERSION='3.0'>
<DISPLAY>
<ACTION TYPE="ACCEPT" TASK="GO" DEST="checkbrows.asp">

<b>Inbox</b>-----
<br><br>
<%

Set pop3 = Server.CreateObject( "Jmail.pop3" )
pop3.Connect uname, pwd, "http://192.168.1.100"


c = 1
if (pop3.Count>0) then
while (c<=pop3.Count)
 Set msg = pop3.Messages.item(c)

 ReTo = ""
 ReCC = ""

 Set Recipients = msg.Recipients
 delimiter = ", "


 For i = 0 To Recipients.Count - 1
  If i = Recipients.Count - 1 Then
   delimiter = ""
  End If
```

```
   Set re = Recipients.item(i)
   If re.ReType = 0 Then
    ReTo = ReTo & re.Name & " (" & re.EMail & ")" & delimiter
   else
    ReCC = ReTo & re.Name & " (" & re.EMail & ")" & delimiter
   End If
 Next

%>
  <b>Subject</b>
  <br>
  <%=msg.Subject %>
 <br>


  <b>From</b>
  <br>
  <%=msg.FromName %>
 <br>


  <b>Message</b>
  <br>
  <%=msg.Body %>

<%

c = c + 1
wend
end if

pop3.Disconnect

%>

</DISPLAY>
</HDML>
```

In Listing 7-13, you create an object on the server to connect to the e-mail sever and receive mail from it. These mail messages are then displayed in HDML format by generating HDML on-the-fly using the following lines of code:

```
  <b>Subject</b>
  <br>
  <%=msg.Subject %>
 <br>


  <b>From</b>
  <br>
  <%=msg.FromName %>
 <br>


  <b>Message</b>
  <br>
  <%=msg.Body %>
```

This code generates the output in Figure 7-16 on the screen of the HDML browser.

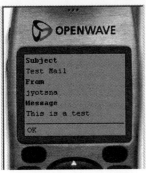

Figure 7-16: Displaying mail on the HDML device (Image of Up.SDK courtesy Openwave Systems Inc.)

Transform the Send Mail Section

In this section, you will focus on sending mail from an HDML device. If the user selects the Send Mail option from the main menu, you call the `loadmailhdml.asp` file with a different value in the `mode` parameter. This time the value you use is `send`. Listing 7-14 shows the code for `loadmailhdml.asp` when it is modified to send mail.

Listing 7-14: loadmailhdml.asp (modified to send mail)

```
<%@ Language=VBScript %>

<%
fname = trim(Request.QueryString("name"))
emode = trim(Request.QueryString("mode"))


'declaring variables
dim client
dim sXML
dim sXSL
dim xmlDocument,xslDocument

'creating the objects for the XMLDOM
set xmlDocument = Server.CreateObject("MSXML2.DOMDocument")
set xslDocument = Server.CreateObject("MSXML2.DOMDocument")
sXML = fname
fname = mid(fname,1,(len(fname) - 4))

if (emode="read") then
 sXSL = "readhdml.xsl"
end if

if (emode="send") then
 sXSL = "sendhdml.xsl"
end if

xmlDocument.async = false
xslDocument.async = false

xmlDocument.load(Server.MapPath(sXML))
xslDocument.load(Server.MapPath(sXSL))

Response.ContentType = "text/vnd.wap.wml"
```

```
Response.Write "<?xml version=""1.0"" ?>"


Response.Write "<!DOCTYPE wml PUBLIC ""-//WAPFORUM//DTD " & _
"WML 1.1//EN"" ""http://www.wapforum.org/DTD/wml_1.1.xml"">"

Response.Write(xmlDocument.transformNode(xslDocument))
%>
```

This time you are setting a new value to the sXSL variable by adding the following lines of code:

```
if (emode="send") then
  sXSL = "sendhdml.xsl"
end if
```

So the code will now load the XML document with a different style-sheet document named sendhdml.xsl, shown in Listing 7-15.

Listing 7-15: sendhdml.xsl

```
<?xml version="1.0" ?>
<xsl:stylesheet xmlns:xsl="http://www.w3.org/1999/XSL/Transform" version="1.0">
<xsl:output method="text"/>


<xsl:template match="/">
<![CDATA[<HDML VERSION='3.0'>]]>

<xsl:apply-templates select="logininfo/username" />

<xsl:apply-templates select="logininfo/password" />

<![CDATA[</HDML>]]>

</xsl:template>

<xsl:template match="logininfo/username">

<xsl:for-each select=".">
  <![CDATA[<ENTRY KEY="username" NAME="user">]]>
    <![CDATA[<ACTION TYPE='ACCEPT' Task="GO" DEST="#pass">]]>
    Enter your login name
  <![CDATA[</ENTRY>]]>
</xsl:for-each>
</xsl:template>


<xsl:template match="logininfo/password">

<xsl:for-each select=".">
 <![CDATA[<ENTRY KEY="password" NAME="pass">]]>
   <![CDATA[<ACTION TYPE='ACCEPT' TASK="GO"
DEST="sendhdml.asp?uname=$username&amp;pwd=$password">]]>
   Enter your password
 <![CDATA[</ENTRY>]]>
</xsl:for-each>

</xsl:template>
```

```
</xsl:stylesheet>
```

This file contains almost the same code that you used in the `receivehdml.xsl` file. There is only one difference, shown in the following lines:

```
<![CDATA[<ACTION TYPE='ACCEPT' TASK="GO"
DEST="sendhdml.asp?uname=$username&amp;pwd=$password">]]>
```

Instead of calling the `receivehdml.asp` file, you are now calling the `sendhdml.asp` file with the same parameters, `username` and `password`, in the query string.

Compose Mail

Now take a look at the `sendhdml.asp` file in Listing 7-16. You use this file to collect all the required data from the user and then post it back to the server so that the mail can be sent. The data is posted in different parameters in the query string.

Listing 7-16: sendhdml.asp

```
<%@LANGUAGE=VBSCRIPT %>

<%
uname = trim(Request.QueryString("uname"))
pwd = trim(Request.QueryString("pwd"))


Response.ContentType = "text/x-hdml"

%>
<HDML VERSION='3.0'>


  <ENTRY KEY="remail" NAME="email">
  <ACTION TYPE='ACCEPT' TASK="GO" DEST="#sub">
  Enter recipient's E-mail address
  </ENTRY>

  <ENTRY KEY="subject" NAME="sub">
  <ACTION TYPE='ACCEPT' TASK="GO" DEST="#mbody">
  Enter the subject of E-mail
  </ENTRY>

  <ENTRY KEY="body" NAME="mbody">
  <ACTION TYPE='ACCEPT' TASK="GO"
DEST="sendmailhdml.asp?sender=jyotsna&amp;recipient=$remail&amp;subject=$subject
&amp;body=$body">
  Enter the body of E-mail
  </ENTRY>


</HDML>
```

The output of Listing 7-16 is shown in Figures 7-17 to 7-19.

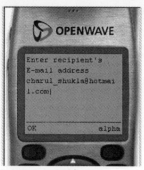

Figure 7-17: Entering the recipient's address (Image of Up.SDK courtesy Openwave Systems Inc.)

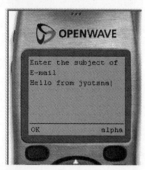

Figure 7-18: Entering the subject of the message (Image of Up.SDK courtesy Openwave Systems Inc.)

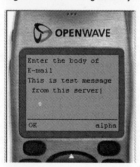

Figure 7-19: Asking for the message to be mailed (Image of Up.SDK courtesy Openwave Systems Inc.)

When the user enters all the data and selects Next on the HDML device, you call the sendmailhdml.asp file (Listing 7-17) to send the mail to the desired person using the mail server on the Web server. This file also displays a message to the user regarding delivery status.

Listing 7-17: sendmailhdml.asp

```
<%@LANGUAGE = VBSCRIPT%>
<%

senderEmail = trim(Request.QueryString("sender"))
subject     = trim(Request.QueryString("subject"))
recipient   = trim(Request.QueryString("recipient"))
body = trim(Request.QueryString("body"))

Response.ContentType = "text/x-hdml"
%>
<HDML VERSION='3.0'>
```

```
<DISPLAY>
<ACTION TYPE="ACCEPT" TASK="GO" DEST="checkbrows.asp">
<%
' Create the JMail message Object
set msg = Server.CreateOBject( "JMail.Message" )

msg.From = senderEmail
msg.AddRecipient recipient
msg.Subject = subject
msg.body = body


if not msg.Send("192.168.1.100") then
    Response.write msg.log
else
    Response.write "Message sent successfully!"
end if

%>
</DISPLAY>
</HDML>
```

The output of the preceding code is shown in Figure 7-20.

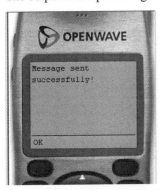

Figure 7-20: The output of sendmailhdml.asp (Image of Up.SDK courtesy Openwave Systems Inc.)

If any problems arise while the program is sending the message, it displays an error log to the user by using the following lines of code:

```
<%
' Create the JMail message Object
set msg = Server.CreateOBject( "JMail.Message" )

msg.From = senderEmail
msg.AddRecipient recipient
msg.Subject = subject
msg.body = body


if not msg.Send("192.168.1.100") then
    Response.write msg.log
else
    Response.write "Message sent successfully!"
end if
```

```
%>
```

After sending a message, the user is again redirected to the application's home page when he selects Next on the HDML device.

Transform the Movie Ticket Booking System

With the movie ticket booking system, the user can browse through the movie listings for the local theaters and book movie tickets online using his HDML-enabled device.

Generate XML Data for Theater Listings

When the user clicks on the Movies section, the system displays names of the all theaters from the database. The user can select any of the theaters listed on the device. To generate XML data for the theater listings, we use the generatexml.asp file in Listing 7-18.

Listing 7-18: generatexml.asp

```
if (section="Movies") then
  sql = "select * from hall"
  objrs.Open sql,conn
  Set fso =Server.CreateObject("Scripting.FileSystemObject")
  fname = "tmovies.xml"
  fullpath = server.MapPath("main")
  Set MyFile=fso.CreateTextFile(fullpath & "/" & fname)
  MyFile.WriteLine("<?xml version='1.0'?>")
  MyFile.WriteLine("<ROOT>")
  while not(objrs.EOF)
    hallid = objrs.Fields("hall_id")
    hallname = objrs.Fields("hall_name")
    add = objrs.Fields("address")
    MyFile.WriteLine("<hall>")
    MyFile.WriteLine("<hall_id>" & hallid & "</hall_id>")
    MyFile.WriteLine("<hall_name>" & hallname & "</hall_name>")
    MyFile.WriteLine("<address>" & add & "</address>")
    MyFile.WriteLine("</hall>")
  objrs.MoveNext
  wend

  MyFile.WriteLine("</ROOT>")
  address = "http://192.168.1.100/loadxsl.asp?name=" & fname
  Response.Redirect(address)
end if
```

In Listing 7-18, you first check for the Movie section and then start fetching the records accordingly from the HALL table. After saving all the records in XML format in the file named tmovies.xml, you transfer control to the loadxml.asp file to load the corresponding XSL document for the transformation of tmovies.xml. Figure 7-21 is a screen shot of tmovies.xml   in Internet Explorer.

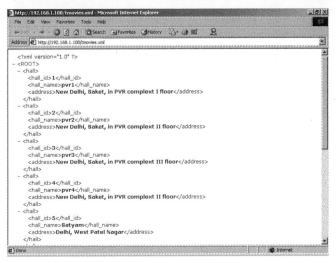

Figure 7-21: tmovies.xml

XSLT for Theater Listings

Listing 7-19 shows the XSLT document you use for the transformation of `tmovies.xml`. This file transforms XML content into HDML content so the user is able to browse it using an HDML-enabled device.

Listing 7-19: tmovieshdml.xsl

© 2001 Dreamtech Software India Inc.
All Rights Reserved

```
<?xml version='1.0'?>
<xsl:stylesheet xmlns:xsl="http://www.w3.org/1999/XSL/Transform"  version="1.0">

<xsl:output method="text"/>
<xsl:template match="/">
<![CDATA[<HDML VERSION='3.0'>]]>
<xsl:apply-templates select="ROOT" />
<![CDATA[</HDML>]]>
</xsl:template>
<xsl:template match="ROOT">
<![CDATA[<CHOICE KEY="hallname">]]>
 <![CDATA[<ACTION TYPE='ACCEPT' Task="GO"
DEST="loadmoviehdml.asp?hall=$hallname">]]>
<xsl:for-each select="hall">
<![CDATA[<CE value=]]>
 <xsl:value-of select="hall_name" />
<![CDATA[>]]>
 <xsl:value-of select="hall_name" />
</xsl:for-each>
<![CDATA[</CHOICE>]]>
</xsl:template>
</xsl:stylesheet>
```

In Listing 7-19, you generate a HDML document by the transformation process of `tmovies.xml`. In the beginning of Listing 7-19 (after the XML version declaration), the first step is to define the output method for the resulting document after the transformation process by using the `<xsl:output>`

element. Later we start generating the HDML contents using the CDATA section and apply a template named ROOT.

In the ROOT template, you are constructing a loop to fetch all the theater names from the XML file and transform them as a list of choices. You can do this easily by using the following lines of code:

```
<![CDATA[<CHOICE KEY="hallname">]]>
 <![CDATA[<ACTION TYPE='ACCEPT' Task="GO"
DEST="loadmoviehdml.asp?hall=$hallname">]]>
 <xsl:for-each select="hall">
<![CDATA[<CE value=]]>
 <xsl:value-of select="hall_name" />
<![CDATA[>]]>
<xsl:value-of select="hall_name" />
</xsl:for-each>
<![CDATA[</CHOICE>]]>
```

When this transformation takes place on the server, it shows the output in Figure 7-22.

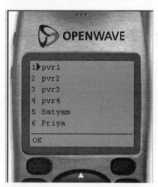

Figure 7-22: The transformation results of tmovies.xml (Image of Up.SDK courtesy Openwave Systems Inc.)

Display the Movie List

When user selects a theater, the titles of the movies that are currently being screened there are displayed. Now you have to generate XML data for the movie listing at the theater which the user selects. For this process, you use the loadmoviehdml.asp file (Listing 7-20).

Listing 7-20: loadmoviehdml.asp

```
<%@ Language=VBScript %>
<%
dim address,section,fname
hallname = trim(Request.QueryString("hall"))
set conn = Server.CreateObject("ADODB.Connection")
conn.ConnectionString = "Provider=SQLOLEDB.1;Persist Security Info=False;User
ID=sa;Initial Catalog=portal;Data Source=developers"
conn.Open
Set objrs = Server.CreateObject("ADODB.Recordset")
objrs.CursorType = adOpenStatic
Set rsmovie = Server.CreateObject("ADODB.Recordset")
rsmovie.CursorType = adOpenStatic
Set rsstatus = Server.CreateObject("ADODB.Recordset")
```

```
rsstatus.CursorType = adOpenStatic
Set rshall = Server.CreateObject("ADODB.Recordset")
rshall.CursorType = adOpenStatic
  sqlhall = "select hall_id from hall where hall_name='" & hallname & "'"
  rshall.Open sqlhall,conn
  while not (rshall.EOF)
    id = rshall.Fields("hall_id")
  rshall.MoveNext
  wend
  sqlmovies = "select DISTINCT(status.movie_id),movie.movie_name from
movie,status where status.movie_id=movie.movie_id and status.hall_id=" & id
  rsmovie.Open sqlmovies,conn
  Set fso =Server.CreateObject("Scripting.FileSystemObject")
  fname = "tmovielist.xml"
  fullpath = server.MapPath("main")
  Set MyFile=fso.CreateTextFile(fullpath & "/" & fname)
  MyFile.WriteLine("<?xml version='1.0'?>")
  MyFile.WriteLine("<ROOT>")
  MyFile.WriteLine("<hall>")
  MyFile.WriteLine("<hall_name>" & hallname & "</hall_name>")
  MyFile.WriteLine("<hall_id>" & id & "</hall_id>")
  while not (rsmovie.EOF)
    MyFile.WriteLine("<movie>")
    moviename = rsmovie.Fields("movie_name")
    MyFile.WriteLine("<movie_name>" & moviename & "</movie_name>")
    movieid = rsmovie.Fields("movie_id")
    MyFile.WriteLine("<movie_id>" & movieid & "</movie_id>")
    sqlstatus = "select
movie_date,showtime,balcony,rear_stall,middle_stall,upper_stall  from status
where movie_id=" & movieid & "and hall_id=" & id
    rsstatus.Open sqlstatus,conn
    while not (rsstatus.EOF)
      MyFile.WriteLine("<status>")
      moviedate = rsstatus.Fields("movie_date")
      showtime = rsstatus.Fields("showtime")
      balcony = rsstatus.Fields("balcony")
      rearstall = rsstatus.Fields("rear_stall")
      middlestall = rsstatus.Fields("middle_stall")
      upperstall = rsstatus.Fields("upper_stall")
      MyFile.WriteLine("<movie_date>" & moviedate & "</movie_date>")
      MyFile.WriteLine("<showtime>" & showtime & "</showtime>")
      MyFile.WriteLine("<balcony>" & balcony & "</balcony>")
      MyFile.WriteLine("<rear_stall>" & rearstall & "</rear_stall>")
      MyFile.WriteLine("<middle_stall>" & middlestall & "</middle_stall>")
      MyFile.WriteLine("<upper_stall>" & upperstall & "</upper_stall>")
      MyFile.WriteLine("</status>")
      rsstatus.MoveNext
    wend
  MyFile.WriteLine("</movie>")
  rsstatus.Close
  rsmovie.MoveNext
  wend
  MyFile.WriteLine("</hall>")
  MyFile.WriteLine("</ROOT>")

  conn.Close
```

```
   address = "http://192.168.1.100/loadxsl.asp?name=" & fname
   Response.Redirect(address)
%>
```

Now let's see how the code in Listing 7-20 transforms the contents Consider the following fragments of
`loadmoviehdml.asp`.

```
dim address,section,fname

hallname = trim(Request.QueryString("hall"))

set conn = Server.CreateObject("ADODB.Connection")
conn.ConnectionString = "Provider=SQLOLEDB.1;Persist Security Info=False;User
ID=sa;Initial Catalog=portal;Data Source=developers"
conn.Open

Set objrs = Server.CreateObject("ADODB.Recordset")
objrs.CursorType = adOpenStatic

Set rsmovie = Server.CreateObject("ADODB.Recordset")
rsmovie.CursorType = adOpenStatic

Set rsstatus = Server.CreateObject("ADODB.Recordset")
rsstatus.CursorType = adOpenStatic
Set rshall = Server.CreateObject("ADODB.Recordset")
rshall.CursorType = adOpenStatic

   sqlhall = "select hall_id from hall where hall_name='" & hallname & "'"
   rshall.Open sqlhall,conn
```

In the second line of the code, you declare a variable named `hallname` to store the name of the theater
passed to you by the previously executed document on the server. Next, you establish a connection with
the database and then select the `hall_id` based on the theater name passed to the file from the HALL
table. In the next lines of code, you select the `movie_name` and `movie_id` of the movies currently on
the screen in that theater by writing the following SQL statement:

```
   sqlmovies = "select DISTINCT(status.movie_id),movie.movie_name from
movie,status where status.movie_id=movie.movie_id and status.hall_id=" & id
   rsmovie.Open sqlmovies,conn
```

After getting all the records from the movie table, save all the data in XML format in a file named
`tmovielist.xml` on the server (using the file system object) with the following lines of code:

```
   Set fso =Server.CreateObject("Scripting.FileSystemObject")
   fname = "tmovielist.xml"

   fullpath = server.MapPath("main")

   Set MyFile=fso.CreateTextFile(fullpath & "/" & fname)

   MyFile.WriteLine("<?xml version='1.0'?>")
   MyFile.WriteLine("<ROOT>")
   MyFile.WriteLine("<hall>")
   MyFile.WriteLine("<hall_name>" & hallname & "</hall_name>")
   MyFile.WriteLine("<hall_id>" & id & "</hall_id>")

   while not (rsmovie.EOF)
```

```
    MyFile.WriteLine("<movie>")
    moviename = rsmovie.Fields("movie_name")
    MyFile.WriteLine("<movie_name>" & moviename & "</movie_name>")
    movieid = rsmovie.Fields("movie_id")
    MyFile.WriteLine("<movie_id>" & movieid & "</movie_id>")
    sqlstatus = "select
movie_date,showtime,balcony,rear_stall,middle_stall,upper_stall  from status
where movie_id=" & movieid & "and hall_id=" & id
    rsstatus.Open sqlstatus,conn
      while not (rsstatus.EOF)
        MyFile.WriteLine("<status>")
        moviedate = rsstatus.Fields("movie_date")
        showtime = rsstatus.Fields("showtime")
        balcony = rsstatus.Fields("balcony")
        rearstall = rsstatus.Fields("rear_stall")
        middlestall = rsstatus.Fields("middle_stall")
        upperstall = rsstatus.Fields("upper_stall")
        MyFile.WriteLine("<movie_date>" & moviedate & "</movie_date>")
        MyFile.WriteLine("<showtime>" & showtime & "</showtime>")
        MyFile.WriteLine("<balcony>" & balcony & "</balcony>")
        MyFile.WriteLine("<rear_stall>" & rearstall & "</rear_stall>")
        MyFile.WriteLine("<middle_stall>" & middlestall & "</middle_stall>")
        MyFile.WriteLine("<upper_stall>" & upperstall & "</upper_stall>")
        MyFile.WriteLine("</status>")
        rsstatus.MoveNext
      wend
    MyFile.WriteLine("</movie>")
    rsstatus.Close
    rsmovie.MoveNext
    wend
    MyFile.WriteLine("</hall>")
    MyFile.WriteLine("</ROOT>")
```

Figure 7-23 is a screen shot of tmovielist.xml.

Figure 7-23: tmovielist.xml

Writing XSLT code for the Movie Listings

Now let's prepare an XSLT document for the transformation of the tmovielist.xml file for HDML clients. This will allow HDML devices to display the list of the movies. Consider the code in Listing 7-21.

Listing 7-21: tmovielisthdml.xsl

```
<?xml version='1.0'?>
<xsl:stylesheet xmlns:xsl="http://www.w3.org/1999/XSL/Transform" version="1.0">
<xsl:output method="text"/>
<xsl:template match="/">
<![CDATA[<HDML VERSION='3.0'>]]>
<xsl:apply-templates select="ROOT" />
<xsl:apply-templates select="ROOT/hall" />
<![CDATA[</HDML>]]>
</xsl:template>
<xsl:template match="ROOT">
<xsl:variable name="hallname" select="hall/hall_name" />
<![CDATA[<CHOICE KEY="hallname">]]>
<![CDATA[<ACTION TYPE='ACCEPT' Task="GO" DEST="#m$hallname">]]>
<xsl:for-each select="hall/movie">
<![CDATA[<CE value=]]>
<xsl:value-of select="./movie_id" />
<![CDATA[>]]>
<xsl:value-of select="./movie_name" />
</xsl:for-each>
<![CDATA[</CHOICE>]]>
</xsl:template>
<xsl:template match="ROOT/hall">
 <xsl:variable name="hname" select="hall_name" />
 <xsl:variable name="hid" select="hall_id" />
 <xsl:for-each select="movie">
  <![CDATA[<DISPLAY NAME=m]]>
  <xsl:value-of select="./movie_id" />
  <![CDATA[>]]>
  <![CDATA[<ACTION TYPE='ACCEPT' Task="GO" DEST="#user" VARS=var1=]]>
  <xsl:value-of select="./movie_id" />
  <![CDATA[>]]>
  Status:--
  <![CDATA[<BR>]]>
  <![CDATA[<B>]]><xsl:value-of select="./movie_name" /><![CDATA[</B>]]>
  <![CDATA[<BR>]]>
  ------------
  <xsl:for-each select="status">
  <![CDATA[<BR>]]>
  <![CDATA[<B>]]>Date<![CDATA[</B>]]>
  <![CDATA[<BR>]]>
  <xsl:value-of select="movie_date" />
  <![CDATA[<BR>]]>
  <![CDATA[<B>]]>Showtime<![CDATA[</B>]]>
  <![CDATA[<BR>]]>
  <xsl:value-of select="showtime" />
  <![CDATA[<BR>]]>
```

```
<![CDATA[<B>]]>Balcony<![CDATA[</B>]]>
<![CDATA[<BR>]]>
<xsl:value-of select="balcony" />
<![CDATA[<BR>]]>
<![CDATA[<B>]]>Rear Stall<![CDATA[</B>]]>
<![CDATA[<BR>]]>
<xsl:value-of select="rear_stall" />
<![CDATA[<BR>]]>
<![CDATA[<B>]]>Middle Stall<![CDATA[</B>]]>
<![CDATA[<BR>]]>
<xsl:value-of select="middle_stall" />
<![CDATA[<BR>]]>
<![CDATA[<B>]]>Upper Stall<![CDATA[</B>]]>
<![CDATA[<BR>]]>
<xsl:value-of select="upper_stall" />
</xsl:for-each>
<![CDATA[<BR>]]>
<![CDATA[</DISPLAY>]]>
</xsl:for-each>
<![CDATA[<ENTRY KEY="username" NAME="user">]]>

    <![CDATA[<ACTION TYPE='ACCEPT' Task="GO" DEST="#mail">]]>
    Enter your name
<![CDATA[</ENTRY>]]>
<![CDATA[<ENTRY KEY="email" NAME="mail">]]>
    <![CDATA[<ACTION TYPE='ACCEPT' Task="GO" DEST="#showdate">]]>
    Enter your E-mail Address
<![CDATA[</ENTRY>]]>
<![CDATA[<ENTRY KEY="sdate" NAME="showdate">]]>
    <![CDATA[<ACTION TYPE='ACCEPT' Task="GO" DEST="#showtime">]]>
    Enter the show date
<![CDATA[</ENTRY>]]>
<![CDATA[<ENTRY KEY="stime" NAME="showtime">]]>
    <![CDATA[<ACTION TYPE='ACCEPT' Task="GO" DEST="#stall">]]>
    Enter the show time
<![CDATA[</ENTRY>]]>
<![CDATA[<ENTRY KEY="stallname" NAME="stall">]]>
    <![CDATA[<ACTION TYPE='ACCEPT' Task="GO" DEST="#tickets">]]>
    Enter stall name
<![CDATA[</ENTRY>]]>
<![CDATA[<ENTRY KEY="ticket" NAME="tickets" FORMAT="nn">]]>
    <![CDATA[<ACTION TYPE='ACCEPT' Task="GO"
DEST="ticketbookinghdml.asp?uname=$username&amp;mail=$email&amp;hallid=$hallname
&amp;movieid=$var1&amp;showdate=$sdate&amp;showtime=$stime&amp;stall=$stallname&
amp;no=$ticket">]]>
    Enter the no. of tickets
<![CDATA[</ENTRY>]]>
</xsl:template>
</xsl:stylesheet>
```

In the beginning of the code, you start generating an HDML document by first defining the
<xsl:output> method and then by using the CDATA section and applying two templates named
"ROOT" and "ROOT/HALL".

In the "ROOT" template, you first declare a variable named "hallname" using the following lines of
code:

```
<xsl:variable name="hallname" select="hall/hall_name" />
```

In next few lines, you construct a loop and fetch the movie id of all the movies from the XML document with the help of the `<xsl:value-of  select>` element on the basis of movie names. The following lines of code perform this task:

```
<xsl:for-each select="hall/movie">
<![CDATA[<CE value=]]>
<xsl:value-of select="./movie_id" />
<![CDATA[>]]>
<xsl:value-of select="./movie_name" />
</xsl:for-each>
```

When this code executes, it shows the output in Figure 7-24.

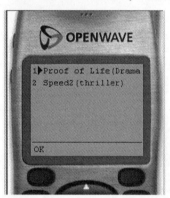

Figure 7-24: The transformation result of tmovielist.xml (Image of Up.SDK courtesy Openwave Systems Inc.)

Display the Movie Status

When the user clicks any of the movie names displayed on the screen, we use our second template in the file named `tmovielisthdml.xsl` (Listing 7-21) to generate the status of reservations at the selected movie.

```
<xsl:for-each select="movie">

<![CDATA[<DISPLAY NAME=m]]>
<xsl:value-of select="./movie_id" />
<![CDATA[>]]>
<![CDATA[<ACTION TYPE='ACCEPT' Task="GO" DEST="#user" VARS=var1=]]>
<xsl:value-of select="./movie_id" />
<![CDATA[>]]>
Status:--
<![CDATA[<BR>]]>
<![CDATA[<B>]]><xsl:value-of select="./movie_name" /><![CDATA[</B>]]>
<![CDATA[<BR>]]>
------------

<xsl:for-each select="status">
<![CDATA[<BR>]]>
<![CDATA[<B>]]>Date<![CDATA[</B>]]>
<![CDATA[<BR>]]>
<xsl:value-of select="movie_date" />
<![CDATA[<BR>]]>
```

```
<![CDATA[<B>]]>Showtime<![CDATA[</B>]]>
<![CDATA[<BR>]]>
<xsl:value-of select="showtime" />
<![CDATA[<BR>]]>

<![CDATA[<B>]]>Balcony<![CDATA[</B>]]>
<![CDATA[<BR>]]>
<xsl:value-of select="balcony" />
<![CDATA[<BR>]]>

<![CDATA[<B>]]>Rear Stall<![CDATA[</B>]]>
<![CDATA[<BR>]]>
<xsl:value-of select="rear_stall" />
<![CDATA[<BR>]]>

<![CDATA[<B>]]>Middle Stall<![CDATA[</B>]]>
<![CDATA[<BR>]]>
<xsl:value-of select="middle_stall" />
<![CDATA[<BR>]]>

<![CDATA[<B>]]>Upper Stall<![CDATA[</B>]]>
<![CDATA[<BR>]]>
<xsl:value-of select="upper_stall" />

</xsl:for-each>

<![CDATA[<BR>]]>
<![CDATA[</DISPLAY>]]>
</xsl:for-each>
```

In the preceding code, you first declare two variables for storing the values of "hallid" and "hallname" and then generate the status in HDML using the CDATA section. When this code executes on the server, it generates the following output shown below in Figure 7-25 and Figure 7-26.

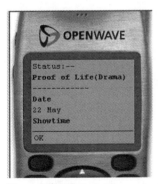

Figure 7-25: Displaying the movie status (Image of Up.SDK courtesy Openwave Systems Inc.)

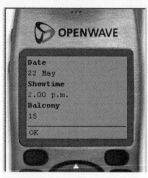

Figure 7-26: Displaying the movie status (Image of Up.SDK courtesy Openwave Systems Inc.)

The next section of the code generates a form to fill in the information about tickets reservation. The following is the code for doing this:

```
<![CDATA[<ENTRY KEY="username" NAME="user">]]>

  <![CDATA[<ACTION TYPE='ACCEPT' Task="GO" DEST="#mail">]]>

  Enter your name

<![CDATA[</ENTRY>]]>

<![CDATA[<ENTRY KEY="email" NAME="mail">]]>

  <![CDATA[<ACTION TYPE='ACCEPT' Task="GO" DEST="#showdate">]]>

  Enter your E-mail Address
<![CDATA[</ENTRY>]]>

<![CDATA[<ENTRY KEY="sdate" NAME="showdate">]]>

  <![CDATA[<ACTION TYPE='ACCEPT' Task="GO" DEST="#showtime">]]>

  Enter the show date
<![CDATA[</ENTRY>]]>

<![CDATA[<ENTRY KEY="stime" NAME="showtime">]]>

  <![CDATA[<ACTION TYPE='ACCEPT' Task="GO" DEST="#stall">]]>
  Enter the show time
<![CDATA[</ENTRY>]]>
<![CDATA[<ENTRY KEY="stallname" NAME="stall">]]>
  <![CDATA[<ACTION TYPE='ACCEPT' Task="GO" DEST="#tickets">]]>
  Enter stall name


<![CDATA[</ENTRY>]]>

<![CDATA[<ENTRY KEY="ticket" NAME="tickets" FORMAT="nn">]]>

  <![CDATA[<ACTION TYPE='ACCEPT' Task="GO"
DEST="ticketbookinghdml.asp?uname=$username&amp;mail=$email&amp;hallid=$hallname
&amp;movieid=$var1&amp;showdate=$sdate&amp;showtime=$stime&amp;stall=$stallname&
amp;no=$ticket">]]>
```

```
    Enter the no. of tickets

<![CDATA[</ENTRY>]]>
```

When this part of the code executes, it shows the output in Figures 7-27 to 7-32.

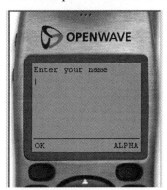

Figure 7-27: The form-filling process (Image of Up.SDK courtesy Openwave Systems Inc.)

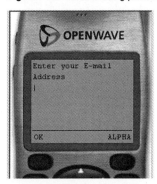

Figure 7-28: The form-filling process (continued) (Image of Up.SDK courtesy Openwave Systems Inc.)

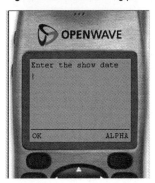

Figure 7-29: The form-filling process (continued) (Image of Up.SDK courtesy Openwave Systems Inc.)

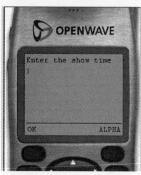

Figure 7-30: The form-filling process (continued) (Image of Up.SDK courtesy Openwave Systems Inc.)

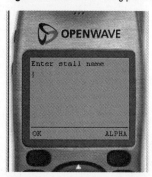

Figure 7-31: The form-filling process (continued)(Image of Up.SDK courtesy Openwave Systems Inc.)

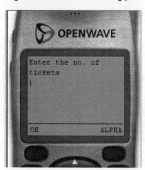

Figure 7-32: The form-filling process (completed) (Image of Up.SDK courtesy Openwave Systems Inc.)

After the user fills in all the requisite information and presses the Submit button on the device, all the data is posted to the server and the `ticketbookinghdml.asp` file is called.

Write Logic for the Booking Process

In this part, let's write the code for ticket reservation and for displaying a message to the user on the process completion. Listing 7-22 is the code for `ticketbookinghdml.asp`, which you use to book the tickets and redirect to the `bookhdml.asp` file to display the status message to the user.

Listing 7-22: ticketbookinghdml.asp

© 2001 Dreamtech Software India Inc.
All Rights Reserved

```
<%@ Language=VBScript %>
<HTML>
```

```
<HEAD>
<META NAME="GENERATOR" Content="Microsoft Visual Studio 6.0">
</HEAD>
<BODY>
<%
uname = trim(Request.QueryString("uname"))
email = trim(Request.QueryString("mail"))
hall = trim(Request.QueryString("hallname"))
hallid = trim(Request.QueryString("hallid"))
movieid = trim(Request.QueryString("movieid"))
movie = trim(Request.QueryString("moviename"))
showdate = trim(Request.QueryString("showdate"))
showtime = trim(Request.QueryString("showtime"))
stall = UCase(trim(Request.QueryString("stall")))
ticket = trim(Request.QueryString("no"))

dim conn,objrs,tsql,balcony,rear,middle,upper,total

set conn = Server.CreateObject("ADODB.Connection")
conn.ConnectionString = "Provider=SQLOLEDB.1;Persist Security Info=False;User
ID=sa;Initial Catalog=portal;Data Source=developers"
conn.Open

Set objrs = Server.CreateObject("ADODB.Recordset")
tsql="select balcony,rear_stall,middle_stall,upper_stall from status where
hall_id=" & hallid & "and movie_id=" & movieid & "and showtime='" & showtime &
"'and movie_date='" & showdate &"'"
objrs.CursorType = adOpenStatic
objrs.Open tsql,conn

while not (objrs.EOF )
 if (stall = "BALCONY") then
  total = UCase(trim(objrs.Fields("Balcony")))
 end if

 if (stall = "REAR_STALL") then
  total = UCase(trim(objrs.Fields("rear_stall")))
 end if

 if (stall = "MIDDLE_STALL") then
  total = UCase(trim(objrs.Fields("middle_stall")))
 end if

 if (stall = "UPPER_STALL") then
  total = UCase(trim(objrs.Fields("upper_stall")))
 end if

 balcony = objrs.Fields("Balcony")
 rear = objrs.Fields("rear_stall")
 middle = objrs.Fields("middle_stall")
 upper = objrs.Fields("upper_stall")

objrs.MoveNext
wend

if (CInt(total) < CInt(ticket) ) then
```

```
Response.Redirect "http://192.168.1.100/bookhdml.asp?name=fail"
else
 if (stall = "BALCONY") then
  balcony = CInt(balcony) - CInt(ticket)
 end if

 if (stall = "MIDDLE_STALL") then
  middle = CInt(middle) - CInt(ticket)
 end if

 if (stall = "REAR_STALL") then
  rear = CInt(rear) - CInt(ticket)
 end if

 if (stall = "UPPER_STALL") then
  upper = CInt(upper) - CInt(ticket)
 end if

 conn.Execute("update status set balcony=" & balcony & ",rear_stall=" & rear &
",middle_stall=" & middle & ",upper_stall=" & upper & "where hall_id=" & hallid
& "and movie_id=" & movieid & "and showtime='" & showtime & "'")
 Response.Redirect "http://192.168.1.100/bookhdml.asp?name=pass"

end if

conn.Close

%>

</BODY>
</HTML>
```

At the beginning of Listing 7-22, you declare some variables for later use and store the values in the variables passed to you by the previous file, as in the following code:

```
uname = trim(Request.QueryString("uname"))
email = trim(Request.QueryString("mail"))
hallid = trim(Request.QueryString("hallid"))
movieid = trim(Request.QueryString("movieid"))
showdate = trim(Request.QueryString("showdate"))
showtime = trim(Request.QueryString("showtime"))
stall = UCase(trim(Request.QueryString("stall")))
ticket = trim(Request.QueryString("no"))
```

The program then checks the database for the number of available tickets for the specific category the user selects, The number of tickets the user enters is the input to the system in two `if` conditions. If the number of tickets the user enters is more than the number of tickets available in the database for the selected movie at the selected theatre and date, you redirect control to the `bookhdml.asp` file with the parameter named "NAME" with the value `fail`. If the second condition is true, update the database to reflect the changes and redirect control to the `bookhdml.asp` file, with the parameter "NAME" with the value `pass`.

```
if (stall = "BALCONY") then
  total = UCase(trim(objrs.Fields("Balcony")))
end if

if (stall = "REAR_STALL") then
```

```
   total = UCase(trim(objrs.Fields("rear_stall")))
  end if

  if (stall = "MIDDLE_STALL") then
   total = UCase(trim(objrs.Fields("middle_stall")))
  end if

  if (stall = "UPPER_STALL") then
   total = UCase(trim(objrs.Fields("upper_stall")))
  end if

  balcony = objrs.Fields("Balcony")
  rear = objrs.Fields("rear_stall")
  middle = objrs.Fields("middle_stall")
  upper = objrs.Fields("upper_stall")

objrs.MoveNext
wend


if (CInt(total) < CInt(ticket) ) then
 Response.Redirect "http://192.168.1.100/bookhdml.asp?name=fail"
else
 if (stall = "BALCONY") then
  balcony = CInt(balcony) - CInt(ticket)
 end if

 if (stall = "MIDDLE_STALL") then
  middle = CInt(middle) - CInt(ticket)
 end if

 if (stall = "REAR_STALL") then
  rear = CInt(rear) - CInt(ticket)
 end if

 if (stall = "UPPER_STALL") then
  upper = CInt(upper) - CInt(ticket)
 end if

 conn.Execute("update status set balcony=" & balcony & ",rear_stall=" & rear &
",middle_stall=" & middle & ",upper_stall=" & upper & "where hall_id=" & hallid
& "and movie_id=" & movieid & "and showtime='" & showtime & "'")
 'conn.Execute("Insert user
(firstname,movieid,hallid,stall,movie_date,showtime,totaltickets,emailadd)
values('" & uname & "'," & movieid  & "," & hallid & ",'" & stall & "','" &
showdate & "','" & showtime & "'," & tickets & ",'" & email & "')")
 Response.Redirect "http://192.168.1.100/bookhdml.asp?name=pass"

end if
```

Displaying Status Message

Now take a look at code for bookhdml.asp in Listing 7-23. You use this file to redirect control to display two different messages on process completion on the basis of the parameter passed by ticketbookinghdml.asp.

Listing 7-23: bookhdml.asp

```
<%@ Language=VBScript %>
<%
name = trim(Request.QueryString("name"))
'declaring variables
dim sXML
dim sXSL
dim xmlDocument,xslDocument

'creating the objects for the XMLDOM
set xmlDocument = Server.CreateObject("MICROSOFT.XMLDOM")
set xslDocument = Server.CreateObject("MICROSOFT.XMLDOM")
Response.ContentType = "text/x-hdml"
sXML = "book.xml"
if(name = "fail") then
 sXSL = "bookfailhdml.xsl"
else
 sXSL = "bookpasshdml.xsl"
end if
xmlDocument.async = false
xslDocument.async = false
xmlDocument.load(Server.MapPath(sXML))
xslDocument.load(Server.MapPath(sXSL))
Response.Write(xmlDocument.transformNode(xslDocument))
 %>
```

In the preceding code, you load one of the two XSLT documents by using the `if` condition. This is decided by the value variable `"name"`. The following lines of code can do this quite easily:

```
if(name = "fail") then
 sXSL = "bookfailhdml.xsl"
else
 sXSL = "bookpasshdml.xsl"
end if
```

Now take a look at both of the XSLT documents (Listings 7-24 and 7-25).

Listing 7-24: bookfailhdml.xsl

```
<?xml version="1.0" ?>
<xsl:stylesheet xmlns:xsl="http://www.w3.org/1999/XSL/Transform" version="1.0">
<xsl:template match="/">
<HDML VERSION="3.0">
<DISPLAY>
<a TASK="GO" DEST="checkbrows.asp">Try Again</a>
</DISPLAY>
</HDML>
</xsl:template>
</xsl:stylesheet>
```

Listing 7-25: bookpasshdml.xsl

```
<?xml version="1.0" ?>
<xsl:stylesheet xmlns:xsl="http://www.w3.org/1999/XSL/Transform" version="1.0">
<xsl:template match="/">
<HDML VERSION="3.0">
<DISPLAY>
<a TASK="GO" DEST="checkbrows.asp">Congratulations! Tickets have been booked.
</a>

</DISPLAY>
</HDML>
</xsl:template>

</xsl:stylesheet>
```

Figure 7-33 shows the unsuccessful ticket-reservation message. When the user presses a key to continue, he is again taken to the application's home page.

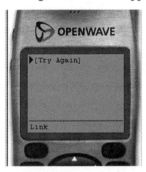

Figure 7-33: Error message in the case of unsuccessful completion (Image of Up.SDK courtesy Openwave Systems Inc.)

The output in Figure 7-34 appears in the event of successful ticket booking.

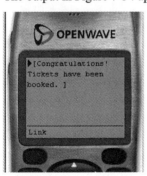

Figure 7-34: Displaying a Congratulations! message to the user (Image of Up.SDK courtesy Openwave Systems Inc.)

This completes the last section on the movie portion. Now the user can access the complete application using an HDML-compliant device.

Summary

In this chapter you have learned how to develop applications for HDML browsers using the XML, XSL, and ASP-based approach. You are using XML files as a database for the application and transforming them using XSL with the help of Active Server Pages.

Chapter 8

Working with i-mode and cHTML

i-mode is a wireless technology that is currently in use only in Japan. This technology is now striving to establish a foothold in Hong Kong, China, the United States, and Europe. As of March 2001, i-mode had an estimated 20 million users. Consequently, there is little doubt that i-mode has emerged as one of the world's most successful wireless technologies, offering Web browsing and e-mail through mobile handsets.

Introduction to i-mode

NTT DoCoMo — ranked as the largest cellular service provider in Japan — developed the i-mode technology. It was launched in Feburary 1999. This wireless technology was popularly called the *i-mode*, which is an abbreviation for *Information-Mode* i-mode allows users to use their mobile phones to access the various Internet services and communicate through e-mail.

The technology behind i-mode is primarily the packet data-transmission technology. Users are charged on the basis of the amount of data transmitted. The technology allows the user to remain connected to the Internet without paying anything. i-mode is perhaps one of the most efficient sources of exchanging e-mail messages with computers, personal digital assistants, and other i-mode-enabled mobile phones. An i-mode-enabled mobile phone generally weighs about 90 grams and contains a large liquid-crystal display for accessing mail, browsing the Web, and so on. The phone contains four basic navigational buttons, which allow the user to move the pointer on the display screen.

In addition to helping the user access the Internet and communicate via e-mail, an i-mode-enabled phone can perform tasks such as making ticket reservations, performing banking transactions, providing news and weather updates, giving stock and trading updates, listing hotel and restaurant information, giving train schedules and flight information, and reporting entertainment news from cinema, music, and so on. I-mode-enabled phones can also download music (in MP3 format), games, and more. For more information on the functions of i-mode, refer to `http://www.nttdocomo.com/imode/`.

Workings of i-mode

i-mode, being a wireless technology, provides continuous Internet access to mobile handsets using a cHTML markup language through the HTTP protocol, which delivers the content to the i-mode center. The i-mode center performs protocol conversions that enable the content to be delivered to the mobile handset. The packet communication network built on DoCoMo's main network allows continuous Internet access.

The essential components required for i-mode service are a mobile handset with an installed browser that supports packet communication, a packet network, an i-mode server, and content providers.

The Future of i-mode

In an effort to expand its operations, NTT DoCoMo is exploring the European and North American markets and has already started to promote i-mode in Europe. The company's investments include an alliance with KPN Communications, Holland, and Telecom Italia Mobile, Italy, to develop its 3G network and eventually roll out mobile Internet applications. The three giants plan to develop applications based around DoCoMo's i-mode mobile Internet service. To enhance its position in the telecommunications sector and strengthen its relationships with wireless telecom operators, NTT DoCoMo has also established its position in Hong Kong and China.

A major breakthrough for NTT DoCoMo has been the introduction of English content on its i-mode-enabled mobile phones. This would facilitate the use of i-mode among non-Japanese people currently residing in Japan whose first language is not Japanese. In March 2000, NTT DoCoMo and Sun Microsystems announced an alliance to further enhance the i-mode platform by launching phones enabled for the Java Virtual Machine (JVM); these types of phones incorporate Sun's Java, Jini, and Java Card technologies into i-mode mobile handsets. In January 2001, i-mode-enabled phones were launched that are capable of supporting Java programs.

At the time of writing of the this book, Nokia is launching an i-mode phone called NM502i, which includes a dual language interface both in Japanese and English.

Besides these enhancements, Internet surfing speeds using i-mode technology will virtually double in the near future with the assistance of 3G technologies that utilize wide-band CDMA. Speeds of more than 300 Kbps would be possible, in contrast to the current maximum speeds of 9.6 Kbps. Moreover, by 2003, maximum speeds could go as high as 2 Mbps, which would greatly improve Internet download times.

i-mode and WAP

The WAP Forum (`http://www.wapforum.org/`) states "The Wireless Application Protocol (WAP) is an open, global specification that empowers mobile users with wireless devices to easily access and interact with information and services instantly." Although i-mode and WAP have technical differences, they support the same market for mobile data services.

An understanding of the vital differences between i-mode and WAP is imperative here. Table 8-1 outlines these basic differences.

Table 8-1: Differences between i-mode and WAP

i-mode	*WAP*
i-mode was developed by NTT DoCoMo.	WAP was developed by European and North American wireless phone companies.
Delivery of i-mode services is through compact HTML (cHTML) pages. i-mode utilizes a packet network for direct communications.	WAP uses Wireless Markup Language (WML) for communication between a WAP Gateway and Internet gateways around the world.
An i-mode phone is always connected to the Internet and the user pays only for the amount of data downloaded or submitted.	The billing mechanism is completely different in the case of WAP.
il-mode devices display information from cHTML and display multicolor images. Some i-mode devices have 256-color capabilities for an enchanced look and feel.	WAP devices support a WAP browser and display only text information. WAP does not have 256-color capability.
Navigation through hyperlinks.	Navigation between layered menus.
Bigger screen space than WAP devices.	Relatively small screen space.

Because cHTML is a subset of HTML (2.0, 3.2, and 4.0) specifications, it has advantages over WML. The future of the Web lies solely with XML because of its adaptability and compatibility with cHTML as compared to WML.

WAP and i-mode are both of vital significance. However, if a comparison is drawn between the two, i-mode definitely has an edge over WAP for the following reasons:

♦ i-mode transmission rates are more or less the same as for WAP; in other words about 9.6 Kbps. Because i-mode service has constant connectivity, however, the user saves dial-up time, which leads to faster access.

♦ Even though both WAP and i-mode are capable of communicating via e-mail, WAP has an edge because e-mail in i-mode is restricted to just 500 bytes.

♦ i-mode can be uploaded to a regular server, whereas WAP requires a special server.

♦ i-mode is HTML-based, and HTML programmers find it easier to handle.

♦ WAP services provide limited content primarily targeted to the business user. In contrast, i-mode conveniently targets the general public.

HTML, WML, HDML, and cHTML Defined

HTML originated from Standard Generalized Markup Language (SGML). Wireless Markup Language (WML) is the markup language for developing content and user interfaces for WAP devices.

Handheld Device Markup Language (HDML) is yet another markup language that enables Web sites to be presented through small display screens on mobile handheld devices such as mobile phones and wireless personal digital assistants (PDAs). HDML was not derived from HTML specifications and was launched before the WAP specifications were developed.

Compact HTML (cHTML) is derived from HTML; in other words, it is a subset of HTML (2.0, 3.2, and 4.0) specifications. It is explained in more detail in the next section.

Introduction to cHTML

HyperText Markup Language (HTML) is the markup language that defines a variety of tags and attributes for developing documents for the Web. cHTML, which stands for compact HTML, is derived from HTML; in other words, it is a subset of HTML (2.0, 3.2, and 4.0) specifications. cHTML is designed for small information appliances, such as mobile devices, wireless personal digital assistants (PDAs), and so on that generally possess the following features:

♦ Limited memory, generally 128 to 512K RAM and 512K to 1M ROM.

♦ Low-power processor, generally 1 to 10 MIPS class CPU for embedded systems.

♦ Small display screens, generally 50 _ 30 dots, 100 _ 72 dots, or 150 _ 100 dots.

♦ Restricted colors, generally monocolor.

♦ Restricted character fonts, generally single-character fonts.

♦ Restricted input method, generally several control buttons and numeric buttons (0 to 9).

Limitations and Guidelines of cHTML

Despite the fact that cHTML is derived from HTML (2.0, 3.2, and 4.0), it does not provide even marginal support for JPEG images, image maps, multiple-character fonts and styles, background color and images, and style sheets. Due to memory constraints in cHTML, buffer size is restricted to 512 bytes and 4,096 bytes, respectively, for the INPUT and SELECT functions.

cHTML works on the assumption that it has to be implemented with limited memory and a low-powered processor. cHTML is most likely to inherit most of the features from the regular HTML because it is a derivative of HTML (2.0, 3.2, and 4.0) specifications. Again, cHTML assumes a single-character font for display on a small, monocolor display screen. Because cHTML does not require two-dimensional cursor movement, the basic operations can be carried out using the four basic buttons. However, scrolling using the four buttons is not required if the content occupies the screen space efficiently. An additional factor is the support for direct selection of anchors using number buttons (0 to 9).

Elements of cHTML

Table 8-2 shows the list of cHTML elements given by W3C. For more information, refer to `http://www.w3.org/TR/1998/NOTE-compactHTML-19980209`.

Table 8-2: cHTML Elements

Elements	Attributes	HTML Version	Notes
! –	–	2.0	
!DOCTYPE	–	2.0	
&xxx;	–	2.0	&, ©, >, <, ", ®, &#0;`&#127;
A	Name= href="URL" rel= rev= title= urn=(deleted from HTML3.2) methods=(deleted from HTML3.2)	2.0	
ABBR	–	4.0	
ACRONYM	–	4.0	
ADDRESS	–	2.0	Single font
APPLET	–	3.2	Deprecated in HTML 4.0
AREA	shape= coords= href="URL" alt= nohref	3.2	
B	–	2.0	Single font
BASE	href="URL"	2.0	
BASEFONT	size=	3.2	Single font Deprecated in HTML 4.0
BDO	–	4.0	
BIG	–	3.2	Single font

Elements	Attributes	HTML Version	Notes
BLOCKQUOTE	–	3.2	
BODY	– bgcolor= background= text= link= vlink= alink=	2.0 3.2 3.2 3.2 3.2 3.2 3.2	Colors other than white are drawn black
BR	– clear=all/left/right	2.0 3.2	
BUTTON	–	4.0	
CAPTION	–	3.2	
CENTER	–	3.2	Deprecated in HTML 4.0
CITE	–	2.0	Single font
CODE	–	2.0	Single font
COL	–	4.0	
COLGROUP	–	4.0	
DD	–	2.0	
DEL	–	4.0	
DFN	–	3.2	
DIR	– compact	2.0	Deprecated in HTML 4.0
DIV	– align=left/center/right	3.2	
DL	– compact	2.0	
DT	–	2.0	
EM	–	2.0	Single font
FIELDSET	–	4.0	
FONT	size=n size=+n/-n color=	3.2	Single font Deprecated in HTML 4.0
FORM	action= method=get/post enctype=	2.0	
FRAME	–	4.0	Frameset DTD
FRAMESET	–	4.0	Frameset DTD
HEAD	–	2.0	

Elements	Attributes	HTML Version	Notes
Hn	– align=left/center/right	2.0 3.2	
HR	– align=left/center/right size= width= noshade	2.0 3.2 3.2 3.2 3.2	
HTML	– version=	2.0 3.2	version="C-HTML 1.0"
I	–	2.0	Single font
IFRAME	–	4.0	Frameset DTD
IMG	src= align=top/middle/bottom align=left/right width= height= hspace= vspace= alt= border= usemap= ismap=	2.0 2.0 3.2 3.2 3.2 3.2 3.2 2.0 3.2 3.2 2.0	Large images compressed automatically
INPUT	type=text name= size= maxlength= value=	2.0	512 bytes (maximum character buffer)
	type=password name= size= maxlength= value=	2.0	
	type=checkbox name= value= checked	2.0	
	type=radio name= value= checked	2.0	
	type=hidden name= value=	2.0	

	type=image name= src= align=top/middle/bot tom/left/right	2.0 2.0 2.0 3.2	
	type=submit name= value=	2.0	
	type=reset name= value=	2.0	
	type=file name= value=	3.2	
INS	–	4.0	
ISINDEX	– prompt=	2.0 3.2	Deprecated in HTML 4.0
KBD	–	2.0	Single font
LABEL	–	4.0	
LEGEND	–	4.0	
LI	– type=1/A/a/I/i type=circle/disc/squ are value=	2.0 3.2 3.2 3.2	
LINK	href="URL" rel= rev= urn= methods= title= id=	2.0	
LISTING	–	2.0	Single font Deprecated in HTML 4.0
MAP	name=	3.2	
MENU	– compact	2.0	Deprecated in HTML 4.0
META	name= http-equiv= content=	2.0	http-equiv="refresh" only
NEXTID	N=	2.0	Deprecated in HTML 3.2
NOFRAMES	–	4.0	Frameset DTD
NOSCRIPT	–	4.0	

Elements	Attributes	HTML Version	Notes
OBJECT	–	4.0	
OL	– type=1/A/a/I/i start= compact	2.0 3.2 3.2 2.0	
OPTGROUP	–	4.0	
OPTION	– selected value=	2.0	
P	– align=left/center/ri ght	2.0 3.2	
PARAM	–	4.0	
PLAINTEXT	–	2.0	Deprecated in HTML 4.0
PRE	– width=	2.0 3.2	
Q	–	T4.0	
S	–	2.0	Deprecated in HTML 4.0
SAMP	–	2.0	Single font
SCRIPT	–	3.2	
SELECT	name= size= multiple	2.0	512 bytes (maximum character buffer)
SMALL	–	3.2	Single font
SPAN	–	4.0	
STRIKE	–	2.0	Deprecated in HTML 4.0
STRONG	–	2.0	Single font
STYLE	–	2.0	
SUB	–	3.2	
SUP	–	3.2	
TABLE	– align=left/center/ri ght etc. border= width= cellspacing= cellpadding=	3.2	
TBODY	–	4.0	

Elements	Attributes	HTML Version	Notes
TD	– align=left/center/right valign=top/middle/bottom/baseline rowspan= colspan= width= height= nowrap	3.2	
TEXTAREA	name= rows= cols=	2.0	512 bytes (maximum character buffer)
TFOOT	–	4.0	
TH	– align=left/center/right valign=top/middle/bottom/baseline rowspan= colspan= width= height= nowrap	3.2	
THEAD	–	4.0	
TITLE	–	2.0	
TR	– align=left/center/right valign=top/middle/bottom/baseline	3.2	
TT	–	2.0	Single font
U	–	3.2	Deprecated in HTML 4.0
UL	– type=disk/circle/square compact	2.0 3.2 2.0	
VAR	–	2.0	Single font
XMP	–	2.0	Single font Deprecated in HTML 4.0

Listing 8-1 shows a "Hello World" program that can be displayed on an i-mode browser.

Listing 8-1: example1.html

```
<html>
   <head>
     <META http-equiv="Content-Type" content="text/html; charset=utf-8" />
     <META name="CHTML" content="yes" />
```

```
     <META name="description" content="cHTML document" />

        <title>Example1</title>
    </head>
    <body>
       <h1>Hello World..</h1>
    </body>
</html>
```

The execution of Listing 8-1 requires an i-mode emulator (which is available for free download at `http://www.wapprofit.com`) and a Web server. The examples covered in this chapter have been successfully tested using the Internet Information Services on Windows 2000. The files are placed in the `drive:\inetpub\wwwroot` directory. Also note that Windows 9x with Personal Web Server 4.0 can also be used for testing the examples.

The output of Listing 8-1, as rendered by the `wapprofit.com` i-mode Emulator, 1.11 is shown in Figure 8-1.

Figure 8-1: The output of example1.html on an i-mode-enabled device

The cHTML file in Listing 8-2, `example2.html`, makes use of the `<font>` tag and its supported attributes: `size` and `color`.

Listing 8-2: example2.html

```
<html>
  <head>
<META http-equiv="Content-Type" content="text/html; charset=utf-8" />
<META name="CHTML" content="yes" />
<META name="description" content="cHTML document" />
    <title>Example2</title>
  </head>
  <body>
     <center>
       <font size=2 color="green">
  <b>Font</b><hr>
             Size = 2<br>
  Color = Red
       </font>
     </center>
  </body>
```

```
</html>
```

The output of Listing 8-2, as rendered by the `wapprofit.com` i-mode Emulator 1.11, is shown in Figure 8-2.

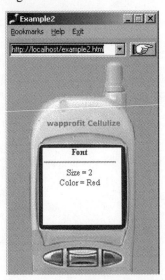

Figure 8-2: The output of example2.html on an i-mode-enabled device

The cHTML file in Listing 8-3, `example3_1.html`, invokes the cHTML file shown in Listing 8-4, `example3_2.html`, using the anchor <a> tag.

Listing 8-3: example3_1.html

```
<html>
  <head>
<META http-equiv="Content-Type" content="text/html; charset=utf-8" />
<META name="CHTML" content="yes" />
<META name="description" content="cHTML document" />
    <title>Example3_1</title>
  </head>
  <body>
   <center>
     <font size=+1 color="Red">
 <br><hr>
 <a href="http://localhost/example3_2.html">
               Next Page
          </a><hr>
     </font>
   </center>
</body>
</html>
```

Listing 8-4: example3_2.html

```
<html>
  <head>
<META http-equiv="Content-Type" content="text/html; charset=utf-8" />
<META name="CHTML" content="yes" />
<META name="description" content="cHTML document" />
    <title>Example3_2</title>
```

```
  </head>
  <body>
   <center>
      <font size=+2 color="Red">
         Welcome<br><br><hr>
      </font>
      <font size=+1>
         <a href="http://localhost/example3_1.html">
            Back
         </a><hr>
      </font>
   </center>
</body>
</html>
```

Figure 8-3 shows the simulation, as rendered by the wapprofit.com i-mode Emulator 1.11.

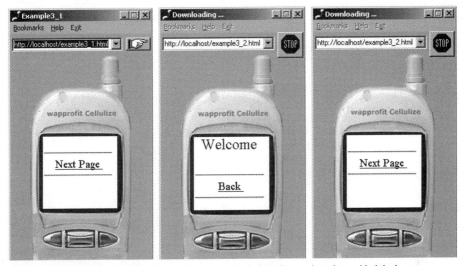

Figure 8-3: The output of example3_1.html and example3_2.html on an i-mode-enabled device

The cHTML file in Listing 8-5, example4.html, makes use of the <img> tag for inserting images.

Listing 8-5: example4.html

```
<html>
  <head>
<META http-equiv="Content-Type" content="text/html; charset=utf-8" />
<META name="CHTML" content="yes" />
<META name="description" content="cHTML document" />
    <title>Example4</title>
  </head>
    <body>
      <center>
         <font size=-1 color="green">Earth</font><hr>
         <img src="c:\ Solar Eclipse.jpg" height=80 width=80 border=1>
      </center>
    </body>
</html>
```

Figure 8-4 shows the simulation of Listing 8-5, as rendered by the `wapprofit.com` i-mode Emulator 1.11.

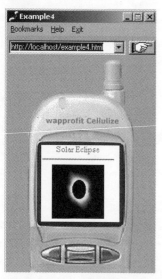

Figure 8-4: The output of example4.html on an i-mode-enabled device

The cHTML file in Listing 8-6, `example5.html`, demonstrates how a table is made using the `<table>` tag; the tags `<td>` and `<tr>` represent the table columns and rows, respectively.

Listing 8-6: example5.html

```
<html>
  <head>
<META http-equiv="Content-Type" content="text/html; charset=utf-8" />
<META name="CHTML" content="yes" />
<META name="description" content="cHTML document" />
    <title>Example5</title>
  </head>
  <body>
    <center>
    <font size=-1><b>DELHI</b></font>
    </center>
    <table border=1 align=center>
      <tr><th colspan=2>Temperature</th></tr>
      <tr>
        <td><font size=-1>Max.</font></td>
        <td><font size=-1>35<sup>o</sup>C</font></td>
      </tr>
      <tr>
        <td><font size=-1>Min.</font></td>
        <td><font size=-1>20<sup>o</sup>C</font></td>
      </tr>
    </table>
  </body>
</html>
```

Figure 8-5 shows the simulation, as rendered by the `wapprofit.com` i-mode Emulator 1.11.

Figure 8-5: The output of example4.html on an i-mode-enabled device

Case Study

This case study presents a vivid description of the various cHTML-supported tags. It's a mini-project in which the main document, main.html (Listing 8-7), provides information about weather, stock updates, and the latest movies. The main.html file offers these three options for the user and invokes weather.html, stocks.html, or movies.html, depending on the user's selection.

Listing 8-7: main.html

```
<html>
  <head>
<META http-equiv="Content-Type" content="text/html; charset=utf-8" />
<META name="CHTML" content="yes" />
<META name="description" content="cHTML document" />
    <title>Case Study</title>
  </head>
  <body>
    <center>
      <font size=4 color=red><b>Information Center</b></font><hr>
      <a href="http://localhost/weather.html">Weather</a><br>
      <a href="http://localhost/stocks.html">Stocks</a><br>
      <a href="http://localhost/movies.html">Movies</a>
    </center>
  </body>
</html>
```

When the user selects Weather, main.html invokes weather.html (Listing 8-8), which, in this example, provides the user with the weather information about four metropolitan cities in India. When the user selects Delhi, weather.html displays the current temperature and forecast for Delhi by invoking the file delhi_weather.html (Listing 8-9). Likewise, if the user selects Bombay, Chennai, or Calcutta, weather.html invokes bombay_weather.html (Listing 8-10), chennai_weather.html (Listing 8-11), or calcutta_weather.html (Listing 8-12), respectively.

Listing 8-8: weather.html

```
<html>
```

```
  <head>
<META http-equiv="Content-Type" content="text/html; charset=utf-8" />
<META name="CHTML" content="yes" />
<META name="description" content="cHTML document" />
    <title>Weather</title>
  </head>
  <body>
   <center>
      <font size=1 color="Red"><b>WEATHER</b></font>
      <hr>
      <table>
          <tr>
            <td align=left>
               <a href="http://localhost/delhi_weather.html">
                  <font size=1>Delhi</font>
               </a>
            </td>
            <td align=right>
               <a href="http://localhost/chennai_weather.html">
                  <font size=1>Chennai</font>
               </a>
            </td>
          </tr>
          <tr>
            <td align=left>
               <a href="http://localhost/bombay_weather.html">
                  <font size=1>Bombay</font>
               </a>
            </td>
            <td align=right>
               <a href="http://localhost/calcutta_weather.html">
                  <font size=1>Calcutta</font>
               </a>
            </td>
          </tr>
      </table>
      <hr>
      <font size=1>
         <a href="http://localhost/main.html">Back</a>
      </font>
   </center>
</body>
</html>
```

Listing 8-9: delhi_weather.html

```
<html>
  <head>
<META http-equiv="Content-Type" content="text/html; charset=utf-8" />
<META name="CHTML" content="yes" />
<META name="description" content="cHTML document" />
    <title>Delhi Weather</title>
  </head>
  <body>
   <center>
     <font size=1 color="Red"><b>DELHI</b></font>
       <table>
```

```
      <tr>
         <th colspan=2><font size=1>Temperature</font></th>
      </tr>
      <tr>
         <td align=center>
            <font size=-1>Max. 35<sup>o</sup>C</font>
         </td>
         <td align=center>
            <font size=-1>Min. 20<sup>o</sup>C</font>
         </td>
      </tr>
    </table>
    <font size=-2><b>Forecast</b>: Clear Sky<br>
      <a href="http://localhost/weather.html">Back</a>
    </font>
  </center>
  </body>
</html>
```

Listing 8-10: bombay_weather.html

```
<html>
  <head>
<META http-equiv="Content-Type" content="text/html; charset=utf-8" />
<META name="CHTML" content="yes" />
<META name="description" content="cHTML document" />
    <title>Bombay Weather</title>
  </head>
  <body>
  <center>
    <font size=1 color="Red"><b>BOMBAY</b></font>
      <table>
         <tr>
            <th colspan=2><font size=1>Temperature</font></th>
         </tr>
         <tr>
           <td align=center>
              <font size=-1>Max. 40<sup>o</sup>C</font>
           </td>
           <td align=center>
              <font size=-1>Min. 25<sup>o</sup>C</font>
           </td>
         </tr>
      </table>
      <font size=-2><b>Forecast</b>: Humidity<br>
        <a href="http://localhost/weather.html">Back</a>
      </font>
  </center>
  </body>
</html>
```

Listing 8-11: chennai_weather.html

```
<html>
  <head>
<META http-equiv="Content-Type" content="text/html; charset=utf-8" />
<META name="CHTML" content="yes" />
```

```
<META name="description" content="cHTML document" />
    <title>Chennai Weather</title>
  </head>
  <body>
  <center>
    <font size=1 color="Red"><b>CHENNAI</b></font>
      <table>
        <tr>
          <th colspan=2><font size=1>Temperature</font></th>
        </tr>
        <tr>
          <td align=center>
            <font size=-1>Max. 40<sup>o</sup>C</font>
          </td>
          <td align=center>
            <font size=-1>Min. 25<sup>o</sup>C</font>
          </td>
        </tr>
      </table>
      <font size=-2><b>Forecast</b>: Sunny<br>
        <a href="http://localhost/weather.html">Back</a>
      </font>
  </center>
  </body>
</html>
```

Listing 8-12: calcutta_weather.html

```
<html>
  <head>
<META http-equiv="Content-Type" content="text/html; charset=utf-8" />
<META name="CHTML" content="yes" />
<META name="description" content="cHTML document" />
    <title>Calcutta Weather</title>
  </head>
  <body>
  <center>
    <font size=1 color="Red"><b>CALCUTTA</b></font>
      <table>
        <tr>
          <th colspan=2><font size=1>Temperature</font></th>
        </tr>
        <tr>
          <td align=center>
            <font size=-1>Max. 30<sup>o</sup>C</font>
          </td>
          <td align=center>
            <font size=-1>Min. 20<sup>o</sup>C</font>
          </td>
        </tr>
      </table>
      <font size=-2><b>Forecast</b>: Rain<br>
        <a href="http://localhost/weather.html">Back</a>
      </font>
  </center>
  </body>
</html>
```

When the user selects Stocks from the main-page options, `main.html` invokes `stocks.html` (Listing 8-13), which gives the user a list of four Indian stock exchanges: DSE (Delhi Stock Exchange), BSE (Bombay Stock Exchange), NSE (National Stock Exchange), and CSE (Calcutta Stock Exchange). After selecting a particular stock exchange, the user is presented with the previous and present stock indexes and the index flow. `stocks.html` can invoke `delhi_stock_exchange.html` (Listing 8-14), `bombay_stock_exchange.html` (Listing 8-15), `national_stock_exchange.html` (Listing 8-16), or `calcutta_stock_exchange.html` (Listing 8-17), depending on the user's choice.

Listing 8-13: stocks.html

```html
<html>
  <head>
<META http-equiv="Content-Type" content="text/html; charset=utf-8" />
<META name="CHTML" content="yes" />
<META name="description" content="cHTML document" />
    <title>Stocks</title>
  </head>
  <body>
   <center>
      <font size=1 color="Red"><b>STOCKS</b></font>
      <hr>
      <table>
        <tr>
          <td align=left>
            <a href="http://localhost/delhi_stock_exchange.html">
              <font size=1>DSE</font>
            </a>
          </td>
          <td align=right>
            <a href="http://localhost/national_stock_exchange.html">
              <font size=1>NSE</font>
            </a>
          </td>
        </tr>
        <tr>
          <td align=left>
            <a href="http://localhost/bombay_stock_exchange.html">
              <font size=1>BSE</font>
            </a>
          </td>
          <td align=right>
            <a href="http://localhost/calcutta_stock_exchange.html">
              <font size=1>CSE</font>
            </a>
          </td>
        </tr>
      </table>
      <hr>
      <font size=1>
        <a href="http://localhost/main.html">Back</a>
      </font>
   </center>
  </body>
</html>
```

Listing 8-14: delhi_stock_exchange.html

```html
<html>
  <head>
<META http-equiv="Content-Type" content="text/html; charset=utf-8" />
<META name="CHTML" content="yes" />
<META name="description" content="cHTML document" />
    <title>Delhi Stock Exchange</title>
  </head>
  <body>
  <center>
    <font size=-2><b>DELHI <br>Stock Exchange</b></font><hr>
    <font size=-2>
       Previous Index : 3990<br>
       Current Index : 4000<br>
       Up by 10 points.
    </font>
        <a href="http://localhost/stocks.html">Back</a>
  </center>
</body>
</html>
```

Listing 8-15: bombay_stock_exchange.html

```html
<html>
  <head>
<META http-equiv="Content-Type" content="text/html; charset=utf-8" />
<META name="CHTML" content="yes" />
<META name="description" content="cHTML document" />
    <title>Bombay Stock Exchange</title>
  </head>
  <body>
  <center>
    <font size=-2><b>BOMBAY<br>Stock Exchange</b></font><hr>
    <font size=-2>
       Previous Index : 4190<br>
       Current Index : 4000<br>
       Down by 190 points.
    </font>
        <a href="http://localhost/stocks.html">Back</a>
  </center>
</body>
</html>
```

Listing 8-16: national_stock_exchange.html

```html
<html>
  <head>
<META http-equiv="Content-Type" content="text/html; charset=utf-8" />
<META name="CHTML" content="yes" />
<META name="description" content="cHTML document" />
    <title>National Stock Exchange</title>
  </head>
  <body>
  <center>
    <font size=-2><b>NATIONAL<br>Stock Exchange</b></font><hr>
    <font size=-2>
       Previous Index : 3550<br>
```

```
      Current Index : 3700<br>
      Up by 150 points.
   </font>
      <a href="http://localhost/stocks.html">Back</a>
   </center>
</body>
</html>
```

Listing 8-17: calcutta_stock_exchange.html

```
<html>
  <head>
<META http-equiv="Content-Type" content="text/html; charset=utf-8" />
<META name="CHTML" content="yes" />
<META name="description" content="cHTML document" />
    <title>Calcutta Stock Exchange</title>
  </head>
  <body>
  <center>
    <font size=-2><b>CALCUTTA<br>Stock Exchange</b></font><hr>
    <font size=-2>
      Previous Index : 4500<br>
      Current Index : 4000<br>
      Down by 500 points.
    </font>
        <a href="http://localhost/stocks.html">Back</a>
   </center>
</body>
</html>
```

When the user selects movies from the options on the main page, main.html invokes movies.html (Listing 8-18), which provides the user with a list of four theaters: Plaza, Regal, Odeon, and Eros. On selecting a particular theater, the user is presented with the details of the movie currently playing at that particular theater. movies.html can invoke plaza.html (Listing 8-19), regal.html (Listing 8-20), odeon.html (Listing 8-21), or eros.html (Listing 8-22), depending on the user's selection.

Listing 8-18: movies.html

```
<html>
  <head>
<META http-equiv="Content-Type" content="text/html; charset=utf-8" />
<META name="CHTML" content="yes" />
<META name="description" content="cHTML document" />
    <title>Movies</title>
  </head>
  <body>
   <center>
     <font size=1 color="Red"><b>MOVIES</b></font>
     <hr>
     <table>
        <tr>
           <td align=left>
              <a href="http://localhost/plaza.html">
                 <font size=1>Plaza</font>
              </a>
           </td>
           <td align=right>
```

```
                <a href="http://localhost/odeon.html">
                    <font size=1>Odeon</font>
                </a>
            </td>
        </tr>
        <tr>
            <td align=left>
                <a href="http://localhost/regal.html">
                    <font size=1>Regal</font>
                </a>
            </td>
            <td align=right>
                <a href="http://localhost/eros.html">
                    <font size=1>Eros</font>
                </a>
            </td>
        </tr>
    </table>
    <hr>
    <font size=1>
        <a href="http://localhost/main.html">Back</a>
    </font>
    </center>
</body>
</html>
```

Listing 8-19: plaza.html

```
<html>
  <head>
<META http-equiv="Content-Type" content="text/html; charset=utf-8" />
<META name="CHTML" content="yes" />
<META name="description" content="cHTML document" />
    <title>Plaza</title>
  </head>
  <body>
  <center>
    <font size=2>
        <b>Almost Famous<br> 12:30,3:30,6:30</b><hr>
    </font>
        A fun film set in the rock and roll years.<hr>
    <a href="http://localhost/movies.html">Back</a>
  </center>
  </body>
</html>
```

In this example we are using web server named HTTP://QUEST, while running the code on your machine please change this to HTTP://YOURSERVERNAME to run the code.

Listing 8-20: regal.html

```
<html>
  <head>
<META http-equiv="Content-Type" content="text/html; charset=utf-8" />
<META name="CHTML" content="yes" />
<META name="description" content="cHTML document" />
    <title>Regal</title>
  </head>
```

```
<body>
<center>
  <font size=2>
     <b>Proof of Life<br> 12:30,3:30,6:30</b><hr>
  </font>
  A kidnap and ransom thriller.<hr>
  <a href="http://localhost/movies.html">Back</a>
</center>
</body>
</html>
```

Listing 8-21: odeon.html

```
<html>
  <head>
<META http-equiv="Content-Type" content="text/html; charset=utf-8" />
<META name="CHTML" content="yes" />
<META name="description" content="cHTML document" />
    <title>Odeon</title>
  </head>
  <body>
  <center>
    <font size=2>
       <b>Terminator II<br> 12:30,3:30,6:30</b><hr>
    </font>
       An action film.<hr>
    <a href="http://localhost/movies.html">Back</a>
  </center>
  </body>
</html>
```

Listing 8-22: eros.html

```
<html>
  <head>
<META http-equiv="Content-Type" content="text/html; charset=utf-8" />
<META name="CHTML" content="yes" />
<META name="description" content="cHTML document" />
    <title>Eros</title>
  </head>
  <body>
  <center>
    <font size=2>
       <b>Vertical Limit<br> 12:30,3:30,6:30</b><hr>
    </font>
       An adventure film.<hr>
    <a href="http://localhost/movies.html">Back</a>
  </center>
  </body>
</html>
```

The output of the preceding application, as rendered by the wapprofit.com i-mode Emulator 1.11, is shown in Figures 8-6 to 8-9.

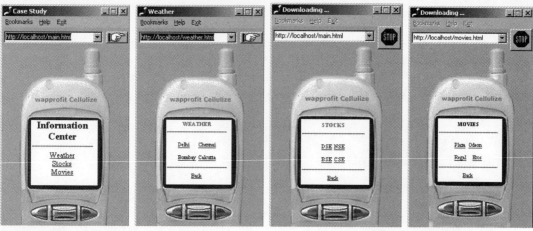

Figure 8-6: Main-page choices and subsequent screens

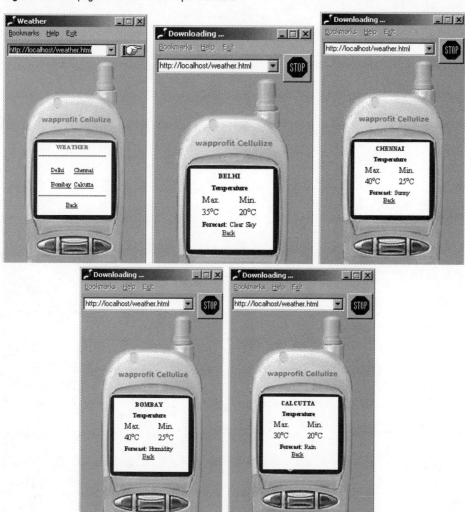

Figure 8-7: The weather section of the application

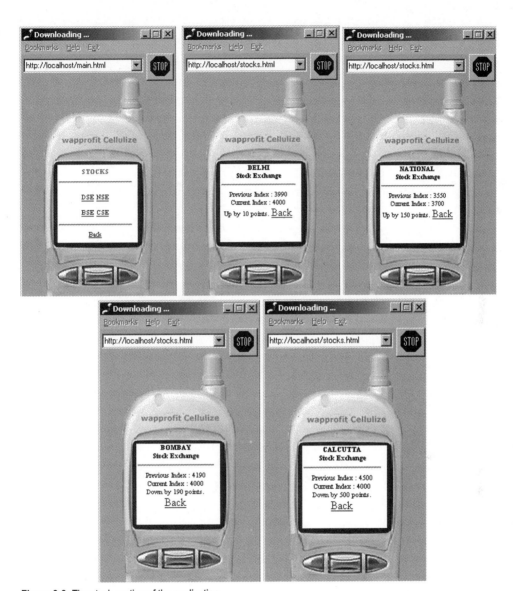

Figure 8-8: The stock section of the application

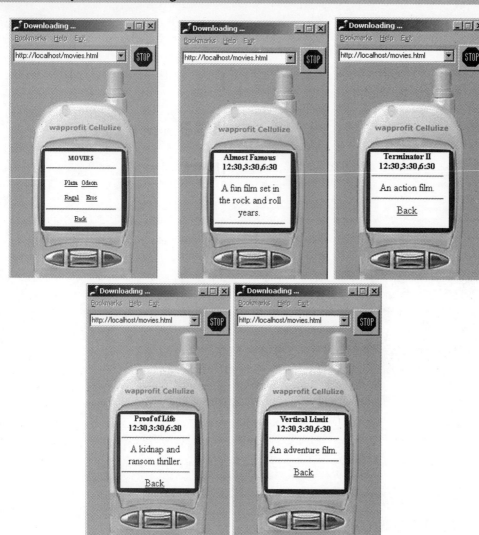

Figure 8-9: The movie portion of the application

Summary

Wireless Web technology is becoming extremely popular, occupying a major share of the Japanese market as compared to what it occupies in the American and European markets. Recent studies have revealed that Japan makes up more than 80 percent of the wireless Web market share, thereby making it the largest wireless market in the world. In Japan, NTT DoCoMo is the largest provider of wireless services. It uses cHTML as a language for serving content over its network. Consequently, it is imperative to learn cHTML if you are planning to develop an i-mode-compatible application. NTT DoCoMo is on its way to expanding its operation in Europe and Asia. Thus in the future, cHTML will become a leading language for serving content in the wireless information arena.

In this chapter, we have analyzed some of the facts about cHTML and i-mode technologies. Moreover, we also undertook a detailed discussion of cHTML with working examples and a sample case study. In the following chapter, we will undertake a complete, full-fledged application for an i-mode-compatible browser using an XML- and XSLT-based approach.

Chapter 9

Making the Web Application i-mode Compatible Using XSLT

In this chapter, you are again going to transform the Web application for use on i-mode-compliant phones and devices. As you learned in the preceding chapter, i-mode technology and its services were invented and operated by NTT DoCoMo, a company based in Japan. In a relatively short time, i-mode technology has had a large impact on the Japanese mobile Internet market. Today, it is poised to spread to the rest of the world. This technology provides access to Internet content, such as e-mail and Web browsing, over a mobile phone network. To display content over phones, i-mode technology uses the cHTML language, which is a scaled-down version of HTML 4.0. Chapter 8 contains a detailed discussion of cHTML.

Advantages of the i-mode Environment

i-mode provides many advantages by virtue of the technology and its unique architecture as compared to wireless application protocol. These advantages are as follows:

- For its browser language, i-mode technology uses cHTML that is a scaled-down version of HTML 4.0. It is very easy to adopt. Web developers can start working on this after a little practice.

- One of the main features of i-mode-enabled phones is the capacity to display color and images on the browser. Some of the new i-mode devices are capable of displaying images with more than 256 colors.

- i-mode technology is based on the "always on" technology and so the user's device is always connected to the Internet. He/she only have to pay for the data transmited over the network; not for the minutes he/she is connected to server.

Framework of an i-mode-Enabled Application

While working with the i-mode transformation process, you use the same framework design that you have used in the previous transformations in this book. This time, you are adding only one set of XSLT documents to the application, so that they become i-mode-platform-enabled. The application's design framework is shown in Figure 9-1.

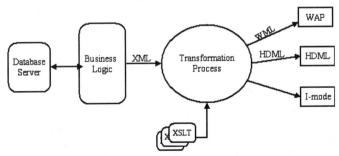

Figure 9-1: The framework design of an i-mode application

As you can see in Figure 9-1, you add one more set of XSLT documents to make the application i-mode compliant. You are using the same set of Active Server Pages for generating the XML data and then using a different set of XSLT documents for transforming the XML data into i-mode-compliant format.

Detect the i-mode Browser

This area is a little confusing because there is no official data on the NTT DoCoMo Web site about the HTTP_USER_AGENT string used with i-mode-enabled phones. These examples use the wapprofit i-mode simulator to test the application. This product uses Internet Explorer to display the content. Because of this, you first test the application using this simulator and then change the browser-detection string to "DoCoMo". It is assumed that most i-mode-enabled browsers will return this string in the HTTP_USER_AGENT string while accessing the i-mode service.

This time you also append some lines of code in the checkbrows.asp file (Listing 9-1) to detect i-mode-enabled browsers and devices. After appending these lines of code, the checkbrows.asp file takes its final shape, containing the detection process for the devices that can access the Internet using any data-transfer medium, including

♦ PC browsers

♦ WAP-enabled devices and phones

♦ HDML-compliant devices and phones

♦ i-mode-enabled devices and phones

Listing 9-1: checkbrows.asp (final version)

© 2001 Dreamtech Software India Inc.
All Rights Reserved

```
<%@ Language=VBScript %>

<%
'declaring variables
dim sXML
dim sXSL
dim xmlDocument,xslDocument
dim browser,bt

'creating the objects for the XMLDOM
set xmlDocument = Server.CreateObject("MICROSOFT.XMLDOM")
set xslDocument = Server.CreateObject("MICROSOFT.XMLDOM")

acc=Request.ServerVariables("HTTP_ACCEPT")

Set bt = Server.CreateObject("MSWC.BrowserType")
browser = bt.browser

uagent = Request.ServerVariables("HTTP_USER_AGENT")
browserinfo = split("/",uagent,5)
browsertype = browserinfo(0)

'check for the browser then according to that browser load the xml files
if (InStr(acc,"text/vnd.wap.wml")) then

  session("browsmode")="wml"
```

```
sXML = "main.xml"
sXSL = "wml.xsl"


xmlDocument.async = false
xslDocument.async = false


xmlDocument.load(Server.MapPath(sXML))
xslDocument.load(Server.MapPath(sXSL))

Response.ContentType = "text/vnd.wap.wml"
Response.Write "<!DOCTYPE wml PUBLIC ""-//WAPFORUM//DTD " & _
  "WML 1.1//EN"" ""http://www.wapforum.org/DTD/wml_1.1.xml"">"

  Response.Write(xmlDocument.transformNode(xslDocument))

else
if (InStr(acc,"hdml")) then

  session("browsmode")="hdml"
  Response.ContentType = "text/x-hdml"

  sXML = "main.xml"
  sXSL = "hdml.xsl"


  xmlDocument.async = false
  xslDocument.async = false


  xmlDocument.load(Server.MapPath(sXML))
  xslDocument.load(Server.MapPath(sXSL))


  Response.Write(xmlDocument.transformNode(xslDocument))

else
  if (browsertype="DoCoMo") then
    session("browsmode")="imode"
    Response.ContentType = "text/html"

    sXML = "main.xml"
    sXSL = "imode.xsl"


    xmlDocument.async = false
    xslDocument.async = false


    xmlDocument.load(Server.MapPath(sXML))
    xslDocument.load(Server.MapPath(sXSL))


  Response.Write(xmlDocument.transformNode(xslDocument))
```

```
      end if
    end if
    end if


if (browser="IE") then
session("browsmode")="pc"
sXML = "main.xml"
sXSL = "main.xsl"


  xmlDocument.async = false
  xslDocument.async = false


  xmlDocument.load(Server.MapPath(sXML))
  xslDocument.load(Server.MapPath(sXSL))
  Response.Write(xmlDocument.transformNode(xslDocument))
end if


%>
```

In Listing 9-1, you add a new `if` condition block to detect i-mode-compliant devices and phones:

```
if (browsertype="DoCoMo") then
  session("browsmode")="imode"
  Response.ContentType = "text/html"


  sXML = "main.xml"
  sXSL = "imode.xsl"


  xmlDocument.async = false
  xslDocument.async = false


  xmlDocument.load(Server.MapPath(sXML))
  xslDocument.load(Server.MapPath(sXSL))


  Response.Write(xmlDocument.transformNode(xslDocument))
end if
```

In the preceding code, you are searching for a specific string, "DoCoMo", in the HTTP_USER_AGENT file. If you find it, you load the corresponding XSLT document to transform the XML data to i-mode-compliant format. This document loads the main XML document, named main.xml, with a different XSLT document named imode.xsl for performing the transformation operation. You also set the content type to text/html for i-mode-enabled browsers.

If you are using the wapprofit i-mode simulator, change the searchable string to "PJPEG" instead of "DoCoMo".

Transform the Home Page

Listing 9-2 shows the code of imode.xsl, which transforms the home page into an i-mode-compliant home page.

Listing 9-2: imode.xsl

```
<?xml version='1.0'?>
<xsl:stylesheet xmlns:xsl="http://www.w3.org/1999/XSL/Transform" version="1.0">
<xsl:output method="html" />

<xsl:template match="/">
<html>
<head>
<META http-equiv="Content-Type" content="text/html; charset=utf-8" />
<META name="CHTML" content="yes" />
<META name="description" content="CHTML document" />
<title>Sample cHTML document</title>
</head>
<body>
Menu <br />
<a href="generatexml.asp?sec=News">1.</a> News Section <br />
<a href="generatexml.asp?sec=Weather">2.</a> Weather Report <br />
<a href="generatexml.asp?sec=E-mail">3.</a> E-mail <br />
<a href="generatexml.asp?sec=Movies">4.</a> Movie Ticket Booking<br />
</body>
</html>
</xsl:template>
</xsl:stylesheet>
```

This code is very similar to HTML. There is one noticeable change, however: You add some META tags while generating the cHTML document, as follows:

```
<META name="CHTML" content="yes" />
<META name="description" content="CHTML document" />
```

These META tags indicate the type of document created. They must be included while you are generating the cHTML document. After defining the META tags, you create a series of links to display all the choices to the user by using the <href> element.

```
<a href="generatexml.asp?sec=News">1.</a> News Section <br />
<a href="generatexml.asp?sec=Weather">2.</a> Weather Report <br />
<a href="generatexml.asp?sec=E-mail">3.</a> E-mail <br />
<a href="generatexml.asp?sec=Movies">4.</a> Movie Ticket Booking<br />
```

As you can see, this is a very simple generation of cHTML data using XSLT and XML for i-mode-enabled browsers.

Test the Home Page with the Simulator

As indicated earlier, we're using the WAPPROFIT i-mode simulator to test the application. When you open the checkbrows.asp file on the Web server, it performs the transformation and shows the output in Figure 9-2.

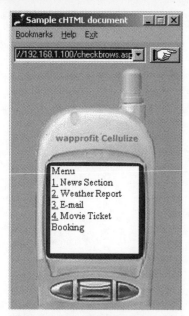

Figure 9-2: The application's home page

Figure 9-2 displays the application's home page. It contains four sections from which the user can select. The following options are available:

♦ News Section

♦ Weather Report

♦ E-mail`

♦ Movie Ticket Booking

The following sections show you how to perform transformations on each of these sections to make them i-mode compatible.

Transform the Weather Section

This section discusses the transformation of the application's weather section to make it i-mode compatible. For this you first generate the XML data for the weather section with the generatexml.asp file in Listing 9-3. Then you will transform that data into an i-mode-compatible format.

Listing 9-3: generatexml.asp

© 2001 Dreamtech Software India Inc.
All Rights Reserved

```
<%@ Language=VBScript %>

<%
dim address,section,fname,ctg,city,emode
ctg =" "
city=" "
emode =" "
section = " "
section = trim(Request.QueryString("sec"))
emode = trim(Request.QueryString("mode"))
```

```
ctg = trim(Request.QueryString("category"))
city = trim(Request.QueryString("city"))


set conn = Server.CreateObject("ADODB.Connection")
conn.ConnectionString = "Provider=SQLOLEDB.1;Persist Security Info=False;User
ID=sa;Initial Catalog=portal;Data Source=developers"
conn.Open

Set objrs = Server.CreateObject("ADODB.Recordset")
objrs.CursorType = adOpenStatic

if (section="Weather")  then
  sql = "select city from weather"
  objrs.Open sql,conn

  Set fso =Server.CreateObject("Scripting.FileSystemObject")
  fname = "tweather.xml"

  fullpath = server.MapPath("main")

  Set MyFile=fso.CreateTextFile(fullpath & "/" & fname)

  MyFile.WriteLine("<?xml version='1.0'?>")
  MyFile.WriteLine("<ROOT>")


  while not(objrs.EOF)
    city = objrs.Fields("city")
    MyFile.WriteLine("<city>" & city & "</city>")

  objrs.MoveNext
  wend

  MyFile.WriteLine("</ROOT>")
  address = "http://192.168.1.100/loadxsl.asp?name=" & fname
  Response.Redirect(address)

end if


if ((city="Sydney") OR (city="NewDelhi") OR (city="Miami") OR (city="Tokyo") OR
(city="Phoenix") OR (city="NewYork"))  then
  sql = "select * from weather where city = '" & city & "'"
  objrs.Open sql,conn

  Set fso =Server.CreateObject("Scripting.FileSystemObject")
  fname = "tweatherdetail.xml"

  fullpath = server.MapPath("main")

  Set MyFile=fso.CreateTextFile(fullpath & "/" & fname)

  MyFile.WriteLine("<?xml version='1.0'?>")
  MyFile.WriteLine("<ROOT>")
```

```
while not(objrs.EOF)
   lowtemp = objrs.Fields("low_temp")
   hightemp = objrs.Fields("high_temp")
   srise = objrs.Fields("sunrise")
   sset = objrs.Fields("sunset")
   mrise = objrs.Fields("moonrise")
   mset = objrs.Fields("moonset")
   dhumid = objrs.Fields("dayhumidity")
   nhumid = objrs.Fields("nighthumidity")
   rain = objrs.Fields("rainfall")
   fig = objrs.Fields("figure")

   MyFile.WriteLine("<weather>")
   MyFile.WriteLine("<low_temp>" & lowtemp & "</low_temp>")
   MyFile.WriteLine("<high_temp>" & hightemp & "</high_temp>")
   MyFile.WriteLine("<sunrise>" & srise & "</sunrise>")
   MyFile.WriteLine("<sunset>" & sset & "</sunset>")
   MyFile.WriteLine("<moonrise>" & mrise & "</moonrise>")
   MyFile.WriteLine("<moonset>" & mset & "</moonset>")
   MyFile.WriteLine("<dayhumidity>" & dhumid & "</dayhumidity>")
   MyFile.WriteLine("<nighthumidity>" & nhumid & "</nighthumidity>")
   MyFile.WriteLine("<rainfall>" & rain & "</rainfall>")
   MyFile.WriteLine("<figure>" & fig & "</figure>")
   MyFile.WriteLine("</weather>")
 objrs.MoveNext
 wend

 MyFile.WriteLine("</ROOT>")
 address = "http://192.168.1.100/loadxsl.asp?name=" & fname
 Response.Redirect(address)

end if


objrs.Close
conn.Close

%>
```

This `generatexml.asp` code is the same as the code you have used in the previous chapters to generate XML for the application's weather section. Here, you are collecting the data from the database and then saving it in an XML file named `tweather.xml` on the server. After saving the XML file, you redirect control to the `loadxsl.asp` file to load the XSL document for making this data i-mode compliant. Figure 9-3 is a screen shot of `tweather.xml`.

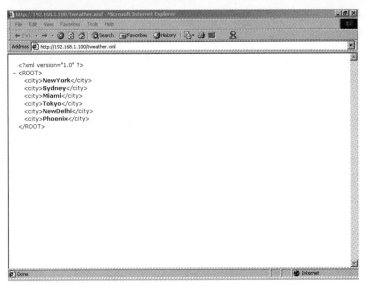

Figure 9-3: tweather.xml

You use the `loadxsl.asp` file (Listing 9-4) to load the XML and XSL documents and then perform the data transformation.

Listing 9-4: loadxsl.asp (final version)

© 2001 Dreamtech Software India Inc.
All Rights Reserved

```
<%@ Language=VBScript %>

<%
fname = trim(Request.QueryString("name"))


'declaring variables
dim sXML
dim sXSL
dim xmlDocument,xslDocument

'creating the objects for the XMLDOM
set xmlDocument = Server.CreateObject("MSXML2.DOMDocument")
set xslDocument = Server.CreateObject("MSXML2.DOMDocument")

sXML = fname
fname = mid(fname,1,(len(fname) - 4))


if (session.Contents(1)="wml") then
 sXSL = fname & "wml.xsl"

 xmlDocument.async = false
 xslDocument.async = false

 xmlDocument.load(Server.MapPath(sXML))
 xslDocument.load(Server.MapPath(sXSL))
```

```
  Response.ContentType = "text/vnd.wap.wml"
  Response.Write "<?xml version=""1.0"" ?>"


  Response.Write "<!DOCTYPE wml PUBLIC ""-//WAPFORUM//DTD " & _
  "WML 1.1//EN"" ""http://www.wapforum.org/DTD/wml_1.1.xml"">"

  Response.Write(xmlDocument.transformNode(xslDocument))

end if


if (session.Contents(1)="imode") then
  sXSL = fname & "imode.xsl"

  xmlDocument.async = false
  xslDocument.async = false

  xmlDocument.load(Server.MapPath(sXML))
  xslDocument.load(Server.MapPath(sXSL))

  Response.ContentType = "text/html"
  Response.Write(xmlDocument.transformNode(xslDocument))

end if


if (session.Contents(1)="hdml") then
  sXSL = fname & "hdml.xsl"

  xmlDocument.async = false
  xslDocument.async = false

  Response.ContentType = "text/x-hdml"

  xmlDocument.load(Server.MapPath(sXML))
  xslDocument.load(Server.MapPath(sXSL))


  Response.Write(xmlDocument.transformNode(xslDocument))
end if

%>
```

Listing 9-4 is the final version of the `loadxsl.asp` file. It includes a section for i-mode clients. In this code, you are checking for the user's mode so that you can detect the type of device he has. By adding the following section for i-mode clients, you can perform the transformation for i-mode devices:

```
if (session.Contents(1)="imode") then
  sXSL = fname & "imode.xsl"

  xmlDocument.async = false
  xslDocument.async = false

  xmlDocument.load(Server.MapPath(sXML))
  xslDocument.load(Server.MapPath(sXSL))
```

```
Response.ContentType = "text/html"
Response.Write(xmlDocument.transformNode(xslDocument))

end if
```

As you can see in the preceding code, you are setting up a different content type for i-mode browsers as well as setting the value of the sXSL variable in the same way you did in earlier chapters. After setting the content type and the value of the sXSL variable, you carry out the transformation process.

Now look at the file named tweatherimode.xsl (Listing 9-5), which you will use to transform the tweather.xml file for i-mode devices.

Listing 9-5: tweatherimode.xsl

```xml
<?xml version='1.0'?>
<xsl:stylesheet xmlns:xsl="http://www.w3.org/1999/XSL/Transform" version="1.0">
<xsl:output method="html" />

<xsl:template match="/">
<html>

<head>
<META http-equiv="Content-Type" content="text/html; charset=utf-8" />
<META name="CHTML" content="yes" />
<META name="description" content="cHTML document" />
<title>Sample cHTML document</title>
</head>


<body>
 <a href="checkbrows.asp">Back </a>
 <br />
 <xsl:apply-templates select="ROOT" />

</body>
</html>
</xsl:template>


<xsl:template match="ROOT">
Cities
<br />
<xsl:for-each select="city">
<a>
   <xsl:attribute name="href">
   generatexml.asp?city=<xsl:value-of select="." />
   </xsl:attribute>
   <xsl:value-of select="." />
</a>
<br />
</xsl:for-each>
</xsl:template>
</xsl:stylesheet>
```

You are using this relatively simple XSL document for the transformation. The first template provides a link for the application's home page. It then applies a template that contains the code for fetching the records from the XML document. So when this XSL code executes, it will display the names of the cities on the screen. The user can choose a city from the list and browse the weather details for it.

Figure 9-4 is a screen shot of the `tweather.xml` file after the transformation process with the `tweatherimode.xsl` file.

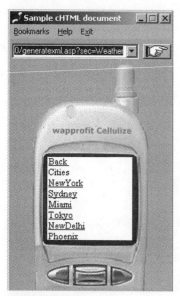

Figure 9-4: The city names

The user selects any city from the list. After the user makes a selection, you again call the `generatexml.asp` file to generate the weather report for that particular city. The `generatexml.asp` file contains the code for generating the weather report for the city the user chooses. You use the following code to generate the weather report for any particular city the user selects in the `generatexml.asp` file:

```
if ((city="Sydney") OR (city="NewDelhi") OR (city="Miami") OR (city="Tokyo") OR
(city="Phoenix") OR (city="NewYork"))   then
  sql = "select * from weather where city = '" & city & "'"
  objrs.Open sql,conn

  Set fso =Server.CreateObject("Scripting.FileSystemObject")
  fname = "tweatherdetail.xml"

  fullpath = server.MapPath("main")

  Set MyFile=fso.CreateTextFile(fullpath & "/" & fname)

  MyFile.WriteLine("<?xml version='1.0'?>")
  MyFile.WriteLine("<ROOT>")


  while not(objrs.EOF)
     lowtemp = objrs.Fields("low_temp")
     hightemp = objrs.Fields("high_temp")
     srise = objrs.Fields("sunrise")
```

```
        sset = objrs.Fields("sunset")
        mrise = objrs.Fields("moonrise")
        mset = objrs.Fields("moonset")
        dhumid = objrs.Fields("dayhumidity")
        nhumid = objrs.Fields("nighthumidity")
        rain = objrs.Fields("rainfall")
        fig = objrs.Fields("figure")

        MyFile.WriteLine("<weather>")
        MyFile.WriteLine("<low_temp>" & lowtemp & "</low_temp>")
        MyFile.WriteLine("<high_temp>" & hightemp & "</high_temp>")
        MyFile.WriteLine("<sunrise>" & srise & "</sunrise>")
        MyFile.WriteLine("<sunset>" & sset & "</sunset>")
        MyFile.WriteLine("<moonrise>" & mrise & "</moonrise>")
        MyFile.WriteLine("<moonset>" & mset & "</moonset>")
        MyFile.WriteLine("<dayhumidity>" & dhumid & "</dayhumidity>")
        MyFile.WriteLine("<nighthumidity>" & nhumid & "</nighthumidity>")
        MyFile.WriteLine("<rainfall>" & rain & "</rainfall>")
        MyFile.WriteLine("<figure>" & fig & "</figure>")
        MyFile.WriteLine("</weather>")
    objrs.MoveNext
    wend

    MyFile.WriteLine("</ROOT>")
    address = "http://192.168.1.100/loadxsl.asp?name=" & fname
    Response.Redirect(address)

end if
```

In the beginning of the code, you check the value in the parameter the device passes to the program and then generate the XML data. You save all the data in a file named `tweatherdetail.xml` on the server using the file system object (FSO). After saving the file on the server, you again call the `loadxsl.asp` file to perform the transformation for displaying the weather report. Figure 9-5 is a screen shot of the `tweatherdetail.xml` file.

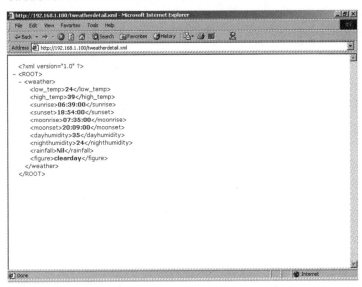

Figure 9-5: tweatherdetail.xml

Listing 9-6 (tweatherdetailimode.xsl) shows the style sheet that you use to transform the tweatherdetail.xml file so that it is i-mode compatible.

Listing 9-6: tweatherdetailimode.xsl

```xml
<?xml version='1.0'?>
<xsl:stylesheet xmlns:xsl="http://www.w3.org/1999/XSL/Transform" version="1.0">
<xsl:output method="html" />

<xsl:template match="/">
<html>

<head>
<META http-equiv="Content-Type" content="text/html; charset=utf-8" />
<META name="CHTML" content="yes" />
<META name="description" content="cHTML document" />
<title>Sample cHTML document</title>
</head>


<body>
 <a href="checkbrows.asp">Back </a>
 <br />
 <xsl:apply-templates select="ROOT" />

</body>
</html>
</xsl:template>

<xsl:template match="ROOT">

  <xsl:for-each select="weather">

 <b>Low Temp-</b>
 <xsl:value-of select="low_temp" />
 <br />
 <b>High Temp-</b>
 <xsl:value-of select="high_temp" />
 <br />
 <b>Sunrise-</b>
 <xsl:value-of select="sunrise" />
 <br />
  <b>Sunset-</b>
 <xsl:value-of select="sunset" />
 <br />
 <b>Moonrise-</b>
 <xsl:value-of select="moonrise" />
 <br />
  <b>Moonset-</b>
 <xsl:value-of select="moonset" />
 <br />
 <b>DayHumidity-</b>
 <xsl:value-of select="dayhumidity" />
```

```
<br />
  <b>NightHumidity-</b>
<xsl:value-of select="nighthumidity" />
<br />
    <b>Rainfall-</b>
<xsl:value-of select="rainfall" />
<br />


</xsl:for-each>
</xsl:template>
</xsl:stylesheet>
```

Let's examine Listing 9-6 step by step so that you understand it completely. While you are starting the document, you use the <XSL:output> element to define the method of the output by using the following line of code:

```
<xsl:output method="html" />
```

After defining the output method, you start the first template in the XSL document. You begin by providing a link for the application's home page and then using <XSL:apply template> to call the template named ROOT.

In the ROOT template, you are fetching all the records from the XML document and formatting them to display on i-mode-compatible browsers. You do so by using the following lines of code:

```
<xsl:template match="ROOT">

  <xsl:for-each select="weather">

<b>Low Temp-</b>
<xsl:value-of select="low_temp" />
<br />
<b>High Temp-</b>
<xsl:value-of select="high_temp" />
<br />
<b>Sunrise-</b>
<xsl:value-of select="sunrise" />
<br />
  <b>Sunset-</b>
<xsl:value-of select="sunset" />
<br />
<b>Moonrise-</b>
<xsl:value-of select="moonrise" />
<br />
  <b>Moonset-</b>
<xsl:value-of select="moonset" />
<br />
<b>DayHumidity-</b>
<xsl:value-of select="dayhumidity" />
<br />
  <b>NightHumidity-</b>
<xsl:value-of select="nighthumidity" />
<br />
    <b>Rainfall-</b>
<xsl:value-of select="rainfall" />
<br />
```

```
</xsl:for-each>
</xsl:template>
```

In this code, you first construct a loop for fetching all the records. You then display the records using the `<XSL:value-of select>` statement. When this code executes using the `loadxsl.asp file`, it shows the output in Figure 9-6 in the i-mode browser.

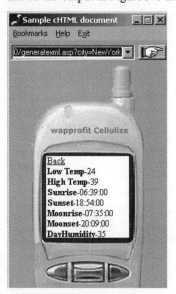

Figure 9-6: The weather report

Transform the News Section

In this section, you transform the news section of the application for i-mode clients. In the news section there are four subsections, called *categories* The user can select any of these and read the news for that category. Figure 9-7 is a diagram of the application's news section's navigation flow.

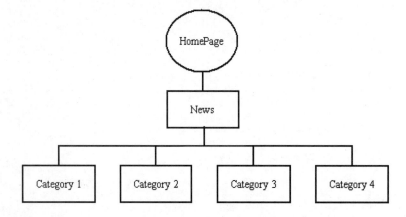

Figure 9-7: The navigation flow of the news section

Generate XML Data for the News Section

First, you have to generate the XML data for the news section's main page to display all the categories to the user so that he can choose one of them. The code for the generatexml.asp file is shown in Listing 9-7.

Listing 9-7: generatexml.asp

```
if (section="News")   then
  sql = "select DISTINCT category from news"
  objrs.Open sql,conn

  Set fso =Server.CreateObject("Scripting.FileSystemObject")
  fname = "tnews.xml"

  fullpath = server.MapPath("main")

  Set MyFile=fso.CreateTextFile(fullpath & "/" & fname)

  MyFile.WriteLine("<?xml version='1.0'?>")
  MyFile.WriteLine("<ROOT>")


  while not(objrs.EOF)

     category = objrs.Fields("category")

     MyFile.WriteLine("<category>" & category & "</category>")

  objrs.MoveNext
  wend

  MyFile.WriteLine("</ROOT>")
  address = "http://192.168.1.100/loadxsl.asp?name=" & fname
  Response.Redirect(address)

end if
```

In Listing 9-7, you select all the unique categories from the database and then save all the data in XML format in a file named tnews.xml on the server, using the file system object (FSO). Having saved all the data, you can use it in the transformation for i-mode clients. Figure 9-8 is a screen shot of tnews.xml.

Write XSL for the News Category Section

After saving all the data in the tnews.xml file, you call the loadxsl.asp file to perform the transformation for i-mode devices. The file you use for the transformation is tnewsimode.xsl (Listing 9-8).

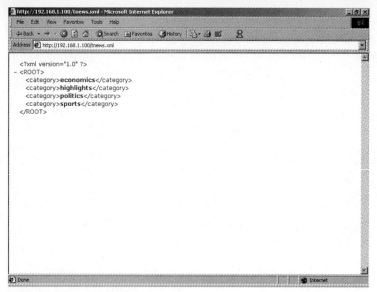

Figure 9-8: tnews.xml

Listing 9-8: tnewsimode.xsl

```
<?xml version='1.0'?>
<xsl:stylesheet xmlns:xsl="http://www.w3.org/1999/XSL/Transform" version="1.0">
<xsl:output method="html" />

<xsl:template match="/">
<html>

<head>
<META http-equiv="Content-Type" content="text/html; charset=utf-8" />
<META name="CHTML" content="yes" />
<META name="description" content="sample cHTML document" />
<title>Sample cHTML document</title>
</head>
<body>
 <a href="checkbrows.asp">Back </a>
 <br />

 <xsl:apply-templates select="ROOT" />
   </body>

</html>
</xsl:template>


<xsl:template match="ROOT">
Sections
<br />
```

```
<xsl:for-each select="category">
<a>
    <xsl:attribute name="href">
    generatexml.asp?category=<xsl:value-of select="." />
    </xsl:attribute>
    <xsl:value-of select="." />
</a>
<br />

</xsl:for-each>
</xsl:template>

</xsl:stylesheet>
```

In Listing 9-8, you first provide a link for the application's home page. Then, by using the ROOT template, you display all the categories in the news section to the user for selection. When the user selects any category, you again call the generatexml.asp file to generate the news for that category. After the transformation, the i-mode browser shows the output in Figure 9-9.

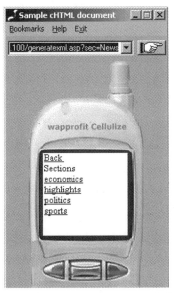

Figure 9-9: The transformation result of tnews.xml

Generate XML for the News Details Section

After the user selects a category of news to read, you generate XML data using the generatexml.asp file. The code for this task is as follows:

```
if ((ctg="highlights") OR (ctg="politics") OR (ctg="economics") OR
(ctg="sports"))   then
  sql = "select * from news where category = '" & ctg & "'"
  objrs.Open sql,conn

  Set fso =Server.CreateObject("Scripting.FileSystemObject")
  fname = "tnewsdetail.xml"
  fullpath = server.MapPath("main")
  Set MyFile=fso.CreateTextFile(fullpath & "/" & fname)
```

```
MyFile.WriteLine("<?xml version='1.0'?>")
MyFile.WriteLine("<ROOT>")

  while not(objrs.EOF)
    heading = objrs.Fields("heading")
    category = objrs.Fields("category")
    description = objrs.Fields("description")
    newsdate = objrs.Fields("news_date")
    MyFile.WriteLine("<news>")
    MyFile.WriteLine("<category>" & category & "</category>")
    MyFile.WriteLine("<heading>" & heading & "</heading>")
    MyFile.WriteLine("<description>" & description & "</description>")
    MyFile.WriteLine("<news_date>" & newsdate & "</news_date>")
    MyFile.WriteLine("</news>")
  objrs.MoveNext
  wend


  MyFile.WriteLine("</ROOT>")
  address = "http://192.168.1.100/loadxsl.asp?name=" & fname
  Response.Redirect(address)

end if
```

In this code, you select all the news from the database in the category the user selects. You then save the data in XML format using the file system object (FSO) on the server in the file named tnewsdetail.xml. After saving the data in an XML file, you now call the loadxsl.asp file to transform the tnewsdetail.xml file (Figure 9-10).

Figure 9-10: tnewsdetail.xml

Preparing XSLT for the News Details Section

In Listing 9-9, you first generate a simple cHTML document and apply a template named ROOT within the <body> tag. In the ROOT template, you construct a loop for getting all the headings and their descriptions from the XML document one by one. You then display all the news on the i-mode simulator (Figure 9-11).

Listing 9-9: tnewsdetailimode.xsl

```
<?xml version='1.0'?>
<xsl:stylesheet xmlns:xsl="http://www.w3.org/1999/XSL/Transform" version="1.0">
<xsl:output method="html" />

<xsl:template match="/">
<html>
<head>
<META http-equiv="Content-Type" content="text/html; charset=utf-8" />
<META name="CHTML" content="yes" />
<META name="description" content="sample cHTML document" />
<title>Sample cHTML document</title>
</head>
<body>
 <a href="checkbrows.asp">Back </a>
 <br />
 <xsl:apply-templates select="ROOT" />
   </body>
</html>
</xsl:template>

<xsl:template match="ROOT">
  <xsl:for-each select="news">

     <b><xsl:value-of select="heading" /></b><br />
       <xsl:value-of select="description" /><br />


  </xsl:for-each>
</xsl:template>

</xsl:stylesheet>
```

Transform the E-mail Section

In this section, we will discuss transforming the e-mail section of the Web portal so that it is i-mode compatible. There are two categories in the e-mail section:

- Read Mail
- Send Mail

By transforming this section, you enable the user to access e-mail messages with his i-mode browser.

Design the Login Form for the E-mail Section

First take a look at the login form for the e-mail section. This section's main page has two options: Read Mail and Send Mail. To start the transformation process and generate XML data, you go back to the generatexml.asp file in Listing 9-10.

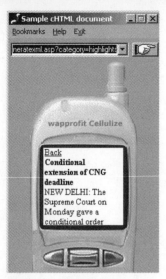

Figure 9-11: Displaying the news

Listing 9-10: generatexml.asp

© 2001 Dreamtech Software India Inc.
All Rights Reserved

```
if (section="E-mail") then

   Set fso =Server.CreateObject("Scripting.FileSystemObject")
   fname = "tlogin.xml"

   fullpath = server.MapPath("main")
   Set MyFile=fso.CreateTextFile(fullpath & "/" & fname)

   MyFile.WriteLine("<?xml version='1.0'?>")
   MyFile.WriteLine("<logininfo>")
   MyFile.WriteLine("<username />")
   MyFile.WriteLine("<password />")
   MyFile.WriteLine("<login />")
   MyFile.WriteLine("</logininfo>")

   address = "http://192.168.1.100/loadxsl.asp?name=" & fname
   Response.Redirect(address)
end if
```

In Listing 9-10, you create a file on the server using the file system object named `tlogin.xml` and then write some XML data in that file. Figure 9-12 is a screen shot of `tlogin.xml`.

After creating the file on the server, you redirect control to the `loadxsl.asp` file to load the corresponding XSL file for the transformation for i-mode clients.

Prepare an XSLT E-mail Home Page

To transform the e-mail section's main page, you use the `tloginimode.xsl` file (Listing 9-11).

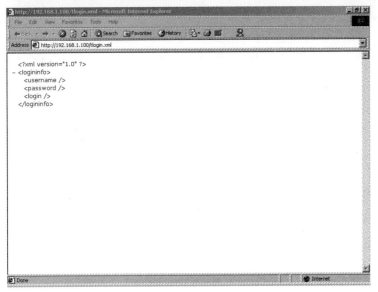

Figure 9-12: tlogin.xml

Listing 9-11: tloginimode.xsl

© 2001 Dreamtech Software India Inc.
All Rights Reserved

```xml
<?xml version='1.0'?>
<xsl:stylesheet xmlns:xsl="http://www.w3.org/1999/XSL/Transform" version="1.0">
<xsl:output method="html" />

<xsl:template match="/">
<html>

<head>
<META http-equiv="Content-Type" content="text/html; charset=utf-8" />
<META name="CHTML" content="yes" />
<META name="description" content="sample cHTML document" />
<title>Sample cHTML document</title>
</head>
    <body>
 <a href="checkbrows.asp">Back </a>
 <br />

  <a  href="generatexml.asp?mode=read">Read Mail</a>
  <br />
  <a  href="generatexml.asp?mode=send">Send Mail</a>
   </body>

</html>
</xsl:template>
</xsl:stylesheet>
```

In Listing 9-11, you generate the simple cHTML document on-the-fly as a result of the transformation process. You use some <META> elements in the beginning of the code so that the browser can identify the document's content type.

After defining the <META> elements, you generate the body of the document and use the <href> element to provide two links for both of the categories in the e-mail section.

When this code executes, the output in Figure 9-13 is shown on the i-mode browser.

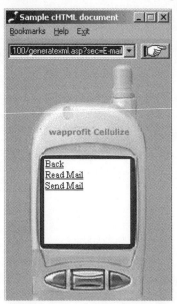

Figure 9-13: The transformation result of tlogin.xml

When the user selects either of the two available options, you again call the generatexml.asp file with the MODE parameter, which contains the value of the user's choice.

Read Mail Using i-mode Clients

When the user selects the Read Mail option from the available list, you again generate XML data using the generatexml.asp file to ask for the username and password for the authentication process. You use the following piece of code in the generatexml.asp file:

```
if (emode <> " ") then

   Set fso =Server.CreateObject("Scripting.FileSystemObject")
   fname = "tlogin.xml"

   fullpath = server.MapPath("main")
   Set MyFile=fso.CreateTextFile(fullpath & "/" & fname)

   MyFile.WriteLine("<?xml version='1.0'?>")
   MyFile.WriteLine("<logininfo>")
   MyFile.WriteLine("<username />")
   MyFile.WriteLine("<password />")
   MyFile.WriteLine("<login />")
   MyFile.WriteLine("</logininfo>")

   if (session.Contents(1)="wml") then
   address = "http://192.168.1.100/loadmailwml.asp?name=" & fname & "&mode=" &
emode
   end if
   if (session.Contents(1)="imode") then
```

```
       address = "http://192.168.1.100/loadmailimode.asp?name=" & fname & "&mode=" &
emode
  end if
  if (session.Contents(1)="hdml") then
    address = "http://192.168.1.100/loadmailhdml.asp?name=" & fname & "&mode=" &
emode
  end if
  Response.Redirect(address)
end if
```

In the preceding code, you again create the `tlogin.xml` file on the server with the same XML data structure. Then you redirect control to the `loadmailimode.asp` file with the filename and mode name as the parameter in the query string.

```
  if (session.Contents(1)="imode") then
    address = "http://192.168.1.100/loadmailimode.asp?name=" & fname & "&mode=" &
emode
  end if
```

Collect User Information

The `loadmailimode.asp` file (Listing 9-12) checks for the option the user selects. On the basis of the selected option, it then loads the XSL file. If the user selects read mode, you load the `readimode.xsl` file to transform the `tlogin.xml` file on the server.

Listing 9-12: loadmailimode.asp

© 2001 Dreamtech Software India Inc.
All Rights Reserved

```
<%@ Language=VBScript %>
<%
fname = trim(Request.QueryString("name"))
emode = trim(Request.QueryString("mode"))
'declaring variables
dim client
dim sXML
dim sXSL
dim xmlDocument,xslDocument
'creating the objects for the XMLDOM
set xmlDocument = Server.CreateObject("MSXML2.DOMDocument")
set xslDocument = Server.CreateObject("MSXML2.DOMDocument")

sXML = fname
fname = mid(fname,1,(len(fname) - 4))

if (emode="read") then
 sXSL = "readimode.xsl"
end if

if (emode="send") then
 sXSL = "sendimode.xsl"
end if
xmlDocument.async = false
xslDocument.async = false
xmlDocument.load(Server.MapPath(sXML))
xslDocument.load(Server.MapPath(sXSL))
```

```
Response.ContentType = "text/html"
Response.Write(xmlDocument.transformNode(xslDocument))
%>
```

Prepare XSLT for the Login Form

Now consider the code in the `readimode.xsl` file (Listing 9-13). In this file, you generate the form for submitting the username and password and then post the data to the server for the authentication process.

Listing 9-13: readimode.xsl

```xml
<?xml version='1.0'?>
<xsl:stylesheet xmlns:xsl="http://www.w3.org/1999/XSL/Transform" version="1.0">
<xsl:output method="html" />

<xsl:template match="/">
<html>

<head>
<META http-equiv="Content-Type" content="text/html; charset=utf-8" />
<META name="CHTML" content="yes" />
<META name="description" content="cHTML document" />
<title>Email document</title>
</head>
<body bgcolor="#ffffff" text="#000000">
<p>
<xsl:apply-templates select="logininfo" />
</p>
</body>

</html>
</xsl:template>


<xsl:template match="logininfo">
<FORM method="get" action="http://192.168.1.100/receiveimode.asp">
<xsl:for-each select="username">
 Enter your User Name
  <INPUT type="text" name="username" size="14" maxlength="20" /><BR />
</xsl:for-each>
<xsl:for-each select="password">
  Enter your Password
  <INPUT type="password" name="password" size="14" maxlength="20" /><BR />

</xsl:for-each>
<INPUT type="submit" name="submit" value="Submit" /><BR />
</FORM>

</xsl:template>

</xsl:stylesheet>
```

In Listing 9-13, you generate a cHTML document that contains the form for collecting the username and password with the `<form>` element. This data is then posted to the server using the `<submit>` element, as in the following lines:

```
<xsl:template match="logininfo">
<FORM method="get" action="http://192.168.1.100/receiveimode.asp">
<xsl:for-each select="username">
 Enter your User Name
   <INPUT type="text" name="username" size="14" maxlength="20" /><BR />
</xsl:for-each>
<xsl:for-each select="password">
   Enter your Password
   <INPUT type="password" name="password" size="14" maxlength="20" /><BR />

</xsl:for-each>
<INPUT type="submit" name="submit" value="Submit" /><BR />
</FORM>

</xsl:template>
```

When the user fills in this form and presses the Submit button, you call the `receiveimode.asp` file and post all the data to the `receiveimode.asp` file using the GET method.

When this transformation takes place, it shows the output in Figure 9-14.

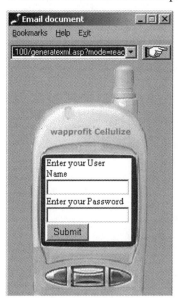

Figure 9-14: The transformation result of readimode.xsl

Display the Inbox on i-mode Devices

When the user logs in and presses the Submit button to proceed, you call the `receiveimode.asp` file (Listing 9-14) on the server.

Listing 9-14: receiveimode.asp

```
<%@LANGUAGE=VBSCRIPT %>
<%
uname = trim(Request.QueryString("username"))
```

```
pwd = trim(Request.QueryString("password"))
Response.ContentType = "text/html"

%>
<html>

<head>
<META http-equiv="Content-Type" content="text/html; charset=utf-8" />
<META name="CHTML" content="yes" />
<META name="description" content="CHTML document" />
<title>CHTML document</title>
</head>
<body bgcolor="#ffffff" text="#000000">
<a href="checkbrows.asp">Back</a>
<p>
<b><i>Inbox</i></b><br />--------<br />
<%
Set pop3 = Server.CreateObject("Jmail.pop3")
pop3.Connect uname, pwd, "http://192.168.1.100"

c = 1
if (pop3.Count>0) then
while (c<=pop3.Count)
 Set msg = pop3.Messages.item(c)

 ReTo = ""
 ReCC = ""

 Set Recipients = msg.Recipients
 delimiter = ", "

 For i = 0 To Recipients.Count - 1
  If i = Recipients.Count - 1 Then
   delimiter = ""
  End If

  Set re = Recipients.item(i)
  If re.ReType = 0 Then
   ReTo = ReTo & re.Name & " (" & re.EMail & ")" & delimiter
  else
   ReCC = ReTo & re.Name & " (" & re.EMail & ")" & delimiter
  End If
 Next

%>

<b>Subject</b><br />
<%=msg.Subject %> <br />

  <b>From</b><br />
  <%=msg.FromName %><br />

 <b>Body</b><br />
  <%=msg.Body %><br />
 <%
```

```
c = c + 1
wend
end if

pop3.Disconnect

%>
</p>
</body>
</html>
```

In the beginning of Listing 9-14, you declare two variables for holding the values that you get from the calling file. After declaring the variables, you start generating the cHTML document and define some <META> elements in it. You then connect to the mail server by providing the username and password. Then you fetch all the records from the inbox using the loop and display them on the browser using the following lines of code:

```
Set pop3 = Server.CreateObject("Jmail.pop3")
pop3.Connect uname, pwd, "http://192.168.1.100"

c = 1
if (pop3.Count>0) then
while (c<=pop3.Count)
 Set msg = pop3.Messages.item(c)

 ReTo = ""
 ReCC = ""

 Set Recipients = msg.Recipients
 delimiter = ", "

 For i = 0 To Recipients.Count - 1
  If i = Recipients.Count - 1 Then
    delimiter = ""
  End If

  Set re = Recipients.item(i)
  If re.ReType = 0 Then
    ReTo = ReTo & re.Name & " (" & re.EMail & ")" & delimiter
  else
    ReCC = ReTo & re.Name & " (" & re.EMail & ")" & delimiter
  End If
 Next

%>

<b>Subject</b><br />
<%=msg.Subject %> <br />

  <b>From</b><br />
  <%=msg.FromName %><br />

 <b>Body</b><br />
  <%=msg.Body %><br />
 <%

c = c + 1
```

```
wend
end if
```

When you finish displaying the messages, you disconnect from the mail server using the following line of code:

```
pop3.Disconnect
```

When this code runs, it shows the output in Figure 9-15.

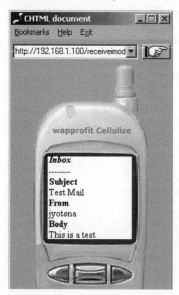

Figure 9-15: The ouptut of receiveimode.asp

Transform the Compose Mail Section

When you transform the compose mail section of the application, the user can compose mail messages on i-mode-compatible browsers and devices. When the user selects the Send Mail option on the e-mail system's main page, the following piece of code in `loadmailimode.asp` comes into action:

```
if (emode="send") then
 sXSL = "sendimode.xsl"
end if
```

This code loads the `sendimode.xsl` file for generating the login form that collects the username and password from the user for the authentication process.

Write XSLT for the User-Authentication Form

Now you write an XSLT document (`sendimode.xsl`, in Listing 9-15) for generating the login for the send mail section of the e-mail system.

Listing 9-15: sendimode.xsl

```
<?xml version='1.0'?>
<xsl:stylesheet xmlns:xsl="http://www.w3.org/1999/XSL/Transform" version="1.0">
<xsl:output method="html" />
```

```
<xsl:template match="/">
<html>
<head>
<META http-equiv="Content-Type" content="text/html; charset=utf-8" />
<META name="CHTML" content="yes" />
<META name="description" content="cHTML document" />
<title>Email document</title>
</head>
<body bgcolor="#ffffff" text="#000000">
<p>
<xsl:apply-templates select="logininfo" />
</p>
</body>
</html>
</xsl:template>

<xsl:template match="logininfo">
<FORM method="get" action="http://192.168.1.100/sendimode.asp">

<xsl:for-each select="username">
  Enter your User Name
  <INPUT type="text" name="username" /><BR />
</xsl:for-each>

<xsl:for-each select="password">
 Enter your Password
 <INPUT type="password" name="password" /><BR />
</xsl:for-each>
<INPUT type="submit" value="Submit" /><BR />

</FORM>
</xsl:template>
</xsl:stylesheet>
```

In Listing 9-15, you generate a simple cHTML document containing a simple form for collecting the username and password. You then post all the data on the server using the cHTML `<submit>` element. You can do this easily using the following lines of code:

```
<xsl:template match="logininfo">
<FORM method="get" action="http://192.168.1.100/sendimode.asp">

<xsl:for-each select="username">
  Enter yourUser Name
  <INPUT type="text" name="username" /><BR />
</xsl:for-each>

<xsl:for-each select="password">
 Enter your Password
 <INPUT type="password" name="password" /><BR />
</xsl:for-each>
<INPUT type="submit" value="Submit" /><BR />

</FORM>
</xsl:template>
```

When the user fills in all the required information and presses the Submit button to proceed, you call the `sendimode.asp` file (Listing 9-16) and post all the data on the server.

Listing 9-16: sendimode.asp

```
<%@LANGUAGE=VBSCRIPT %>

<%
uname = trim(Request.QueryString("username"))
pwd = trim(Request.QueryString("password"))

Response.ContentType = "text/html"

%>
<html>
<head>
<META http-equiv="Content-Type" content="text/html; charset=utf-8" />
<META name="CHTML" content="yes" />
<META name="description" content="CHTML document" />
<title>CHTML document</title>
</head>
<body bgcolor="#ffffff" text="#000000">

  <p>
  <form method="get" action="http://192.168.1.100/sendmailimode.asp">

  Enter recipient's E-mail address<BR />
  <input type="text" name="remail" /><BR />

  Enter the subject of E-mail <BR />
  <input type="text" name="subject" /><BR />

  Enter the body of E-mail<BR />
  <input type="text" name="body" /><BR />
  <input type="submit" value="Submit" /><BR />
  <input type="hidden" name="sender" value="<%=uname%>" />
  </form>
  </p>

</body>
</html>
```

In Listing 9-16, you generate a cHTML document containing a `<form>` element and some `<input>` elements for collecting the information about the message from the user. You ask the user for the subject, the recipient's address, and the message by using the following lines of code:

```
  <form method="get" action="http://192.168.1.100/sendmailimode.asp">

  Enter recipient's E-mail address<BR />
  <input type="text" name="remail" /><BR />

  Enter the subject of E-mail <BR />
  <input type="text" name="subject" /><BR />

  Enter the body of E-mail<BR />
  <input type="text" name="body" /><BR />
  <input type="submit" value="Submit" /><BR />
```

```
    <input type="hidden" name="sender" value="<%=uname%>" />
    </form>
```

When the user fills in all the information and presses Submit, you call the `sendmailimode.asp` file (shown in the next section) to send the mail message using the mail server. You then display a confirmation to the user. When this code executes, it shows the output in Figure 9-16.

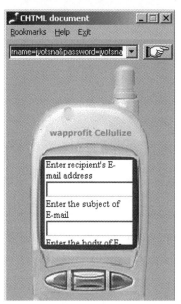

Figure 9-16: The output of sendimode.asp

Send Mail and Display Confirmation

Listing 9-17 shows the code for `sendmailimode.asp`.

Listing 9-17: sendmailimode.asp

```asp
<%@LANGUAGE = VBSCRIPT%>
<%

senderEmail = trim(Request.QueryString("sender"))
subject     = trim(Request.QueryString("subject"))
recipient   = trim(Request.QueryString("remail"))
body = trim(Request.QueryString("body"))

Response.ContentType = "text/html"
 %>

<html>
<head>
<META http-equiv="Content-Type" content="text/html; charset=utf-8" />
<META name="CHTML" content="yes" />
<META name="description" content="CHTML document" />
<title>Sample cHTML document</title>
</head>
<body bgcolor="#ffffff" text="#000000">
```

```
<p>
<%
' Create the JMail message Object
set msg = Server.CreateOBject( "JMail.Message" )


msg.From = senderEmail
msg.AddRecipient recipient
msg.Subject = subject
msg.body = body


if not msg.Send("192.168.1.100") then
    Response.write msg.log
else
    Response.write "Message sent successfully!"
end if


%>
</p>
<a href="checkbrows.asp">Back</a>
</body>
 </html>
```

In the beginning of Listing 9-17, you declare some variables and store the values that you get from the `sendimode.asp` file as follows:

```
senderEmail = trim(Request.QueryString("sender"))
subject     = trim(Request.QueryString("subject"))
recipient   = trim(Request.QueryString("remail"))
body = trim(Request.QueryString("body"))
```

In the next part of the code, you start generating the cHTML document. For this, you create a "`Jmail.Message`" object on the server and use it to send the message. If there is an error while sending the mail message, you display a message to the user; otherwise, you display a confirmation message that the message was sent successfully. The following lines of code do this:

```
set msg = Server.CreateOBject( "JMail.Message" )
msg.From = senderEmail
msg.AddRecipient recipient
msg.Subject = subject
msg.body = body


if not msg.Send("192.168.1.100") then
    Response.write msg.log
else
    Response.write "Message sent successfully!"
end if
```

When this code executes, it shows the output in Figure 9-17.

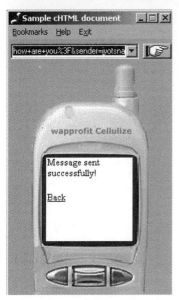

Figure 9-17: Confirmation of successful message sending

There is only one more section of the application that you need to transform for compatibility with i-mode clients: the movie ticket booking system.

Transform the Movie Ticket Booking System

By using the movie ticket booking system, the user can browse through the movie listings for the local theaters, as well as book movie tickets online using any kind of i-mode-enabled device.

Generate XML Data for Theater Listings

When the user accesses the movies section of the application, the system begins by displaying all the names of the theaters in the database. This enables the user to choose any of the theaters listed in the browser. To generate XML data for the theater listings, you use the code from the generatexml.asp file as shown in Listing 9-18.

Listing 9-18: generatexml.asp

© 2001 Dreamtech Software India Inc.
All Rights Reserved

```
if (section="Movies") then

    sql = "select * from hall"
    objrs.Open sql,conn

    Set fso =Server.CreateObject("Scripting.FileSystemObject")
    fname = "tmovies.xml"

    fullpath = server.MapPath("main")

    Set MyFile=fso.CreateTextFile(fullpath & "/" & fname)

    MyFile.WriteLine("<?xml version='1.0'?>")
    MyFile.WriteLine("<ROOT>")
```

```
while not(objrs.EOF)
   hallid = objrs.Fields("hall_id")
   hallname = objrs.Fields("hall_name")
   add = objrs.Fields("address")
   MyFile.WriteLine("<hall>")
   MyFile.WriteLine("<hall_id>" & hallid & "</hall_id>")
   MyFile.WriteLine("<hall_name>" & hallname & "</hall_name>")
   MyFile.WriteLine("<address>" & add & "</address>")
   MyFile.WriteLine("</hall>")
objrs.MoveNext
wend

MyFile.WriteLine("</ROOT>")
address = "http://192.168.1.100/loadxsl.asp?name=" & fname
Response.Redirect(address)

end if
```

In Listing 9-18, you first check for the section the user selects and then fetch the records accordingly from the HALL table. You save all the records in XML format in the file named tmovies.xml using the file system object. In the next step, you transfer control to the loadxml.asp file to load the corresponding XSL document for the transformation of tmovies.xml. Figure 9-18 is a screen shot of tmovies.xml.

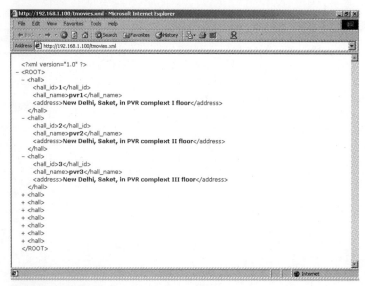

Figure 9-18: tmovies.xml

Write the XSLT Document for Theater Listings

Listing 9-19 shows the XSLT document for the transformation of tmovies.xml, which makes the data accessible to i-mode users.

Listing 9-19: tmoviesimode.xsl

```
<?xml version='1.0'?>
<xsl:stylesheet xmlns:xsl="http://www.w3.org/1999/XSL/Transform" version="1.0">
<xsl:output method="html" />
<xsl:template match="/">
<html>

<head>
<META http-equiv="Content-Type" content="text/html; charset=utf-8" />
<META name="CHTML" content="yes" />
<META name="description" content="cHTML document" />
<title>Movie document</title>
</head>
<body>

<p>
<xsl:apply-templates select="ROOT" />
</p>

</body>
</html>
</xsl:template>


<xsl:template match="ROOT">
<b>select a theatre name</b>
<br />

<xsl:for-each select="hall">
   <a>
 <xsl:attribute name="href">
 loadmovieimode.asp?hall=<xsl:value-of select="hall_name" />
 </xsl:attribute>
 <xsl:value-of select="hall_name" />

   </a>
   <br />

</xsl:for-each>
</xsl:template>
</xsl:stylesheet>
```

In Listing 9-19, you generate a cHTML document because of the transformation process of
tmovies.xml. At the start of the listing, you define some <META> elements and then apply them in a
template named "ROOT" using the <xsl:template> element.

In the "ROOT" template, you are constructing a loop to fetch all the values from the XML document, as
well as the names of all the theaters. You can do this easily by using the following lines of code:

```
<xsl:for-each select="hall">
   <a>
 <xsl:attribute name="href">
 loadmovieimode.asp?hall=<xsl:value-of select="hall_name" />
 </xsl:attribute>
 <xsl:value-of select="hall_name" />

   </a>
   <br />
```

```
</xsl:for-each>
```

When this transformation takes place on the server, it shows the output in Figure 9-19.

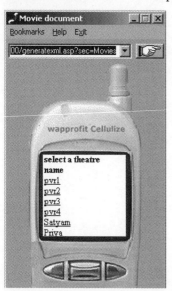

Figure 9-19: The transformation results of tmovies.xml

Display the Movie List

The user selects the theater of his choice to browse the movies that are currently being screened there. Now you have to generate XML data for the movie listing at the theater the user selected. For this task, you use the `loadmovieimode.asp` file (Listing 9-20).

Listing 9-20: loadmovieimode.asp

© 2001 Dreamtech Software India Inc.
All Rights Reserved

```
<%@ Language=VBScript %>

<%
dim address,section,fname

hallname = trim(Request.QueryString("hall"))

set conn = Server.CreateObject("ADODB.Connection")
conn.ConnectionString = "Provider=SQLOLEDB.1;Persist Security Info=False;User
ID=sa;Initial Catalog=portal;Data Source=developers"
conn.Open

Set objrs = Server.CreateObject("ADODB.Recordset")
objrs.CursorType = adOpenStatic

Set rsmovie = Server.CreateObject("ADODB.Recordset")
rsmovie.CursorType = adOpenStatic

Set rsstatus = Server.CreateObject("ADODB.Recordset")
```

```
rsstatus.CursorType = adOpenStatic
Set rshall = Server.CreateObject("ADODB.Recordset")
rshall.CursorType = adOpenStatic

  sqlhall = "select hall_id from hall where hall_name='" & hallname & "'"
  rshall.Open sqlhall,conn

  while not (rshall.EOF)
   id = rshall.Fields("hall_id")

  rshall.MoveNext
  wend
  sqlmovies = "select DISTINCT(status.movie_id),movie.movie_name from
movie,status where status.movie_id=movie.movie_id and status.hall_id=" & id
  rsmovie.Open sqlmovies,conn

  Set fso =Server.CreateObject("Scripting.FileSystemObject")
  fname = "tmovielist.xml"

  fullpath = server.MapPath("main")

  Set MyFile=fso.CreateTextFile(fullpath & "/" & fname)

  MyFile.WriteLine("<?xml version='1.0'?>")
  MyFile.WriteLine("<ROOT>")
  MyFile.WriteLine("<hall>")
  MyFile.WriteLine("<hall_name>" & hallname & "</hall_name>")
  MyFile.WriteLine("<hall_id>" & id & "</hall_id>")

  while not (rsmovie.EOF)
   MyFile.WriteLine("<movie>")
   moviename = rsmovie.Fields("movie_name")
   MyFile.WriteLine("<movie_name>" & moviename & "</movie_name>")
   movieid = rsmovie.Fields("movie_id")
   MyFile.WriteLine("<movie_id>" & movieid & "</movie_id>")
   sqlstatus = "select
movie_date,showtime,balcony,rear_stall,middle_stall,upper_stall from status
where movie_id=" & movieid & "and hall_id=" & id
   rsstatus.Open sqlstatus,conn
     while not (rsstatus.EOF)
      MyFile.WriteLine("<status>")
      moviedate = rsstatus.Fields("movie_date")
      showtime = rsstatus.Fields("showtime")
      balcony = rsstatus.Fields("balcony")
      rearstall = rsstatus.Fields("rear_stall")
      middlestall = rsstatus.Fields("middle_stall")
      upperstall = rsstatus.Fields("upper_stall")
      MyFile.WriteLine("<movie_date>" & moviedate & "</movie_date>")
      MyFile.WriteLine("<showtime>" & showtime & "</showtime>")
      MyFile.WriteLine("<balcony>" & balcony & "</balcony>")
      MyFile.WriteLine("<rear_stall>" & rearstall & "</rear_stall>")
      MyFile.WriteLine("<middle_stall>" & middlestall & "</middle_stall>")
      MyFile.WriteLine("<upper_stall>" & upperstall & "</upper_stall>")
      MyFile.WriteLine("</status>")
      rsstatus.MoveNext
     wend
```

```
MyFile.WriteLine("</movie>")
rsstatus.Close
rsmovie.MoveNext
wend
MyFile.WriteLine("</hall>")
MyFile.WriteLine("</ROOT>")


conn.Close

address = "http://192.168.1.100/loadxsl.asp?name=" & fname
Response.Redirect(address)
%>
```

Now let's see how Listing 9-20 performs in a real-world context by examining it fully. Consider the following segment of `loadmovieimode.asp`.

```
dim address,section,fname

hallname = trim(Request.QueryString("hall"))

set conn = Server.CreateObject("ADODB.Connection")
conn.ConnectionString = "Provider=SQLOLEDB.1;Persist Security Info=False;User
ID=sa;Initial Catalog=portal;Data Source=developers"
conn.Open

Set objrs = Server.CreateObject("ADODB.Recordset")
objrs.CursorType = adOpenStatic

Set rsmovie = Server.CreateObject("ADODB.Recordset")
rsmovie.CursorType = adOpenStatic

Set rsstatus = Server.CreateObject("ADODB.Recordset")
rsstatus.CursorType = adOpenStatic
Set rshall = Server.CreateObject("ADODB.Recordset")
rshall.CursorType = adOpenStatic

  sqlhall = "select hall_id from hall where hall_name='" & hallname & "'"
  rshall.Open sqlhall,conn
```

In the first line of the code, you declare a variable named `hallname` to store the name of the theater passed to you by the previous file. In the next step, you establish a connection with the database. In the last few lines, you select the `hall_id` based on the theater name passed to you from the HALL table. In the next few lines, you select the `movie_name` and `movie_id` of the movies currently on the screen in that theater by providing the following SQL statement:

```
  sqlmovies = "select DISTINCT(status.movie_id),movie.movie_name  from
movie,status where status.movie_id=movie.movie_id and status.hall_id=" & id
  rsmovie.Open sqlmovies,conn
```

After collecting all the records from the database, you save all the data in XML format in a file named `tmovielist.xml` on the server (with the help of the file system object) by using the following lines of code:

```
  Set fso =Server.CreateObject("Scripting.FileSystemObject")
  fname = "tmovielist.xml"

  fullpath = server.MapPath("main")
```

```
Set MyFile=fso.CreateTextFile(fullpath & "/" & fname)

MyFile.WriteLine("<?xml version='1.0'?>")
MyFile.WriteLine("<ROOT>")
MyFile.WriteLine("<hall>")
MyFile.WriteLine("<hall_name>" & hallname & "</hall_name>")
MyFile.WriteLine("<hall_id>" & id & "</hall_id>")

while not (rsmovie.EOF)
  MyFile.WriteLine("<movie>")
  moviename = rsmovie.Fields("movie_name")
  MyFile.WriteLine("<movie_name>" & moviename & "</movie_name>")
  movieid = rsmovie.Fields("movie_id")
  MyFile.WriteLine("<movie_id>" & movieid & "</movie_id>")
  sqlstatus = "select
movie_date,showtime,balcony,rear_stall,middle_stall,upper_stall  from status
where movie_id=" & movieid & "and hall_id=" & id
    rsstatus.Open sqlstatus,conn
    while not (rsstatus.EOF)
      MyFile.WriteLine("<status>")
      moviedate = rsstatus.Fields("movie_date")
      showtime = rsstatus.Fields("showtime")
      balcony = rsstatus.Fields("balcony")
      rearstall = rsstatus.Fields("rear_stall")
      middlestall = rsstatus.Fields("middle_stall")
      upperstall = rsstatus.Fields("upper_stall")
      MyFile.WriteLine("<movie_date>" & moviedate & "</movie_date>")
      MyFile.WriteLine("<showtime>" & showtime & "</showtime>")
      MyFile.WriteLine("<balcony>" & balcony & "</balcony>")
      MyFile.WriteLine("<rear_stall>" & rearstall & "</rear_stall>")
      MyFile.WriteLine("<middle_stall>" & middlestall & "</middle_stall>")
      MyFile.WriteLine("<upper_stall>" & upperstall & "</upper_stall>")
      MyFile.WriteLine("</status>")
      rsstatus.MoveNext
    wend
  MyFile.WriteLine("</movie>")
  rsstatus.Close
  rsmovie.MoveNext
wend
MyFile.WriteLine("</hall>")
MyFile.WriteLine("</ROOT>")
```

Figure 9-20 is a screen shot of `tmovielist.xml`.

Write XSLT for the Movie Listings

Now you will prepare an XSLT document for the transformation of the `tmovielist.xml` file for i-mode clients. This will allow i-mode browsers to display the list of the movies on their screens. Consider Listing 9-21.

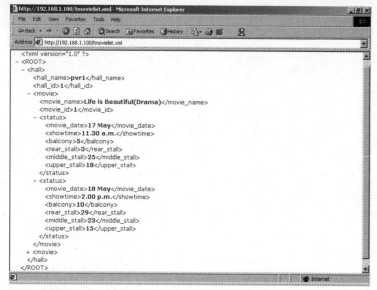

Figure 9-20: tmovielist.xml

Listing 9-21: tmovielistimode.xsl

© 2001 Dreamtech Software India Inc.
All Rights Reserved

```xml
<?xml version='1.0'?>
<xsl:stylesheet xmlns:xsl="http://www.w3.org/1999/XSL/Transform" version="1.0">
<xsl:output method="html" />


<xsl:template match="/">
<html>


<head>
<META http-equiv="Content-Type" content="text/html; charset=utf-8" />
<META name="CHTML" content="yes" />
<META name="description" content="cHTML document" />
<title>Movie document</title>
</head>
 <body>
 <a href="browsimode.asp">Back </a>
 <br />
 <p>
  <xsl:apply-templates select="ROOT" />
 </p>
</body>
</html>
</xsl:template>
<xsl:template match="ROOT">
<xsl:variable name="hallid" select="hall/hall_id" />
<b>List of movies on <xsl:value-of select="hall/hall_name" /></b><br />
<xsl:for-each select="hall/movie">
<br />
<a>
```

```
<xsl:attribute name="href">tmoviestatusimode.asp?hallid=<xsl:value-of
select="$hallid" />&amp;movieid=<xsl:value-of select="./movie_id"
/></xsl:attribute>
<xsl:value-of select="./movie_name" />
</a>
</xsl:for-each>
</xsl:template>
</xsl:stylesheet>
```

In the beginning of the code, you generate a cHTML document. In the body, you apply a template named "ROOT" using the <xsl:apply-template> element.

In the "ROOT" template, you first declare a variable named "hallid" using the following line of code:

```
<xsl:variable name="hallid" select="hall/hall_id" />
```

In next few lines, you construct a loop and fetch the names of all the movies from the XML document with the help of the <xsl:value-of select> element on the basis of the "hallid". The following lines of code help you do so:

```
 <xsl:for-each select="hall/movie">
<br />
<a>
<xsl:attribute name="href">tmoviestatusimode.asp?hallid=<xsl:value-of
select="$hallid" />&amp;movieid=<xsl:value-of select="./movie_id"
/></xsl:attribute>
<xsl:value-of select="./movie_name" />
</a>
</xsl:for-each>
```

When this code executes, it shows the output in Figure 9-21.

Figure 9-21: The transformation result of tmovielist.xml

Display the Movie Status

When the user clicks any of the movie names displayed on the screen, the program calls the tmoviestatusimode.asp file (Listing 9-22) to check the status of reservations at the selected theater.

Listing 9-22: tmoviestatusimode.asp

```
<%@ Language=VBScript %>

<%
dim address,section,fname

hallid = trim(Request.QueryString("hallid"))
movieid = trim(Request.QueryString("movieid"))

set conn = Server.CreateObject("ADODB.Connection")
conn.ConnectionString = "Provider=SQLOLEDB.1;Persist Security Info=False;User
ID=sa;Initial Catalog=portal;Data Source=developers"
conn.Open

Set objrs = Server.CreateObject("ADODB.Recordset")
objrs.CursorType = adOpenStatic


Set rsstatus = Server.CreateObject("ADODB.Recordset")
rsstatus.CursorType = adOpenStatic

  sqlmovies = "select
movie_date,showtime,balcony,rear_stall,middle_stall,upper_stall  from status
where movie_id=" & movieid & "and hall_id=" & hallid
  rsstatus.Open sqlmovies,conn

  Set fso =Server.CreateObject("Scripting.FileSystemObject")
  fname = "tmoviestatus.xml"

  fullpath = server.MapPath("main")

  Set MyFile=fso.CreateTextFile(fullpath & "/" & fname)

  MyFile.WriteLine("<?xml version='1.0'?>")
  MyFile.WriteLine("<ROOT>")
  MyFile.WriteLine("<hall_id>" & hallid & "</hall_id>")
  MyFile.WriteLine("<movie_id>" & movieid & "</movie_id>")

  while not (rsstatus.EOF)

      MyFile.WriteLine("<status>")
      moviedate = rsstatus.Fields("movie_date")
      showtime = rsstatus.Fields("showtime")
      balcony = rsstatus.Fields("balcony")
      rearstall = rsstatus.Fields("rear_stall")
      middlestall = rsstatus.Fields("middle_stall")
      upperstall = rsstatus.Fields("upper_stall")
      MyFile.WriteLine("<movie_date>" & moviedate & "</movie_date>")
      MyFile.WriteLine("<showtime>" & showtime & "</showtime>")
      MyFile.WriteLine("<balcony>" & balcony & "</balcony>")
      MyFile.WriteLine("<rear_stall>" & rearstall & "</rear_stall>")
      MyFile.WriteLine("<middle_stall>" & middlestall & "</middle_stall>")
```

```
      MyFile.WriteLine("<upper_stall>" & upperstall & "</upper_stall>")
      MyFile.WriteLine("</status>")
      rsstatus.MoveNext
   wend

   rsstatus.Close
   MyFile.WriteLine("</ROOT>")
   conn.Close


   address = "http://192.168.1.100/loadxsl.asp?name=" & fname
   Response.Redirect(address)
%>
```

The preceding code is divided into two main parts. In the first part, you declare two variables for storing the values of "hallid" and "movieid" that you get from the previous file. After declaring the variables, you establish a connection with the database and fetch the records from the STATUS table using the following lines of code:

```
dim address,section,fname

hallid = trim(Request.QueryString("hallid"))
movieid = trim(Request.QueryString("movieid"))

set conn = Server.CreateObject("ADODB.Connection")
conn.ConnectionString = "Provider=SQLOLEDB.1;Persist Security Info=False;User
ID=sa;Initial Catalog=portal;Data Source=developers"
conn.Open

Set objrs = Server.CreateObject("ADODB.Recordset")
objrs.CursorType = adOpenStatic
Set rsstatus = Server.CreateObject("ADODB.Recordset")
rsstatus.CursorType = adOpenStatic

  sqlmovies = "select
movie_date,showtime,balcony,rear_stall,middle_stall,upper_stall  from status
where movie_id=" & movieid & "and hall_id=" & hallid
  rsstatus.Open sqlmovies,conn
```

In the second part of the code, you store all the collected data in XML format in the file named tmoviestatus.xml by using the following code. In the last few lines of code, you transfer control to the loadxsl.asp file to load the corresponding XSLT document for the transformation.

```
   Set fso =Server.CreateObject("Scripting.FileSystemObject")
   fname = "tmoviestatus.xml"
   fullpath = server.MapPath("main")
   Set MyFile=fso.CreateTextFile(fullpath & "/" & fname)
   MyFile.WriteLine("<?xml version='1.0'?>")
   MyFile.WriteLine("<ROOT>")
   MyFile.WriteLine("<hall_id>" & hallid & "</hall_id>")
   MyFile.WriteLine("<movie_id>" & movieid & "</movie_id>")
   while not (rsstatus.EOF)
      MyFile.WriteLine("<status>")
      moviedate = rsstatus.Fields("movie_date")
      showtime = rsstatus.Fields("showtime")
      balcony = rsstatus.Fields("balcony")
      rearstall = rsstatus.Fields("rear_stall")
```

```
        middlestall = rsstatus.Fields("middle_stall")
        upperstall = rsstatus.Fields("upper_stall")
        MyFile.WriteLine("<movie_date>" & moviedate & "</movie_date>")
        MyFile.WriteLine("<showtime>" & showtime & "</showtime>")
        MyFile.WriteLine("<balcony>" & balcony & "</balcony>")
        MyFile.WriteLine("<rear_stall>" & rearstall & "</rear_stall>")
        MyFile.WriteLine("<middle_stall>" & middlestall & "</middle_stall>")
        MyFile.WriteLine("<upper_stall>" & upperstall & "</upper_stall>")
        MyFile.WriteLine("</status>")
        rsstatus.MoveNext
    wend
  rsstatus.Close
  MyFile.WriteLine("</ROOT>")
  conn.Close
```

Figure 9-22 is a screen shot of `tmoviestatus.xml`.

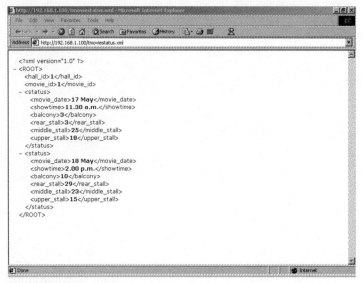

Figure 9-22: tmoviestatus.xml

Write XSLT for Displaying Movie Status

This time you are preparing an XSLT document for the transformation of `tmoviestatus.xml`, so that you can display the movie status for all days on i-mode clients. Listing 9-23 contains the code of `tmoviestatusimode.xsl`.

Listing 9-23: tmoviestatusimode.xsl

```
<?xml version='1.0'?>
<xsl:stylesheet xmlns:xsl="http://www.w3.org/1999/XSL/Transform" version="1.0">
<xsl:output method="html" />
<xsl:template match="/">


<html>


<head>
<META http-equiv="Content-Type" content="text/html; charset=utf-8" />
```

```
<META name="CHTML" content="yes" />
<META name="description" content="cHTML document" />
<title>Movie document</title>
</head>
<body>
 <a href="checkbrows.asp">Back </a>
 <br />
 <p>
 <xsl:apply-templates select="ROOT" />
 </p>

</body>
</html>
</xsl:template>

<xsl:template match="ROOT">

<xsl:variable name="hid" select="hall_id" />
<xsl:variable name="mid" select="movie_id" />

<p>
<b>Status:--</b>
<br /><br />

------------<br />

<xsl:for-each select="status">

  <b>Date</b><br />
  <xsl:value-of select="movie_date" /><br />

  <b>Showtime</b><br />
  <xsl:value-of select="showtime" /><br />

  <b>Balcony</b><br />
  <xsl:value-of select="balcony" /><br />

  <b>Rear Stall</b><br />
  <xsl:value-of select="rear_stall" /><br />

  <b>Middle Stall</b><br />
  <xsl:value-of select="middle_stall" /><br />

  <b>Upper Stall</b><br />
  <xsl:value-of select="upper_stall" /><br />
</xsl:for-each>

<br />
<b>Enter the information for ticket booking</b><br />

<FORM method="get" action="http://192.168.1.100/ticketbookingimode.asp">

Enter your name
<input name="uname" />
Enter your E-mail address
<input name="mail" />
```

```
Enter the showdate
<input name="showdate" />
Enter the showtime
<input name="showtime" />
Enter the stall name
<input name="stall" />
Enter the no. of tickets
<input name="no" maxlength="2" />

<input>
<xsl:attribute name="type">hidden</xsl:attribute>
<xsl:attribute name="name">hallid</xsl:attribute>
<xsl:attribute name="value"><xsl:value-of select="$hid" /></xsl:attribute>
</input>
<input>
<xsl:attribute name="type">hidden</xsl:attribute>
<xsl:attribute name="name">movieid</xsl:attribute>
<xsl:attribute name="value"><xsl:value-of select="$mid" /></xsl:attribute>

</input>

<input type="submit" value="Book" />
</FORM>

</p>

</xsl:template>
</xsl:stylesheet>
```

This is one of the longest XSLT documents that you will see in this chapter; hence, a brief discussion of the code is essential here.

In the beginning of the code, you generate a simple cHTML document and apply a template named ROOT by using the `<xsl:apply-templates>` element as follows:

```
<body>
 <a href="checkbrows.asp">Back </a>
 <br />
 <p>
 <xsl:apply-templates select="ROOT" />
 </p>
</body>
```

There are two main parts of the "ROOT" template to discuss. In the first part, you declare two variables for later use in the code listing, as shown in the following lines of code:

```
<xsl:variable name="hid" select="hall_id" />
<xsl:variable name="mid" select="movie_id" />
```

After declaring the variables, you start building the interface for displaying the movie status on i-mode clients by using the various cHTML formatting tags in a looping statement. After starting the loop, you display the contents by fetching them from the XML document by using the following lines of code:

```
<p>
<b>Status:--</b>
<br /><br />

------------<br />
```

```
<xsl:for-each select="status">

  <b>Date</b><br />
  <xsl:value-of select="movie_date" /><br />

  <b>Showtime</b><br />
  <xsl:value-of select="showtime" /><br />

  <b>Balcony</b><br />
  <xsl:value-of select="balcony" /><br />

  <b>Rear Stall</b><br />
  <xsl:value-of select="rear_stall" /><br />

  <b>Middle Stall</b><br />
  <xsl:value-of select="middle_stall" /><br />

  <b>Upper Stall</b><br />
  <xsl:value-of select="upper_stall" /><br />
</xsl:for-each>
```

When this part of the code executes, it shows the output in Figure 9-23.

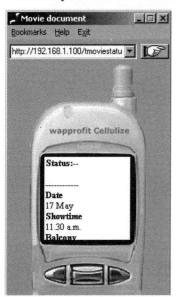

Figure 9-23: Displaying the movie status

In the next part of the code, you build a form for the user to fill in the information about booking tickets. The following is the code for doing this:

```
<b>Enter the information for ticket booking</b><br />

<FORM method="get" action="http://192.168.1.100/ticketbookingimode.asp">

Enter your name
<input name="uname" />
Enter your E-mail address
<input name="mail" />
Enter the showdate
```

```
<input name="showdate" />
Enter the showtime
<input name="showtime" />
Enter the stall name
<input name="stall" />
Enter the no. of tickets
<input name="no" maxlength="2" />


<input>
<xsl:attribute name="type">hidden</xsl:attribute>
<xsl:attribute name="name">hallid</xsl:attribute>
<xsl:attribute name="value"><xsl:value-of select="$hid" /></xsl:attribute>
</input>
<input>
<xsl:attribute name="type">hidden</xsl:attribute>
<xsl:attribute name="name">movieid</xsl:attribute>
<xsl:attribute name="value"><xsl:value-of select="$mid" /></xsl:attribute>

</input>

<input type="submit" value="Book" />
</FORM>

</p>
```

When this part of the code executes, it shows the output in Figure 9-24.

Figure 9-24: The form-filling process

After the user fills in all the requisite information and presses the Submit button on the device, you post all the data to the server and call the `ticketbookingimode.asp` file on the server.

Write Logic for the Booking Process

In this section, you undertake the ticket reservation and display a message to the user after the process is completed. Listing 9-24 is the code for `ticketbookingimode.asp`, which you use to book the tickets and redirect to the `bookimode.asp` file to display the status message to the user.

Listing 9-24: ticketbookingimode.asp

© 2001 Dreamtech Software India Inc.
All Rights Reserved

```asp
<%@ Language=VBScript %>
<HTML>
<HEAD>
<META NAME="GENERATOR" Content="Microsoft Visual Studio 6.0">
</HEAD>
<BODY>

<%
uname = trim(Request.QueryString("uname"))
email = trim(Request.QueryString("mail"))
hallid = trim(Request.QueryString("hallid"))
movieid = trim(Request.QueryString("movieid"))
showdate = trim(Request.QueryString("showdate"))
showtime = trim(Request.QueryString("showtime"))
stall = UCase(trim(Request.QueryString("stall")))
ticket = trim(Request.QueryString("no"))


dim conn,objrs,tsql,balcony,rear,middle,upper,total

set conn = Server.CreateObject("ADODB.Connection")
conn.ConnectionString = "Provider=SQLOLEDB.1;Persist Security Info=False;User
ID=sa;Initial Catalog=portal;Data Source=developers"
conn.Open

Set objrs = Server.CreateObject("ADODB.Recordset")
objrs.CursorType = adOpenStatic

tsql="select balcony,rear_stall,middle_stall,upper_stall from status where
hall_id=" & hallid & "and movie_id=" & movieid & "and showtime='" & showtime &
"'and movie_date='" & showdate &"'"
objrs.Open tsql,conn

while not (objrs.EOF )
 if (stall = "BALCONY") then
  total = UCase(trim(objrs.Fields("Balcony")))
 end if

 if (stall = "REAR_STALL") then
  total = UCase(trim(objrs.Fields("rear_stall")))
 end if

 if (stall = "MIDDLE_STALL") then
  total = UCase(trim(objrs.Fields("middle_stall")))
 end if
```

```
if (stall = "UPPER_STALL") then
  total = UCase(trim(objrs.Fields("upper_stall")))
end if

balcony = objrs.Fields("Balcony")
rear = objrs.Fields("rear_stall")
middle = objrs.Fields("middle_stall")
upper = objrs.Fields("upper_stall")

objrs.MoveNext
wend


if (CInt(total) < CInt(ticket) ) then
 Response.Redirect "http://192.168.1.100/bookimode.asp?name=fail"
else
 if (stall = "BALCONY") then
  balcony = CInt(balcony) - CInt(ticket)
 end if

 if (stall = "MIDDLE_STALL") then
  middle = CInt(middle) - CInt(ticket)
 end if

 if (stall = "REAR_STALL") then
  rear = CInt(rear) - CInt(ticket)
 end if

 if (stall = "UPPER_STALL") then
  upper = CInt(upper) - CInt(ticket)
 end if

 conn.Execute("update status set balcony=" & balcony & ",rear_stall=" & rear &
",middle_stall=" & middle & ",upper_stall=" & upper & "where hall_id=" & hallid
& "and movie_id=" & movieid & "and showtime='" & showtime & "'")
 'conn.Execute("Insert user
(firstname,movieid,hallid,stall,movie_date,showtime,totaltickets,emailadd)
values('" & uname & "'," & movieid  & "," & hallid & ",'" & stall & "','" &
showdate & "','" & showtime & "'," & tickets & ",'" & email & "')")
 Response.Redirect "http://192.168.1.100/bookimode.asp?name=pass"

end if

conn.Close

%>
</BODY>
</HTML>
```

At the top of the preceding code, you declare some variables for later use and store the values in the variables passed to you by the previous file, as in the following code:

```
uname = trim(Request.QueryString("uname"))
email = trim(Request.QueryString("mail"))
hallid = trim(Request.QueryString("hallid"))
movieid = trim(Request.QueryString("movieid"))
showdate = trim(Request.QueryString("showdate"))
```

```
showtime = trim(Request.QueryString("showtime"))
stall = UCase(trim(Request.QueryString("stall")))
ticket = trim(Request.QueryString("no"))
```

In the next step, you check the database for the number of available tickets for the specific category the user selects The number of tickets the user enters is the input to the system, in two if conditions. If the number of tickets the user enters is more than the number of tickets available in the database, you redirect control to the bookimode.asp file with the parameter named "NAME" holding the value fail. If the second condition is true, you update the database to reflect the changes and redirect control to the bookimode.asp file, with the parameter "NAME" holding the value pass.

```
if (stall = "BALCONY") then
  total = UCase(trim(objrs.Fields("Balcony")))
end if

if (stall = "REAR_STALL") then
  total = UCase(trim(objrs.Fields("rear_stall")))
end if

if (stall = "MIDDLE_STALL") then
  total = UCase(trim(objrs.Fields("middle_stall")))
end if

if (stall = "UPPER_STALL") then
  total = UCase(trim(objrs.Fields("upper_stall")))
end if

balcony = objrs.Fields("Balcony")
rear = objrs.Fields("rear_stall")
middle = objrs.Fields("middle_stall")
upper = objrs.Fields("upper_stall")

objrs.MoveNext
wend


if (CInt(total) < CInt(ticket) ) then
  Response.Redirect "http://192.168.1.100/bookimode.asp?name=fail"
else
  if (stall = "BALCONY") then
    balcony = CInt(balcony) - CInt(ticket)
  end if

  if (stall = "MIDDLE_STALL") then
    middle = CInt(middle) - CInt(ticket)
  end if

  if (stall = "REAR_STALL") then
    rear = CInt(rear) - CInt(ticket)
  end if

  if (stall = "UPPER_STALL") then
    upper = CInt(upper) - CInt(ticket)
  end if
```

```
 conn.Execute("update status set balcony=" & balcony & ",rear_stall=" & rear &
",middle_stall=" & middle & ",upper_stall=" & upper & "where hall_id=" & hallid
& "and movie_id=" & movieid & "and showtime='" & showtime & "'")
 'conn.Execute("Insert user
(firstname,movieid,hallid,stall,movie_date,showtime,totaltickets,emailadd)
values('" & uname & "'," & movieid  & "," & hallid & ",'" & stall & "','" &
showdate & "','" & showtime & "'," & tickets & ",'" & email & "')")
 Response.Redirect "http://192.168.1.100/bookimode.asp?name=pass"

end if
```

Display a Status Message When the Process Is Complete

Now take a look at the code for bookimode.asp in Listing 9-25. You use this file to redirect control to display two different messages when the process is complete on the basis of the parameter passed to you by the ticketbookingimode.asp file.

Listing 9-25: bookimode.asp

© 2001 Dreamtech Software India Inc.
All Rights Reserved

```
<%@ Language=VBScript %>

<%
name = trim(Request.QueryString("name"))

'declaring variables
dim sXML
dim sXSL
dim xmlDocument,xslDocument

'creating the objects for the XMLDOM
set xmlDocument = Server.CreateObject("MICROSOFT.XMLDOM")
set xslDocument = Server.CreateObject("MICROSOFT.XMLDOM")

sXML = "book.xml"

if(name = "fail") then
 sXSL = "bookfailimode.xsl"
else
 sXSL = "bookpassimode.xsl"
end if

xmlDocument.async = false
xslDocument.async = false

xmlDocument.load(Server.MapPath(sXML))
xslDocument.load(Server.MapPath(sXSL))
Response.ContentType = "text/html"
Response.Write(xmlDocument.transformNode(xslDocument))

%>
```

This is very simple ASP code. With the help of the `if` condition, you load one of the two XSLT documents in this file. This is done on the basis of the value held by the variable `"name"`. The following lines of code can do this quite easily:

```
if(name = "fail") then
 sXSL = "bookfailimode.xsl"
else
 sXSL = "bookpassimode.xsl"
end if
```

Now take a look at both of the XSLT documents (Listings 9-26 and 9-27) that you are using in this code.

Listing 9-26: bookfailimode.xsl

© 2001 Dreamtech Software India Inc.
All Rights Reserved

```
<?xml version='1.0'?>
<xsl:stylesheet xmlns:xsl="http://www.w3.org/1999/XSL/Transform" version="1.0">
<xsl:output method="html" />
<xsl:template match="/">
<html>
<head>
<META http-equiv="Content-Type" content="text/html; charset=utf-8" />
<META name="CHTML" content="yes" />
<META name="description" content="cHTML document" />
<title>Movie document</title>
</head>
 <body>
 <a href="checkbrows.asp">Back </a>
 <br />
  <p>
   <xsl:apply-templates select="ROOT" />
  </p>
   </body>
</html>
</xsl:template>
<xsl:template match="ROOT">
Try again!
</xsl:template>
</xsl:stylesheet>
```

Listing 9-27: bookpassimode.xsl

© 2001 Dreamtech Software India Inc.
All Rights Reserved

```
<?xml version='1.0'?>
<xsl:stylesheet xmlns:xsl="http://www.w3.org/1999/XSL/Transform" version="1.0">
<xsl:output method="html" />
<xsl:template match="/">
<html>
<head>
<META http-equiv="Content-Type" content="text/html; charset=utf-8" />
<META name="CHTML" content="yes" />
<META name="description" content="cHTML document" />
<title>Movie document</title>
</head>
```

```
<body>
 <a href="checkbrows.asp">Back </a>
 <br />
  <p>
   <xsl:apply-templates select="ROOT" />
  </p>
   </body>
</html>
</xsl:template>
<xsl:template match="ROOT">
Congratulations! Tickets have been booked.
</xsl:template>
</xsl:stylesheet>
```

Figure 9-25 shows the results of unsuccessful ticket booking on the browser. When the user presses the Back link, he is again taken to the application's home page.

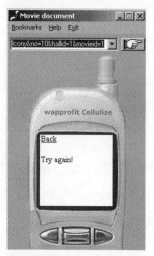

Figure 9-25: Error message in the case of unsuccessful completion

The output in Figure 9-26 appears in the case of successful ticket booking.

Figure 9-26: Displaying a "Congratulations!" message to the user

When the user presses Back, it again takes him to the application's home page; consequently, the whole process that we have covered in this chapter starts all over again.

This completes the last section on the movie area of the application. Now the user can access the complete application using an i-mode-compliant device.

Summary

In this chapter, you have learned the process of transforming an existing application for use by i-mode clients. You did this using the XML- and XSLT-based approach.

Chapter 10

A Preview of VoiceXML

The success of information technology depends solely on networking. After the information technology boom, Internet usage increased rapidly. To make the most of the Internet, Web facilities needed to improve. To this end, AT&T has introduced a revolutionary voice application for use on the Web. Thus begins the story of voice applications.

Introduction to Voice Applications

Voice applications have enormous potential for use in the real world. They provide the user with a number of useful services, as you'll read in the following paragraphs.

Voice applications can help in developing improved calling services, which are poised to be important sources of information and telephone-management capabilities. In contact management, the user can maintain an address book. This information is entered by using a Web surface and is accessible through the telephone. Moreover, this complete system provides a platform for other calling services such as e-mail-by-phone, voice-activated dialing, and so on.

Voice mail management is another area of calling services. This service enables users to maintain voice mailboxes through a telephone line. Voice mailboxes are arranged in sequence. So before the user proceeds to the next mailbox, he interacts with the voice-mail system.

E-mail-by-phone is a service that gives users the ability to listen to and manage e-mail messages over the telephone. The features here consist of reading, deleting, and delivering messages, as well as delivering audio attachments.

Voice-activated dialing is a service that lets a user use his voice to dial a number. Moreover, the user can specify the name or the location from the address book.

Voice applications also provide content services. These services remove the need for people to use a computer or a WAP phone to access day-to-day information. For example, with the weather information service, users get the latest information on the weather conditions for a particular day. Under the traffic reporting service, the user can collect details of traffic conditions in an area and then choose the best route to travel. News service provides users with the options to select news from national, international, or business arenas. The sports service allows users to get information on the latest sports news from around the globe. Transaction services are perhaps the most powerful service of their kind. Here the user can develop services based on a pre-existing e-commerce framework. An example of such a system is the home delivery system for meals, electronic items, and so on.

Background and Scope of Voice Applications

Figure 10-1 illustrates the background of voice applications.

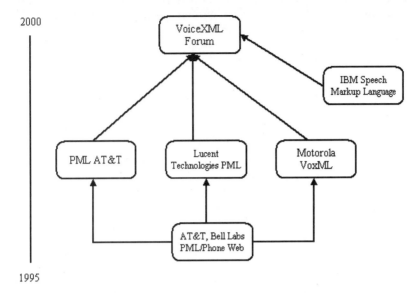

Figure 10-1: The background of voice applications

The process actually began in 1995 when AT&T Bell Labs came up with the dynamic PML/Phone Web application. This application was later split into PML (AT&T) and PML (Lucent Technologies). In 1998, Motorola introduced VoxML using the same application. IBM too stepped in; collectively, these four giants (AT&T, Lucent Technologies, Motorola, and IBM) founded a VoiceXML Forum in March 1999. It paved the way for making Internet content accessible via voice and telephone. The VoiceXML Forum is an industry organization that released two specifications — v0.9 and v1.0 — in August 1999 and March 2000, respectively. This forum has more than 150 supporting companies.

Applications and Possibilities of VoiceXML

VoiceXML, as an enterprise, can find wider use and provide much better results than the contemporary alternatives. The business community has already acknowledged the potency of VoiceXML as a resource generator and service provider. And e-commerce service based on VoiceXML has become almost mandatory for faster dispensation of professional services. Undoubtedly, VoiceXML consolidates all the components that make for a comprehensive communication platform. It is a complete application that contains all the tools needed for fast, accurate, and easy information exchange.

Lately, the ease of uninterrupted business flow has given birth to an altogether new business strategy. Such a strategy can be attributed to the mobile workforce using the VoiceXML platform. Present-day dynamic business is effectively managed using this platform. Employees are able to communicate business decisions more quickly and without delay, irrespective of the receiver's location. This is possible through distributed response statistics, which are available instantly. This access to such real-time information with a pervasive medium has reduced idle time and has paved the way for updated parameters and collective participation. Ultimately, customers are more satisfied. VoiceXML inherits the advantages of almost any application that caters to mobile clients. Such advantages are portability, abundance of tools for upgrading performance, libraries for specific purposes, references, and a satisfactory number of developers.

VoiceXML has interpreters that enable the PC user to take advantage of this application and its facility for running dialogs and extending telephone applications to a PC. For the Internet surfer, it offers the natural medium of voice. It eliminates the need for special skills to surf the net. VoiceXML rejuvenates the network and leverages wireless and Internet connectivity. VoiceXML harbors immense possibilities that have led to the rapid development of speech interfaces. With the maturation of telephone speech recognition technology, the automation of voice response, the provision for speaker verification, inexpensive biometric security, text-to-speech synthesis, and the quantum leap in the user interface, VoiceXML is at the forefront of the technology that dictates the course of the future. It is considered the application that will provide the basis for speech interfaces of the future.

The business world as well as the scientific world stands to benefit enormously from VoiceXML, because it dedicates itself fully to a wide range of proposals involving mass participation. The business world is poised to benefit from the direct seller-buyer interaction caused by VoiceXML. The facilities VoiceXML offers can be reinforced further through implementing accessories or tools in existing applications. Tools such as the XML voice interpreter, browser, speech-recognition engine, and other innovations will undoubtedly redeem what people view as the drawbacks of technology. VoiceXML is a technology of the present as well as the future, and it is heavily armed to face the challenge of time.

Voice Browsers and the W3C's Work on Voice Activity

The W3C is trying to add a new dimension to information technology by providing the means through which users can interact with a Web site via spoken commands and listening to prerecorded speech, music, and synthetic speech. This will enable the user to use telephones to access Web-based services and will be beneficial particularly for visually impaired people and for those who need to use their hands and eyes for other things. The W3C organized a workshop on voice browsers in 1998. In March 1999, an activity proposal was prepared to establish a W3C Voice Browser Activity and Working Group.

The voice browser enables people to access Web-based services through mobile phones regardless of their location. Voice browsers are poised to emerge as extremely useful instruments for the new generation of call centers, which will become voice Web portals to the company's services and related Web sites — regardless of whether they are accessed through the Internet or the phone. Voice facilitates conventional desktop browsers with high-resolution graphic displays, providing an alternative to using the keyboard or screen. Voice overcomes the limitations of keypads and displays as mobile devices become even smaller.

The Web has tremendous potential to increase the opportunities for voice-based applications. The combination of text to speech (TTS) and prerecorded material (audio) can be used to represent the dialogs to the user of the application.

Architecture of a Voice Application

Voice applications are developed and designed using a Web browser and a telephone. After the voice application has been developed, it can be communicated and shared with other voice applications. Figure 10-2 shows the workings of a voice application in a network.

VoiceXML is used to create voice applications, just like HTML is used to create visual applications. A Web browser requests an HTML document to be displayed visually to the user. The VoiceXML interpreter sends a request to the Web server, which then returns a VoiceXML document presented as a voice application via the phone. Technologies such as ASP, JSP, and JavaScript that are used to create Web sites can be used to generate new voice applications, thus making VoiceXML an extremely powerful application.

Implementations of Voice Technology

Today many companies are working on VoiceXML and its related technologies. As a result of this, we have a number of implementations of Voice technology on different platforms and with different extensions. The following is a description of some of the most popular implementations.

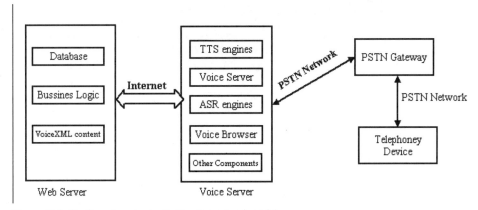

Figure 10-2: The flow of a voice application in a network model

Java Speech Markup Language

Java Speech Markup Language (JSML) improves voice quality and adds a new dimension to it by introducing naturalness in synthesized-speech. A JSML document provides structural knowledge about sentences and paragraphs, enhanced control for the production of synthesized speech that includes pronunciation and emphasis of words, as well as the placement of boundaries and pauses. JSML can even control the pitch and the speaking rate. However, perhaps the best thing about JSML is that it enables markers to be embedded in text and allows synthesizer-specific controls.

JSML supports many types of applications and text markups in different languages by acquiring general information about the text and using cross-language properties. Although JSML can be used in modern languages, a single JSML document contains text in one language. Thus, applications manage and control the speech synthesizers when the output is required in multiple languages.

Consider the following example of JSML used to indicate the start and the end of a sentence:

```
<SENT>Hello World.../SENT>
```

The application converts the source message to JSML text by using cross-language properties as well as general information about the text. For example, a text document can be read by converting messages to JSML.

Figure 10-3 demonstrates the role JSML plays in voice applications. The source message initially goes over to the application part. With help of JSML texts, it enters the speech synthesizer and finally produces the speech output.

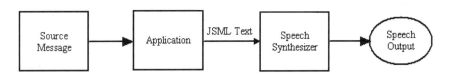

Figure 10-3: The JSML process (courtesy of Sun MicroSystems)

JSML is a subset of XML (Extensible Markup Language)—a simple dialect of SGML. JSML possesses certain advantages. It is fairly easy for computers to read and edit; it can be written using general XML editors; and its markup language is also pretty simple for any synthesizer. If you require more information on JSML, refer to the following Web site: http://java.sun.com/products/java-media/speech/forDevelopers/JSML/.

VoXML

VoXML is derived from Motorola's VoiceXML Markup Language. The VoXML voice browser supports and interprets a small set of reserved words — such as "help," "cancel," "exit," and "goodbye"—which are constant irrespective of any VoXML application. There is a grammar associated with each of these reserved words. For example, the application's response to `help` is specified in VoXML by using the `help` element. In case no `help` element is associated with the form, the current prompt is interrupted and the user hears "No help is available" and remains in the same application position.

Similarly, the application's response to "cancel" is specified in VoXML by using the `cancel` element. If no `cancel` element is associated, the current prompt is interrupted and the user remains in the same application position. If the browser recognizes "exit," the current prompt is interrupted and the user hears "Exiting and returning to the main menu," and the prompt returns to the main menu. If the browser recognizes "goodbye," the current prompt is interrupted and the user hears "goodbye," which results in the phone hanging up.

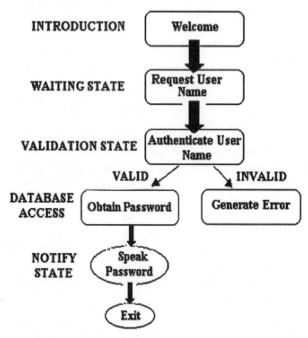

Figure 10-4: State diagram (courtesy of Motorola, Inc.)

VoXML provides a means for building applications that use TTS recorded sounds, navigational controls, and input controls for voice applications. VoXML-based applications are executed and tested by an interactive method provided by UIS Simulator.

The implementation of VoiceGenie is virtually the same as the Tellme implementation of VoiceXML because both are used on the same platform; the only difference lies in that the implementation belongs to two different companies and their own extensions to VoiceXML. For additional information, refer to `http://www.VoiceGenie.com` and `http://www.tellme.com`

VoiceXML

VoiceXML originated from the Extensible Markup Language (XML). One of the goals of using VoiceXML is to extract the maximum possible power from Web development as well as deliver content to voice-response applications. It relieves the developers from the complications involved in low-level programming and resource management. Further, VoiceXML is designed to create audio dialogs that are used in speech, digitized audio, DTMF, key input, recording of spoken input, telephony, and mixed-initiative conversations.

The architecture of VoiceXML

The architecture of VoiceXML is shown in Figure 10-5.

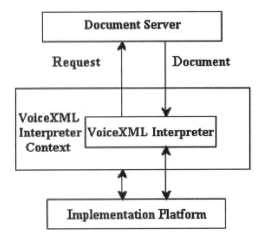

Figure 10-5: VoiceXML architecture

To produce VoiceXML documents, the document server collects the request from the client application through the VoiceXML Interpreter Context. The VoiceXML Interpreter Context here does two things: It monitors the user input and keeps track of the VoiceXML Interpreter. The VoiceXML Interpreter Context and the VoiceXML Interpreter control the platform.

Work flow of VoiceXML

At any particular moment, the user of a VoiceXML document is in a particular conversational state. Thus the VoiceXML document or application can be said to be a *finite-state machine*. For transition between the documents, each dialog is determined by the previous one. Uniform Resource Identifiers (URIs) are used to specify transitions. The present document is assumed if the URI does not refer to a document. The first dialog is assumed if the URI does not refer to a dialog. Execution is terminated if a dialog fails to specify the following one (its successor).

Fundamentally, dialogs can be grouped into two types: forms and menus. Forms define an interaction that collects values for a set of field-item variables that contain the information and the output. They collect the values for a set of field-item variables. Each field specifies the grammar permitted for the input therein. Menus, on the other hand, offer the choices for the next step. They offer a choice of options to the user and take him to another dialog, depending on the option he selects.

A *subdialog* is the mechanism that is used to break down complex sequences of different dialogs into simpler ones. This process enhances the structure by creating components that may be reused later. For example, a passport-inquiry system might require some information from your local police department servers about your identity or your permanent address. Subdialogs are a part of parent dialogs, which work as a function. Subdialogs return results to their parent dialogs when their execution is completed.

The subdialog contains a `<param>` element for accessing the parameters that are passed to that subdialog from the calling document. You must declare all these parameters by using the `<var>` element within the subdialog, as shown in Listing 10-1.

Listing 10-1: example1.vxml

```
<?xml version="1.0"?>
<vxml version="1.0">
    <form>
        <subdialog name="result" src="#getidentity">
            <param name="userid" expr="'007'"/>
            <filled>
                <submit next="http://myserver.com/process"/>
            </filled>
        </subdialog>
    </form>

    <form id="getidentity">
        <var name="userid"/>
        <field name="password">
            <grammar src="http://myserver.com/passgrammar.gram"
type="application/x-jsgf"/>
            <prompt>Please say your password</prompt>
            <filled>
                <if cond="checkvaliduser(userid, password)">
                    <var name="status" expr="true"/>
                <else/>
                    <var name="status" expr="false"/>
                </if>
                <return namelist="status"/>
            </filled>
        </field>
    </form>
</vxml>
```

Here, a password is used to validate the user's identity. The status is returned to the calling dialog, which indicates whether the password is valid or invalid.

When a user interacts with a VoiceXML document, a session begins and continues as documents are loaded and processed.

Grammars and their use in VoiceXML

Grammars define a valid expression from the user when he interacts with a phone application. In system-controlled applications, the grammar comes into action when the user interacts with the application. In multiple-initiative applications—which provide flexibility and power to voice applications—the user and the system take turns in planning the next step.

Grammar is the root document of an application, and it plays an active role in any VoiceXML application. Reference to an external grammar is included by using the <vxml> application attribute; for example:

<vxml-application=http://resourses.tellme .com/lib/universals.vxml>

A grammar consists of a set of expressions and subgrammars that make up a valid vocabulary. A grammar defined within the <vxml> element or the <link> element has a document scope; a grammar defined within the <form> element has a form scope; and a grammar defined within the <field> element has a field scope. The context-free grammar of VoiceXML is categorized as either inline grammar, such as the following:

```
<grammar>
<! [CDATA[
Grammar Description
] ]>
</grammar>
```

where `Grammar Description` defines the vocabulary part; or it is categorized as external grammar, such as the following:

```
<grammar  src = " URL" />
```

Here, the URL is the location of the file on a document server. Java Speech Grammar Format (JSGF) can be used within the VoiceXML `<grammar>` element.

Event Handling and Linking in VoiceXML

VoiceXML has the capacity to handle the user input and any other events that are not covered by the form mechanism. Event handling can be specified at any level. The interpreter throws an event when there is a semantic error in a VoiceXML document. In VoiceXML, tracking of events is possible with the catch elements, which in turn are inherited from enclosing elements. A `<link>` element possesses one or more grammars. Link elements are definitely the child of the `<vxml>` and `<form>` elements or the form item. At the `<vxml>` level, a link contains grammars that are active throughout the document. However, for a `<form>`, the grammar remains active only when the user is present in it. Link, when it is matched, throws an event instead of going to a new document.

Links do support a mixed initiative and are undoubtedly beneficial. A `<link>` element is used to throw events to the destination URI. If the input matches the grammar of the link, the control is transferred to the link's destination URI. Hence, two conditions are possible in this case: The link either goes to the new document or it throws an event.

The following are the attributes associated with a `<link>`:

♦ `next`—specifies the URI to go to (the URI may be a document or a dialog in the present document).

♦ `expr`—similar to `next` except that the URI is dynamically determined by evaluating the given ECMAScript expression.

♦ `event`—specifies the event to throw when any of the link grammars is matched.

♦ `caching`—same as `event`.

♦ `fetchaudio`—specifies the audio clip's URI while the fetch is being displayed.

♦ `fetchtimeout`—specifies the time interval to wait for the content to be returned before an error occurs.

♦ `fetchhint`—specifies when the interpreter context should retrieve content from the server.

VoiceXML contains two types of events. The first are the predefined events, which are thrown by the system. These events can be classified further into plain events and error events. The second type of events are the application-defined events. These are thrown by the voice application and caught by the custom events. All of this is outlined in Figure 10-6.

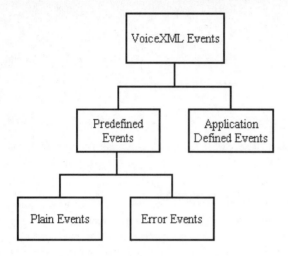

Figure 10-6: VoiceXML events

Predefined normal events include the following:

- ♦ cancel—the user has requested to cancel playing the current prompt.
- ♦ telephone.disconnect.hangup—the user has hung up.
- ♦ telephone.disconnect.transfer—the user has been transferred unconditionally to another line and will not return.
- ♦ exit—the user has asked to exit.
- ♦ help—the user requires help.
- ♦ noinput—this event is thrown within an interactive call state when the user is silent within the timeout period.
- ♦ nomatch—this event is thrown when the user speaks something outside the active grammars.

Predefined error events include the following:

- ♦ error.semantic—this event is thrown when the VoiceXML interpreter finds a semantic error.
- ♦ error.badfetch—this throws the following fails: external script, external grammar, or http request for a VoiceXML document.
- ♦ error.noauthorization—the user is not authorized to perform the operation requested (such as dialing an invalid telephone number or a number the user is not allowed to call).
- ♦ error.unsupported.format—the requested resource's format is not supported by the platform, for example, an unsupported grammar format, an audio file format, an object type, or a MIME type.
- ♦ error.unsupported.element—the platform does not support the given element. For instance, if a platform does not implement <record>, it must throw error.unsupported.record. This allows an author to use event handling to adapt to different platform capabilities.

Application-defined events define behavior in one place that can be called anywhere in the application.

Handling Predefined Events

The <catch> element is used to handle events set by the event attribute; for example:

```
<catch event="nomatch">
 <audio>
  This test show how events can be handled.
 </audio>
 <listen/>
</catch>
```

VoiceXML provides some predefined tags for handling some predefined events such as `noinput`, which plays some audio and replays the prompt, as in the following:

```
<noinput>
   <audio>Hello!</audio>
 </reprompt>
</noinput>
```

Hosting and Executing the Documents

Dialogs are the top-level (root) elements in a VoiceXML document. Now we'll discuss how to execute the documents.

Execution within a document begins with the first dialog by default. Each dialog locates the next one. If the dialog doesn't specify the next dialog, the execution stops. (See Listing 10-2.)

Listing 10-2: Execution with one document (example2.vxml)

© 2001 Dreamtech Software India Inc.
All Rights Reserved

```
<?xml version="1.0"?>
<vxml version="1.0">
   <var name="message" expr="'Hello friends! How are you?'"/>
   <form>
      <block>
         <value expr="message"/>
         <goto next="#say_fine"/>
      </block>
   </form>
   <form id="say_fine">
      <block>
            Fine, Thank you.
      </block>
   </form>
</vxml>
```

As the multiple documents execute as one application, one of these documents can act as a root document and can be referred in all other subdocuments under the application scope. Every time the interpreter loads the application root document, a request is conveyed to the interpretor to load a document in this application. The application document remains loaded unless and until the interpreter is informed that it should load a document belonging to a different application. Two conditions are possible here during interpretation. First, the application root document is already loaded and the user executes it. Second, the root document and the subdocuments are loaded and execute some other document. (See Listings 10-3 and 10-4.)

Listing 10-3: Application root document (root.vxml)

© 2001 Dreamtech Software India Inc.
All Rights Reserved

```
<?xml version="1.0"?>
```

```
<vxml version="1.0">
   <var name="age" expr="'teenager'"/>
   <link next="my.vxml">
      <grammar>
         operator
      </grammar>
   </link>
</vxml>
```

Listing 10-4: Main document (main.vxml)

```
<?xml version="1.0"?>
<vxml version="1.0" application="root.vxml">
   <form id="say_yes_no">
      <field name="response" type="boolean">
         <prompt>
            Are you <value expr="application.age"/>?
         </prompt>
         <filled>
            <if cond="response">
               <exit/>
            </if>
            <clear namelist="response"/>
         </filled>
      </field>
   </form>
</vxml>
```

In the preceding listings, `main.vxml` is loaded first and `root.vxml` is imported (called) as the application root document. So after being imported, `root.vxml` produces the variable `age` and also specifies the link to navigate to `my.vxml`.

Note: In multidocument applications, two documents are loaded at a time and the document (user) also interacts with the application root document. Another possible condition: If the document is a nonexistent application root document or if an application root document refers to itself, a semantic error is thrown.

VoiceXML Forms

Forms are the main components of VoiceXML terminology and play a vital role. Forms are modularized, as shown in Figure 10-7.

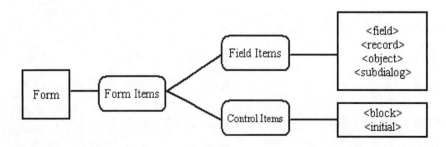

Figure 10-7: Forms in VoiceXML

A form in VoiceXML contains field items and field variables. Field items define the form's form items and are visited in the main loop of the form interpretation algorithm (FIA). This algorithm contains a main loop that repeatedly fetches the form items. The items that don't have satisfactory conditions are selected first lexically. Form items can be classified further into the following subitems:

- ♦ Field item variables—these are classified further into the following:
  - `<field>` contains a value obtained through grammars.
  - `<record>` records an audio file on the server.
  - `<transfer>` transfers the user to a different telephone line.
  - `<object>` is used for calling an object with the required parameters.
  - `<subdialog>` opens a different dialog on the current page, or a different VoiceXML document.

- ♦ Control item variables—these help to control the gathering of the form's field. The various types of control items are the following:
  - `<block>` contains a sequence of procedural statements.
  - `<initial>` is used to control the initial interaction in a mixed-initiative form.
  - `<filled>` actions are blocks of procedural logic that execute when field-item combinations are entered.

Event handlers in VoiceXML forms

A form has the following attributes:

- ♦ `id`—this specifies the form name.
- ♦ `scope`—this attribute represents the scoping status of grammar in a given form. Three kinds of scopes are currently provided by VoiceXML. They are the following:
  - Application specific: If the scope of grammar is limited to the document that acts like a particularly rude document, the grammar of this document is active during the execution of any document in this application.
  - Document specific: If the scope of grammar is limited to the document, the form is active during the execution of this document.
  - Form specific: In this, the scope is limited to a particular form in one document while the other forms of the same document are unable to use the grammar of that particular form.

A form works in the following manner:

- ♦ Selects prompt(s).
- ♦ Plays selected prompt(s).
- ♦ Collects user input.
- ♦ Interprets any `<filled>` action pertaining to the newly filled fields.

The FIA terminates by transferring the control statement (for example, a `<goto>` transfers the control to another dialog or document; or a `<submit>` transfers data to the document server). It also is terminated by an implied `<exit>`, when no form items are available for selection.

Variables and items have an associated form item variable that contains the result of interpreting the form items. This variable is set to `undefined` by default when the form is entered and can be assigned a name using the `name` attribute. An internal name is generated if no name is assigned.

Every form item contains a guard condition, which indicates whether it is possible for the given form item to be selected by the FIA. The default guard condition tests to identify whether the form item variable has a value. If the form item variable contains a value, the form item is not visited. Generally, the variables are not given initial values and the guard conditions are not specified. A form may have a form item variable initially set to hide a field, and later cleared, using `<clear>`, so as to force the field's collection. Form items contain the following attributes:

- `name`—represents a form item variable name that holds the form item value.
- `expr`—represents the initial value of the form item variable. ECMAScript `undefined` is the default value.
- `cond`—evaluates the expression in relation to the form item variable.

Forms can be classified as directed forms and mixed-initiative forms, which we discuss in the following sections.

Directed forms

- Directed forms are some of the most widely used forms in VoiceXML terminology. They are executed row-wise and provide a system-direction dialog. Listing 10-5 demonstrates the process in which a directed form can be used by any organization.

Listing 10-5: example3.vxml

© 2001 Dreamtech Software India Inc.
All Rights Reserved

```xml
<?xml version="1.0"?>
<vxml version="1.0">
   <form id="result_info">
      <block>
         Welcome to the DOEACC Society Result Information Service.
      </block>
      <field name="level">
         <prompt>
            What is your level, either O, A, B, C?
         </prompt>
         <grammar>
            <![CDATA[
            [
                     [ O ] {<option "O">}
                     [ A ] {<option "A">}
                     [ B ] {<option "B">}
                     [ C ] {<option "C">}
            ]
            ]]>
         </grammar>
         <catch event="help">
            Please say your level, either, O, A, B, C?
         </catch>
      </field>
      <field name="rollno" type="number">
         <prompt>
            What is your roll number?
         </prompt>
         <grammar src="rollno.gram" type="application/x-jsgf"/>
         <catch event="help">
            Please say the roll number.
         </catch>
      </field>
      <block>
         <submit next="/servlet/result" namelist="level rollno"/>
```

```
        </block>
    </form>
</vxml>
```

While executing the preceding code, the FIA first selects the initial blocks of the document, which gives the user a welcome message. After selecting the first block, the user is prompted for the level, and the value returned by the user is stored in a variable, and so on. The user is then asked to supply some other information. Finally, when the execution of the block is completed, the transition takes place and the control is passed to the different URL.

Mixed-initiative forms

Mixed-initiative forms provide a more indelible kind of dialog for the VoiceXML user because they often contain more than one form. Containing more than one form provides a better dialog and better utilization of grammars, as well as a better combination of active grammar in one document so that the user can express choices in a better way.

Listing 10-6 demonstrates how a sample pizza information service can effectively use mixed-initiative forms for picking up pizza orders.

Listing 10-6: example4.vxml

```
<?xml version="1.0"?>
<vxml version="1.0">
    <form id="pizza_info">
        <grammar src="size_type.gram" type="application/x-jsgf"/>
        <block>
            <prompt bargein="false">
                Welcome to Sample Pizza Information Service.
                <audio src="http://myserver.com/my.wav"/>
            </prompt>
        </block>
        <initial name="start">
            <prompt>
                what type and size of pizza would you like to order?
            </prompt>
            <help>
                Please say the type and size of the pizza.
            </help>
            <noinput count="1">
                <reprompt/>
            </noinput>
            <noinput count="2">
                <reprompt/>
                <assign name="start" expr="true"/>
            </noinput>
        </initial>
        <field name="ptype">
            <prompt>
                What type?
            </prompt>
            <help>
                Please say the type of pizza you would like to order
            </help>
        </field>
```

```
      <field name="psize">
         <prompt>
            What size of <value expr="ptype"/> pizza you would like to order?
         </prompt>
         <help>
            Please say the size of pizza you would like to order
         </help>
         <filled>
            <if cond="ptype == undefined">
               <assign name="ptype" expr="'Veg'"/>
            </if>
            <if cond="psize == undefined">
               <assign name="psize" expr="'Normal'"/>
            </if>
         </filled>
      </field>
      <field name="go_ahead" type="boolean" modal="true">
         <prompt>
            Are you sure you want to buy <value expr="psize"/>, <value
expr="ptype"/> pizza?
         </prompt>
         <filled>
            <if cond="go_ahead">
               <submit next="/servlet/sample" namelist="psize ptype"/>
            </if>
            <clear namelist="start psize ptype go_ahead"/>
         </filled>
      </field>
   </form>
</vxml>
```

Variables and Expressions in VoiceXML

The naming convention followed in VoiceXML variables is similar to that used for ECMAScript variables. The only difference is that the names start with underscore (_) characters. These are specifically reserved for internal use. To declare the variables, you use the <var> elements present in the form items. The values of these variables are assigned using the expr attribute. Listing 10-7 declares four variables—age, child, teenage, and adult—and assigns a value to each of these.

Listing 10-7: example5.vxml

```
<?xml version="1.0"?>
<vxml version="1.0">
   <form id="test_age">
      <block>
         <var name="age" expr="18"/>
         <var name="child" expr="'you are a child'"/>
         <var name="teenager" expr="'you are a teenager'"/>
         <var name="adult" expr="'you are an adult'"/>
      </block>
   </form>
</vxml>
```

Variable Scope in VoiceXML

It is mandatory to declare variables before they are used. However, if you don't declare an explicit initial value for a variable, that variable is initialized to the ECMAScript `undefined` value. Thus, it is essential to understand the scope of variables in VoiceXML. Variables declared in VoiceXML can have the scopes shown in Figure 10-8.

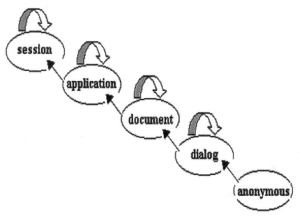

Figure 10-8: The scopes of variables

In Figure 10-8, the curved arrows indicate the presence of a variable in its own scope. The variable name is similar to the name of the scope itself. In order to get a better understanding of this, let's look at each scope individually.

session

♦ Variables with session scope are read-only and have the lifespan of an entire user session. They are declared and set by the interpreter context. It is not possible for VoiceXML documents to declare new session variables. Four standard session variables are possible. They are the following:

♦ `session.telephone.ani`, where `ani` stands for Automatic Number Identification and gives the receiver of the telephone call the caller's phone number.

♦ `session.telephone.dnis`, where `dnis` stands for Dialed Number Identification Service. This variable identifies the receiver and the caller's number.

♦ `session.telephone.iidigits`, where `iidigits` stands for Information Indicator Digit. This variable gives information on where the call originated, which might be a pay phone, a cell phone, and so on.

♦ `session.telephone.uui`, where `uui` stands for User to User Information. This variable returns additional information provided as part of an ISDN call setup from the calling party.

application

♦ Variables with application scope are declared with the `<var>` elements that come under the application root document's `<vxml>` element in the VoiceXML hierarchy. They are initialized when the application root document is loaded and is visible to the root document and other application leaf documents. They exist while the application root document is loaded.

document

♦ Variables with document scope are declared with `<var>` elements that come under the document's `<vxml>` element in the VoiceXML hierarchy. They are initialized when the document is loaded and are visible only within that document. They exist while the document is loaded.

dialog

♦ Dialog scope exists while the user is visiting a dialog (<form> or <menu>). It is visible to the element of that dialog. Variables are declared by <var> elements that come under <form>, by the <var> elements present inside the executable content such as <block> content, <catch> element content, and so on.

anonymous

♦ The anonymous scope contains variables declared in each <block>, <filled>, and <catch> element.

How to Reference Variables

There are two ways to reference a variable–in cond and expr attributes. The expression language these attributes use is ECMAScript. The cond operators include >, <, >=, <=, and &&. Variable references match the closest enclosing scope according to the scope chain. To avoid ambiguity, a reference can be prefixed with a scope name.

Listing 10-8: example6.vxml

© 2001 Dreamtech Software India Inc.
All Rights Reserved

```
<?xml version="1.0"?>
<vxml version="1.0">
   <form id="test_age">
     <block>
     <var name="age" expr="18"/>
     <var name="child" expr="'you are child'"/>
     <var name="teenager" expr="'you are teenager'"/>
     <var name="adult" expr="'you are adult'"/>
     <if cond="age > 19">
        <prompt><value expr="adult"/></prompt>
     <elseif cond="age > 12"/>
        <prompt><value expr="teenager"/></prompt>
     <else/>
        <prompt><value expr="child"/></prompt>
     </if>
     </block>
   </form>
</vxml>
```

Grammars in VoiceXML

Field grammars cannot specify a scope. Link grammars, on the other hand, are providing the scope of the element that contains the link. Consequently, if the link grammars are defined in the application root document, links also activate themselves in any other loaded application document. Grammars within links cannot specify a scope.

By default, form grammars and menu grammars are provided the dialog scope. This is done to facilitate their activation when the user is present in the scope. Likewise, if the form grammars are given a document scope, they are activated only when the user is present in the document. However, it is imperative to note that, if the form grammars are provided the document scope and the document is the application root document, these form documents are activated when the user is positioned in another loaded document in the same application. Document scope is given to a form grammar by specifying either the scope attribute on the <form> element or the scope attribute on the <grammar> element. When both are specified, the form grammar assumes the scope specified by the <grammar> element.

Menu grammars, when given the dialog scope, are active only when the user is located in the menu. And here also, if menu grammars are given the document scope and are active throughout the document, and if their document is the application root document, they are also active in any other loaded document belonging to the application. It is impossible for grammars contained within menu choices to specify a scope.

Subgrammars

A VoiceXML subgrammar is basically used to define a group of expressions that specify a value for each of their items. You can define n number of attributes for a single value. By doing so, you can enhance the application's performance by providing more data for matching the user's input.

Listing 10-9 returns a single value to the system by providing a separate expression for each month.

Listing 10-9: A subgrammar code snippet

© 2001 Dreamtech Software India Inc.
All Rights Reserved

```
months:m {<option ($d)>}

months
[
    january {return (jan)}
    february {return (feb)}
    march {return (mar)}
    april {return (apr)}
    may {return (may)}
    june {return (jun)}
    july {return (jul)}
    august {return (aug)}
    september {return (sep)}
    october {return (oct)}
    november {return (nov)}
    december {return (dec)}
]
```

DTMF expressions

The conventions in Table 10-1 are used to write DTMF expressions.

Table 10-1: DTMF Keys Description

Keypad Symbol	Syntax
0	dtmf-0
1	dtmf-1
2	dtmf-2
3	dtmf-3
4	dtmf-4
5	dtmf-5
6	dtmf-6
7	dtmf-7
8	dtmf-8

Keypad Symbol	Syntax
9	dtmf-9
#	Reserved
*	Reserved

The # key is reserved for the specific setting. These settings possess a unique behavior. While prompts are displayed, entering # makes the platform listen for the user input and interrupts the ongoing dialog. If the user is speaking or entering other DTMF input, pressing # terminates the input and triggers recognition explicitly.

The * key takes the user back to the root document.

Rules for writing a subgrammar

While writing a subgrammar, you should adhere to certain rules. At least one uppercase letter should be used to name the subgrammar you are planning to use. The name length may be extended up to a maximum of 200 characters and may include the dash (-), underscore (_), period (.), single quotes ('), or the "at" sign (@). At the least, two attribute name/value pairs should be defined; if it's used at all, the ParseGrammar() function is used to parse the return value. Further, all name/value pairs attributes should be concatenated into one string within the option directive. The ^ symbol should be utilized to divide each pair.

Rules for writing a grammar description

When writing a grammar description, you should write the expressions in a way that helps the speech-recognition engine compare the possible user responses between brackets [   ]. Describing the subgrammars within the grammar description is also imperative here. However, you need to remember at the outset that a subgrammar can't be referenced in another file, and that an expression that references a subgrammar in another dialog or file is impossible to construct.

Case Study

This case study presents a vivid description of various VoiceXML elements. It is a mini-project wherein the main document, main.vxml (Listing 10-10), provides information about books, weather, and the latest movies. The main.vxml file offers three options for the user, waits for his initial voice response, and accordingly invokes books.vxml, weather.vxml, or movies.vxml. By saying "help," the user gets assistance in performing the appropriate action.

Listing 10-10: main.vxml

```
<?xml version="1.0"?>
<vxml version="1.0">
   <meta name="author" content="vxml author team"/>

   <link next="main.vxml">
      <grammar type="application/x-gsl">
         main menu
      </grammar>
   </link>
   <menu id="top">
   <property name="modes" value="dtmf" />

      <prompt bargein="false">
         Welcome to the voice information center.
```

```
        </prompt>
        <prompt>
           Please choose one of the following options:
           <enumerate/>.
        </prompt>
        <choice dtmf="1" next="books.vxml"> books </choice>
        <choice dtmf="2" next="weather.vxml"> weather </choice>
        <choice dtmf="3" next="movies.vxml"> movies </choice>
        <help>
           For book information, say 'books or press 1 '.
           For weather information, say 'weather or press 2'.
           For movie information, say 'movies or press 3'
           To return to the main menu at any time, say 'main menu'.
           To hear this message again, say 'help'.
        </help>
     </menu>
  </vxml>
```

Output of the above code is displayed in Figure 10-9.

Figure 10-9: Welcome dialog of the application (Image of MADK 2.0.Courtesy of Motorola, Inc.)

When the user loads `main.vxml` into the Motorola ADK browser, it shows the output in Figure 10-10. Here the user can either speak his choice or press the corresponding button for the DTMF detection of the individual choice. In this case, the user has three particular choices to select:

♦ Choice 1: The user can either press 1 or say "book" to choose the books section.

♦ Choice 2: The user can either press 2 or say 2 for invoking the weather section.

♦ Choice 3: The user can either say "movies" or press 3 for the movies section.

If the user requires any sort of help, he can say "help," which starts the help program.

Figure 10-10: Prompting available options (Image of MADK 2.0. Courtesy of Motorola, Inc.)

To make your selection, enter your choice in the response text box in the simulator and press Speak to proceed. You can also select your choice from microphone by pressing the scroll lock key and when the agent is ready to listen, start speaking your choice.

The Books Section

When the user says "books," `main.vxml` invokes `books.vxml`. This document provides various search options to the user; for example, he can search books on the basis of title or the author's name. On the user's response, `books.vxml` performs appropriate action. If the user says "name," `books.vxml` offers the user the available book titles. As per the requirement, the user says the appropriate book title. `books.vxml` then asks the user for a confirmation. If the user confirms that the book is correct, the user can purchase the book by credit card. Likewise, if the user says "author," a similar procedure is undertaken. Listing 10-11 shows the code for `books.vxml`.

Listing 10-11: books.vxml

© Dreamtech Software India Inc.
All Rights Reserved

```
<?xml version="1.0"?>
<vxml version="1.0">
   <meta name="author" content="vxml author team"/>

   <form id="vxml_books">
      <var name="amount" expr="undefined"/>
      <var name="address" expr="undefined"/>

      <link next="main.vxml">
         <grammar type="application/x-jsgf"> main menu </grammar>
      </link>

      <link next="books.vxml">
         <grammar type="application/x-jsgf"> start again</grammar>
      </link>
```

```
    <field name="search_for">

        <prompt>
            Welcome to the book information service.
            To search for books by name, say "name".
            To search for books by author name, say "author".
        </prompt>

        <grammar type="application/x-gsl">
            [ author name   ]
        </grammar>


        <filled>
            <clear namelist="author name "/>
            <clear namelist="r_b_book m_b_book a_b_book"/>
            <clear namelist="client_name credit_card_name credit_card_no
credit_card_exp_dt"/>
            <clear namelist="confirm_delivery amount address"/>

            <if cond="search_for=='name'">
                <goto nextitem="name"/>
            <elseif cond="search_for=='author'"/>
                <goto nextitem="author"/>

            </if>
        </filled>
    </field>
    <field name="name" cond="search_for=='name'">
        <prompt>

            Order by book name
            please say the book name

        </prompt>

        <option>multi platform mobile web portal</option>
        <option>Java 2 Micro Edition</option>
        <option>xml enhancement</option>
        <option>active server pages</option>
        <option>intro to java</option>
        <option>java server pages</option>

        <filled>

            <if cond="name=='multi platform mobile web portal'">
                <assign name="author" expr="'ravinder bhushan'"/>

                <assign name="amount" expr="'$25'"/>

            <elseif cond="name=='Java 2 Micro Edition'"/>
                <assign name="author" expr="'ravinder bhushan'"/>

                <assign name="amount" expr="'$25'"/>
```

```
            <elseif cond="name=='xml enhancement'"/>
                <assign name="author" expr="'madan dhyani'"/>

                <assign name="amount" expr="'$30'"/>

            <elseif cond="name=='active server pages'"/>
                <assign name="author" expr="'madan dhyani'"/>

                <assign name="amount" expr="'$40'"/>

            <elseif cond="name=='intro to java'"/>
                <assign name="author" expr="'ashish baveja'"/>

                <assign name="amount" expr="'$45'"/>

            <elseif cond="name=='java server pages'"/>
                <assign name="author" expr="'ashish baveja'"/>

                <assign name="amount" expr="'$30'"/>


        </if>
        <goto nextitem="client_name"/>
    </filled>
</field>


<field name="author" cond="search_for=='author'">
    <prompt> Order by name of author. please say the author name.
    </prompt>
    <grammar type="application/x-gsl">
        [ (ravinder bhushan) (madan dhyani) (ashish baveja) ]
    </grammar>
    <filled>

        <if cond="author=='ravinder bhushan'">
            <clear namelist="r_b_book"/>
            <goto nextitem="r_b_book"/>

        <elseif cond="author=='madan dhyani'"/>
            <clear namelist="m_b_book"/>
            <goto nextitem="m_b_book"/>

        <elseif cond="author=='ashish baveja'"/>
            <clear namelist="a_b_book"/>
            <goto nextitem="a_b_book"/>
        </if>

    </filled>
</field>

<field name="r_b_book" cond="search_for=='author' &amp;&amp;
author=='ravinder bhushan'">
    <prompt>
        ravinder bhushan srivastava, author of dreamtech software
        please select one of the following names by
```

```
            <value expr="author"/>: <enumerate/>
        </prompt>
        <option>multi platform mobile web portal</option>
        <option>Java 2 Micro Edition</option>

        <filled>
            <assign name="name" expr="r_b_book"/>
            <if cond="r_b_book=='multi platform mobile web portal'">
                <assign name="amount" expr="'$25'"/>
            <elseif cond="r_b_book=='Java 2 Micro Edition'"/>
                <assign name="amount" expr="'$25'"/>
            </if>
            <goto nextitem="client_name"/>
        </filled>
    </field>


    <field name="m_b_book" cond="search_for=='author' &amp;&amp;
author=='madan dhyani'">
        <prompt>
            madan dhyani, author of dreamtech software
            please select one of the following names by
            <value expr="author"/>: <enumerate/>
        </prompt>
        <option>xml enhancement</option>
        <option>active server pages</option>

        <filled>
            <assign name="name" expr="m_b_book"/>
            <if cond="m_b_book=='xml enhancement'">

                <assign name="amount" expr="'$30'"/>
            <elseif cond="m_b__book=='active server pages'"/>

                <assign name="amount" expr="'$40'"/>
            </if>
            <goto nextitem="client_name"/>
        </filled>
    </field>


    <field name="a_b_book" cond="search_for=='author' &amp;&amp;
author=='ashish baveja'">
        <prompt>
            ashish baveja, author of dreamtech software
            please select one of the following names by
            <value expr="author"/>: <enumerate/>
        </prompt>
        <option>intro to java</option>
        <option>java server pages</option>
        <filled>
            <assign name="name" expr="a_b_book"/>
            <if cond="a_b_book=='intro to java'">

                <assign name="amount" expr="'$45'"/>
```

```
            <elseif cond="a_b_book=='java server pages'"/>

                <assign name="amount" expr="'$30'"/>
            </if>
            <goto nextitem="client_name"/>
        </filled>
    </field>


    <field name="client_name">
        <prompt> please say your name. </prompt>
        <grammar type="application/x-gsl">
          [ aparna deepali sunil ]
        </grammar>
        <filled>
          <if cond="client_name=='aparna'">
              <assign name="address" expr="'1, golf links, new delhi 110024'"/>
          <elseif cond="client_name=='deepali'"/>
              <assign name="address" expr="'2, south extension, bombay
220043'"/>
          <elseif cond="client_name=='sunil'"/>
              <assign name="address" expr="'3, cannaught place, new delhi
110001'"/>
          </if>
        </filled>
    </field>
    <field name="credit_card_name">
        <prompt>
            <value expr="client_name"/>, please enter your credit card name?
        </prompt>
        <grammar type="application/x-gsl">
          [ (bank of india) (r b i)(s b i) ]
        </grammar>
    </field>


    <field name="credit_card_no" type="digits">
        <prompt>
            please say your credit card number of . <value
expr="credit_card_name"/>
        </prompt>
    </field>


    <field name="credit_card_exp_dt" type="digits">
        <prompt>
            please say the expiration date of your credit card.
        </prompt>
    </field>


    <field name="confirm_delivery" type="boolean">
        <prompt>
            you have ordered the book <value expr="name"/>,
            written by <value expr="author"/>, priced at
```

```
                <value expr="amount"/>. a final confirmation is required. say yes or
no

        </prompt>
        <filled>
           <if cond="confirm_delivery">
              <prompt>
                 the book you just purchased is named <value expr="name"/>,
                 written by <value expr="author"/>. it is being shipped to
                 your address at <value expr="address"/>. thanks for shopping.
              </prompt>
           <else/>
              <prompt>
                 your order has cancelled now going back to starting point
              </prompt>
           </if>
           <clear namelist="search_for"/>
           <goto nextitem="search_for"/>
        </filled>
     </field>
  </form>
</vxml>
```

Now let's see how the code actually works. When the user simulates this code on the Motorola ADK, it first shows the welcome message in Figure 10-11. In the next few pages, you will see each and every step of the execution process.

Figure 10-11: Welcome dialog of the Books section (Image of MADK 2.0.Courtesy of Motorola, Inc.)

When the execution of the first block is completed, the user has two choices. The choices can be either of the following:

♦ The user can search for books by title by saying "name."

♦ The user can search for books by author by saying "author."

The following are the choices available in this case study for both of these options:

- ◆ Choices for the books options:
  - *XML Enhancement*
  - *Multi-Platform Mobile Web Applications*
  - *Java Server Pages*
  - *Java 2 Micro Edition*
  - *Intro to Java*
  - *Active Server Pages*

- ◆ Choices for the author options:
  - Ashish Baveja
  - Madan Dhyani
  - Ravinder Bhushan

Continuing along, the user first selects the first section and says the book title "*XML Enhancement.*" As soon as the user enters the book name, he is asked to supply the name of the buyer, as shown in Figure 10-12.

Figure 10-12: Prompting the user to enter his name (Image of MADK 2.0.Courtesy of Motorola, Inc.)

Here we have the following choices for the buyer name:

- ◆ Sunil
- ◆ Aparna
- ◆ Deepali

If the user says "Sunil" as the name of the buyer, the system will next process it and ask for the name of the credit card, as shown in Figure 10-13.

Figure 10-13: Prompting the user to enter his credit card name (Image of MADK 2.0.Courtesy of Motorola, Inc.)

Here again, the user has some choices available for the credit card name. These choices are the following:

♦ Bank of India

♦ r b i

♦ s b i

Assume that the user enters "r b i" as the name of the credit card for the buyer "Sunil." He is then asked for the credit card number and the expiration date, as shown in Figure 10-14.

Figure 10-14: Prompting the user to enter the credit card number(Image of MADK 2.0.Courtesy of Motorola, Inc.)

The user can say or enter the five-digit expiration date for the r b i credit card, as shown in Figure 10-15.

Figure 10-15: Prompting the user for the expiration date of his credit card (Image of MADK 2.0.Courtesy of Motorola, Inc.)

After the user enters the expiration date, the program confirms the details of the purchase (title, author, and price). The user then gives a confirmation for the purchase by saying "yes" or "no," as shown in Figure 10-16.

Figure 10-16: Prompting the user for final confirmation (Image of MADK 2.0.Courtesy of Motorola, Inc.)

If the user enters "yes" at the confirmation stage, in the next step the final confirmation message is displayed along with the address to which the book will be shipped, as shown in Figure 10-17.

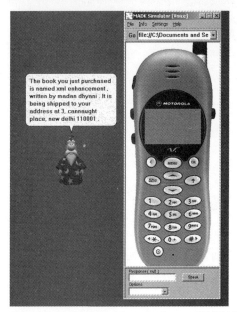

Figure 10-17: Confirming the order (Image of MADK 2.0.Courtesy of Motorola, Inc.)

The system then gives the user a thank-you message and goes back to the main menu, as shown in Figure10-18.

Figure 10-18: Prompting a Thanks message (Image of MADK 2.0.Courtesy of Motorola, Inc.)

If the user canceled the order by saying no after the confirmation dialog, information about the cancellation will be given to the user and the program goes back to the starting point.

Now let's proceed further to take a look at the second process in which the user selects the books based on the author name. When the user says "author" at the first prompt, he sees the message shown in Figure 10-19.

Figure 10-19: Prompting the options to the user (Image of MADK 2.0.Courtesy of Motorola, Inc.)

Here the system asks the user for the name of the author whose book he wants to select. The choices for the author options, in our case, are as follows: Ashish Baveja; Madan Dhyani; Ravinder Bhushan. If the user enters "Ashish Baveja" as the choice, the system provides a list of all the books written by that author. The system then prompts the user for the book name that he or she wants to purchase, as shown in Figure 10-20.

Figure 10-20: Prompting the list of books (Image of MADK 2.0.Courtesy of Motorola, Inc.)

The user has two choices here. Say the user seeks the list of books written by Ashish Baveja. They are the following: *Intro to Java* and Java Server Pages. Say that *"Java Server Pages"* is the user's choice. The system now asks for the buyer's name, as seen in Figure 10-21.

Figure 10-21: Prompting the user to say his name (Image of MADK 2.0.Courtesy of Motorola, Inc.)

The user now says "Sunil" as the name of the buyer. The system then asks for the credit card name and expiration date for processing the order. It asks for a final confirmation for placing the order. By saying "yes," the user can place the order. The program then displays a confirmation screen with the shipping address information. If the user says "no" at the confirmation prompt, the system cancels the order and the user is informed that he will return to the starting point of the program (see Figure 10-22).

Figure 10-22: Prompting the order cancellation message (Image of MADK 2.0.Courtesy of Motorola, Inc.)

The Weather Section

When the user says "weather" at the starting-page prompt, `main.vxml` invokes `weather.vxml` (shown in Listing 10-12). This document provides the user with the weather information for four cities of India (Delhi, Mumbai, Chennai, and Kolkota). When the user says "Delhi," `weather.vxml` offers the current weather information or the forecast for the next day. If the user says "current," `weather.vxml` presents the current weather information. If the user says "forecast," `weather.vxml` provides the forcast for the next day. Likewise, if the user says "Mumbai," "Chennai," or "Kolkota," a similar procedure is followed.

Listing 10-12: weather.vxml

```
<?xml version="1.0"?>
<vxml version="1.0">
   <meta name="author" content="vxml auther team"/>

   <form id="weather">
      <link next="main.vxml">
         <grammar type="application/x-jsgf">
            main menu
         </grammar>
      </link>
      <field name="city">
         <prompt>
            choose a city, or say main menu to return to the main menu.
         </prompt>
   <grammar type="application/x-jsgf">
            delhi | mumbai | chennai | kolkota
         </grammar>
         <nomatch>
            sorry, weather information for the city of is not available
            <value expr="city"/>
         </nomatch>
      </field>
      <field name="when">
         <prompt>
            would you like current conditions or the  forecast. or say help
         </prompt>
         <grammar type="application/x-jsgf">
            current | forecast
         </grammar>
         <help>
            for current weather condition, say 'current'.
            for forecast weather condition, say 'forecast'.
         </help>
      </field>
      <filled namelist="city when">
         weather details for <value expr="city"/>, <value expr="when"/>.
         <if cond="city == 'delhi' &amp;&amp; when == 'current'">
            <prompt>
               it is currently 35 degrees in delhi.
            </prompt>
         <elseif cond="city == 'delhi' &amp;&amp; when == 'forecast'"/>
            <prompt>
               delhi forecast for tuesday. showers. high of 50. low of 44.
```

```
                       wednesday. partly cloudy. high of 39. low of 35.
                  </prompt>
             <elseif cond="city == 'mumbai' &amp;&amp; when == 'current'"/>
                  <prompt>
                       it is currently 31 degrees in dumbai, with showers.
                  </prompt>
             <elseif cond="city == 'mumbai' &amp;&amp; when == 'forecast'"/>
                  <prompt>
                       mumbai forecast for tuesday. cloudy skies with 27 degrees of
high.
                       low of 22. wednesday. showers. high of 27. low of 12.
                  </prompt>
             <elseif cond="city == 'chennai' &amp;&amp; when == 'current'"/>
                  <prompt>
                       it is currently 29 degrees in chennai city, with cloudy skies.
                  </prompt>
             <elseif cond="city == 'chennai' &amp;&amp; when == 'forecast'"/>
                  <prompt>
                       chennai city forecast for tuesday. windy. high of 42. low of 33.
                       wednesday. rain. high of 33. low of 25.
                  </prompt>
             <elseif cond="city == 'kolkota' &amp;&amp; when == 'current'"/>
                  <prompt>
                       it is currently 28 degrees in kolkota, with cloudy skies.
                  </prompt>
             <elseif cond="city == 'kolkota' &amp;&amp; when == 'forecast'"/>
                  <prompt>
                       kolkota forecast for tuesday. partly cloudy. high of 34. low of
28.
                       wednesday. sunny. high of 40. low of 30.
                  </prompt>
             </if>
             <goto next="#weather"/>
        </filled>
   </form>
</vxml>
```

In starting the function, the system asks the user for the name of a city. If the user is not interested in the weather, he can go back to `main.vxml` by saying "main menu" (see Figure 10-23).

Figure 10-23: Prompting the user for the city name (Image of MADK 2.0.Courtesy of Motorola, Inc.)

Let's assume that the user wants to learn about the weather and has selected "Delhi" as the choice. The next dialog system gives the user three options (see Figure 10-24). The options are as follows: current; forecast; help.

Figure 10-24: Prompting the available weather section options (Image of MADK 2.0.Courtesy of Motorola, Inc.)

Here we'll assume that the user says "current" as the choice. Consequently, the system speaks a series of dialogs on the current weather information. All of these dialogs are shown in Figures 10-25 to 10-27.

Figure 10-25: Prompting the weather report for a selected city (continued)

Figure 10-26: Prompting the weather report for a selected city (continued)

Figure 10-27: Prompting the weather report for a selected city (All Images of MADK 2.0.Courtesy of Motorola, Inc.)

Next let's say the user chooses to learn the weather forecast for Delhi. This time, the system gives the user information about the weather forecast for the next two days (Figure 10-28).

Figure 10-28: Prompting weather forcast for a selected city (Image of MADK 2.0.Courtesy of Motorola, Inc.)

The Movies Section

When the user says "movies" at the starting-page prompt, `main.vxml` invokes `movies.vxml` (Listing 10-13). This document provides the user with the latest movie information and reviews. `movies.vxml` initially prompts the user for a six-digit PIN code. If the user enters an invalid PIN code, a `nomatch` event is thrown. If the user enters a valid PIN code, a list of theaters is presented. After choosing a particular theater, the user sees the details of the latest movies running at that particular theater. After saying the name of a particular movie, the user sees a summary of that particular movie.

Listing 10-13: movies.vxml

```xml
<?xml version="1.0"?>
<vxml version="1.0" application="movies_titles.vxml">
 <form id="welcome">


<var name="pin_number"/>

  <help>Please say or type your 6 digit pin number.</help>
  <block>
  <assign name="application.next_dialog" expr="'welcome'"/>
  Welcome to the Voice enabled movie system.
  </block>

  <field name="pin_number" type="digits">
   <prompt>Please say or type your 6 digit pin code number.</prompt>

   <filled>
     <if cond="pin_number.length != 6">
      <clear namelist="pin_number"/>
     <throw event="nomatch"/>
     </if>
     <assign name="application.pin_number" expr="pin_number"/>


     <var name="r" expr="Math.random()"/>
     <if cond="r &lt; 0.30" >
      <assign name="application.next_dialog" expr="'hall_information1'"/>
      <goto next="#hall_information1"/>

     <elseif cond="r &gt;= 0.30 &amp;&amp; r &lt; 0.60"/>
       <assign name="application.next_dialog" expr="'hall_information1'"/>
       <goto next="#hall_information3"/>

     </if>
    </filled>
   </field>
 </form>


<menu id="hall_information1">
  <help> Select a theater for which you want to know the movie listings.
      The choices are Priya, Anupam, and Liberty.
```

```
          Or you may say prompt movie name to hear its review.
          To go back to the main menu, say main menu
    </help>
    <prompt>The following theaters are located within 5 Kms from
    <value expr="application.pin_number"/> pin number.
    </prompt>
    <prompt>
    Pryia, located at 54 Northeast Plaza in Vasant.
    Anupam at Shoping Complex in Rohini.
    Liberty in Indian Mall at second floor in Indian Mall.
    </prompt>
    <prompt>Please say the name of the theater
    to hear its current movie listing. You can also say the movie name
    to hear its review.
    </prompt>

    <choice next="#pryia">priya</choice>
    <choice next="#anupam"> anupam </choice>
    <choice next="#liberty"> liberty </choice>
    </menu>



<menu id="hall_information3">
    <help> Please select a theater for which you want to hear the movie listings.
          The choices are Chankya and Plaza. You can also say
          the movie name to hear its review or say main menu to return to
          the main menu.
    </help>
    <prompt>The following theaters are located within 5 Kms from
    <value expr="application.pin_number"/> pin number.
    </prompt>
    <prompt>
    Chankya, 12/7, Main Road Chankya Puri.
    Plaza, 78 Connnaught Place, New Delhi 110001.
    </prompt>
    <prompt>Please say the name of the theater
    to hear its current movie listing. You can also say the movie name
    to hear its review.
    </prompt>
    <choice next="#chankya"> chankya </choice>
    <choice next="#plaza"> plaza </choice>
    </menu>



<menu id="pryia">
    <help>Say 'another theater' to hear a movie listing for a different theater.
Say
        'another pin number' to locate theaters in a different pin number. Say
'main menu'
        to return to the main menu. You can also say the movie name to hear its
review.
    </help>
    <prompt>AKS at 12.30.
```

```
      AKS at 6:30.
      Daman at 8:30.

   To hear a movie listing for a different theater in <value
expr="application.pin_number"/>
   pin number say another theater. To hear a movie listing for a theater in a
different
   pin number say 'another pin number'. To listen to a movie review say the name
   of the movie. To return to the main menu say 'main menu'.
  </prompt>
  <choice next="#hall_information1"> another theater </choice>
  <choice next="#welcome"> another pin number </choice>
 </menu>

 <menu id="anupam">
  <help>Say 'another theater' to hear a movie listing for a different theater.
Say
    'another pin number' to locate theaters in a different pin number. Say
'main menu'
    to return to the main menu. You can also say the movie name to hear its
review.
  </help>
  <prompt>lagan at 4:00.
     Lagan at 6:30.
     Daman at 8:00.

   To hear a movie listing for a different theater in <value
expr="application.pin_number"/>
   pin number say another theater. To hear a movie listing for a theater in a
different
   pin number say 'another pin number'. To listen to a movie review say the name
   of the movie. To return to the main menu say 'main menu'.
  </prompt>
  <choice next="#hall_information1"> another theater </choice>
  <choice next="#welcome"> another pin number </choice>
 </menu>

 <menu id="liberty">
  <help>Say 'another theater' to hear a movie listing for a different theater.
Say
    'another pin number' to locate theaters in a different pin number. Say
'main menu'
    to return to the main menu. You can also say the movie name to hear its
review.
  </help>
  <prompt>AKS at 12:30.
     AKS at 6:30.
     Lagan at 8:30.

   To hear a movie listing for a different theater in <value
expr="application.pin_number"/>
   pin number say another theater. To hear a movie listing for a theater in a
different
   pin number say 'another pin number'. To listen to a movie review say the name
   of the movie. To return to the main menu say 'main menu'.
  </prompt>
```

```
  <choice next="#hall_information1"> another theater </choice>
  <choice next="#welcome"> another pin number </choice>
 </menu>



 <menu id="chankya">
  <help>Say 'another theater' to hear a movie listing for a different theater.
Say
      'another pin number' to locate theaters in a different pin number. Say
'main menu'
      to return to the main menu. You can also say the movie name to hear its
review.
  </help>
  <prompt>Lagan at 12:30.
       Daman at 7:00.
       Pyar Tune Kya Kiya at 9:30.

    To hear a movie listing for a different theater in <value
expr="application.pin_number"/>
    pin number say another theater. To hear a movie listing for a theater in a
different
    pin number say 'another pin number'. To listen to a movie review say the name
    of the movie. To return to the main menu say 'main menu'.
  </prompt>
  <choice next="#hall_information3"> another theater </choice>
  <choice next="#welcome"> another pin number </choice>
 </menu>



 <menu id="plaza">
  <help>Say 'another theater' to hear a movie listing for a different theater.
Say
      'another pin number' to locate theaters in a different pin number. Say
'main menu'
      to return to the main menu. You can also say the movie name to hear its
review.
  </help>
  <prompt>AKS at 12:30.
       Daman at 7:00.
       Pyar Tune Kya Kiya at 9:30.

    To hear a movie listing for a different theater in <value
expr="application.pin_number"/>
    pin number say another theater. To hear a movie listing for a theater in a
different
    pin number say 'another pin number'. To listen to a movie review say the name
    of the movie. To return to the main menu say 'main menu'.
  </prompt>
  <choice next="#hall_information3"> another theater </choice>
  <choice next="#welcome"> another pin number </choice>

 </menu>


</vxml>
```

When this file is executed, the user first gets a welcome message (Figure 10-29) followed by a request asking for the PIN code for the geographic area in which the user resides (Figure 10-30).

Figure 10-29: Welcome dialog of the movies section (Image of MADK 2.0.Courtesy of Motorola, Inc.)

Figure 10-30: Prompting the user for his pin code number (Image of MADK 2.0.Courtesy of Motorola, Inc.)

When the user enters his PIN code number, the system provides a list of all the theaters in that particular area (Figures 10-31 and 10-32) and asks for a theater name from the list. The user here also has the option to mention the name of a movie to see its review before actually selecting the theater (Figure 10-33).

Figure 10-31: Prompting the theater listings(continued) (Image of MADK 2.0.Courtesy of Motorola, Inc.)

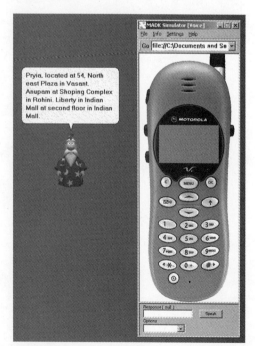

Figure 10-32: Prompting the theater listing (continued) (Image of MADK 2.0.Courtesy of Motorola, Inc.

Figure 10-33: Prompting the user for a theater name and a movie name (Image of MADK 2.0.Courtesy of Motorola, Inc.)

Let's assume that the user asks for "Plaza." The system now gives information about the movies being screened in the plaza theater (Figure 10-34). From this point on, if the user wants to hear the movie listing for a different theater, he can say "another theater." If the user wants to hear the movie listing in a different area, he can say "another PIN number."

Figure 10-34: Prompting movies listing for the selected theater (Image of MADK 2.0.Courtesy of Motorola, Inc.)

Figure 10-35: Prompting available options (Image of MADK 2.0.Courtesy of Motorola, Inc.)

Here the user says "Lagan" as the movie name to hear its review. Immediately, the system reviews the film "Lagan"(see Figure 10-36).

Figure 10-36: Prompting movie analysis (Image of MADK 2.0.Courtesy of Motorola, Inc.)

The movie_titles.vxml file in Listing 10-14 presents the user with the names of the latest movie titles.

Listing 10-14: movies_titles.vxml

```
<?xml version="1.0"?>
<vxml version="1.0">
 <var name="next_dialog" expr="'welcome'"/>
 <var name="pin_number"/>
 <link next="main.vxml">
  <grammar type="application/x-gsl">
   [ (main menu) (main menu please) ]
  </grammar>
 </link>
 <menu scope="document">
  <choice next="movies_analysis.vxml#lagan"> lagan </choice>
  <choice next="movies_analysis.vxml#daman"> daman </choice>
  <choice next="movies_analysis.vxml#pyar_tune_kya_kiya"> pyar tune kya kiya
</choice>
  <choice next="movies_analysis.vxml#AKS"> AKS </choice>
 </menu>
</vxml>
```

The `movie_analysis.vxml` document in Listing 10-15 presents the user with the analysis of the latest movie titles.

Listing 10-15: movie_analysis.vxml

```
<?xml version="1.0"?>
<vxml version="1.0" application="movies_titles.vxml">

 <form id="lagan">
  <block>This is the new one from the amir khan under his home production, he is
hoping for big success with this movie.
   <goto expr="'movies.vxml#'+application.next_dialog"/>
  </block>
 </form>

 <form id="daman">
  <block>Directed by Kalpna Laazmi, this movie is based on violence against
women by their husbands.
   <goto expr="'movies.vxml#'+application.next_dialog"/>
  </block>
 </form>

 <form id="pyar_tune_kya_kiya">
  <block>A multistar cast-based movie. Staring Urmila and Fradeen Khan and
directed by Ram Gopal Verma. Nice to watch.
   <goto expr="'movies.vxml#'+app lication.next_dialog"/>
  </block>
 </form>

 <form id="AKS">
  <block>This is a thriller movie once again starring Amitabh Bachhan.
```

```
  <goto expr="'movies.vxml#'+application.next_dialog"/>
  </block>
 </form>

 <form id="Lagan">
  <block>This is a new one from the amir khan under his home production, he is
hoping a big success from this movie.
  <goto expr="'movies.vxml#'+application.next_dialog"/>
  </block>
 </form>
</vxml>
```

Summary

VoiceXML is up-and-coming and perhaps the most promising technology for building efficient and economical telephone-based speech-recognition applications. The future of VoiceXML lies with the applications that will allow customers to access information in a company database over the phone. The VoiceXML 1.0 specification states the following:

> VoiceXML's main goal is to bring the full power of Web development and content delivery to voice-response applications and to free the authors of such applications from low-level programming and resource management. It enables integration of voice services with data services using the familiar client-server paradigm.

Thus, VoiceXML adds a new dimension in voice applications to a wider audience of developers and simplifies the integration of phone-based interfaces with Web infrastructures and existing applications. Its future is bright and the coming years promise to unleash a whole range of applications for almost everything we do and want to do.

Chapter 11

Transforming Your Site for Voice Clients

This chapter covers one of the most vital topics in the whole spectrum of this book. Throughout this book, you have seen the various transformations that are used for the various devices and platforms available on the market today. In all the previous chapters, while transforming the application for various devices and platforms, you saw some basic similarities among these devices and platforms. Some of the features common to all these devices are the following:

♦ All of these devices are capable of showing text using a visual display.

♦ All of them have some kind of keyboard to enter the character data.

♦ These devices can be connected to the Internet and can download data from the Internet.

By carrying out the transformation for all these devices, you can undoubtedly target many users. However, it is imperative to consider a situation in which a person in a remote area can use the services with an ordinary telephone connection and without a device with the above characteristics. You will appreciate the relevance of such a situation when you consider the following facts:

♦ There are more phones in the world than any other communciation device.

♦ Phones have established themselves as a potent medium for transferring data over networks.

With regard to the growing popularity of phones, let's consider transforming the application for ordinary telephone users. This application is free and is accessible to anyone in the world, provided he has a telephone line. To access online information, all the user needs to do is dial a telephone number (not even with a modem).

This chapter will focus on using VoiceXML to build up the interactive voice-response system for the Mobile Portal application using the XML- and XSLT-based approach. You will learn the processes for transforming all four parts of the application. After you go through this chapter, you will be able to enable the user to access the application using a normal PSTN network. We are using the VoiceXML 1.0 specification for the development process. Several other implementations are available from different companies such as Tellme Networks, Voicegenie, and Nuance. For writing the grammar across the application, we are using the nuance GSL format. For testing the application in a simulated environment, we are using the Motorola Application Development Kit Beta 2.0, which can be downloaded from `http://www.motorola.com`. (Note that VoiceXML 2.0 is scheduled to be released by the end of 2001. It will work in conjunction with many other languages for improved control over speech markup and dialog enhancement. For more information, visit `http://www.w3c.org/voice`.

Develop the Application's Home Page

In this section, you use the `main.xml` file as XML data and the `mainvxml.xsl` file as the XSL document for the transformation process of the home page. Listing 11-1 contains the code for `mainvxml.xsl`.

Listing 11-1: mainvxml.xsl

© 2001 Dreamtech Software India Inc.

All Rights Reserved

```
<?xml version="1.0" ?>
<xsl:stylesheet xmlns:xsl="http://www.w3.org/TR/WD-xsl">
<xsl:template match="/">

<vxml version="1.0">

 <link next="http://192.168.1.100/voice.asp">
  <grammar type="application/x-gsl">
  main menu
  </grammar>
 </link>


 <menu>
 <prompt >Please say or type one of the following sections to
proceed.<enumerate/> Or say help for further assistance.</prompt>

<choice dtmf="1" next="http://192.168.1.100/xml.asp?sec=Weather">
   Weather
</choice>

<choice dtmf="2" next="http://192.168.1.100/xml.asp?sec=News">
   News
</choice>

<choice dtmf="3" next="http://192.168.1.100/xml.asp?sec=Email">
   E-mail
</choice>

<choice dtmf="4" next="http://192.168.1.100/xml.asp?sec=Movies">
   Movies
</choice>

<choice dtmf="5" next="#exit">Exit</choice>

    <help>
        For weather information of various cities, say 'weather or press 1'.
        For the latest news updates from around the globe, say 'News or press
2'.
        To access your mail box, say 'Email or press 3'.
        To book movie tickets, say 'Movies or press 4'.
        To exit from this application, say 'Exit or press 5'.
        To hear this message again, say 'help'.
        To start this application again, say main menu.
    </help>

  <noinput>I didn't hear anything. Please choose any section from the
following<enumerate/></noinput>
  <nomatch> I didn't get that. Please say or type your choice
again<enumerate/></nomatch>
```

```
   <catch event="nomatch noinput" count="4">

    <prompt> You have exceeded the limits of retry. System will now stop the
application. </prompt>
    <throw event="telephone.disconnect.hangup"/>

   </catch>

 </menu>

 <form id="exit">
  <block> Thanks for using this application </block>
  <catch event="exit">
   <throw event="telephone.disconnect.hangup"/>
  </catch>

 </form>

 </vxml>

</xsl:template>
</xsl:stylesheet>
```

Consider the code of the `mainvxml.xsl` file. Here you are not fetching any kind of data from the XML document; instead, you generate a complete VXML document using the XSL document. At the beginning of the code, you create a `<link>` statement and define the grammar for it using the GSL format. During the execution of the document, if the caller says "main menu," the application restarts from the top level. This is achieved by using the following lines of code:

```
<link next="http://192.168.1.100/voice.asp">
  <grammar type="application/x-gsl">
  main menu
  </grammar>
</link>
```

After creating a top-level link and defining the grammar for it, you now start constructing a `<menu>` item. Under the `<menu>` tag, you use the `<choice>` tag to provide the different options to the user by typing the following lines of code:

```
<choice dtmf="1" next="http://192.168.1.100/xml.asp?sec=Weather">
    Weather
</choice>

<choice dtmf="2" next="http://192.168.1.100/xml.asp?sec=News">
    News
</choice>

<choice dtmf="3" next="http://192.168.1.100/xml.asp?sec=Email">
    E-mail
</choice>

<choice dtmf="4" next="http://192.168.1.100/xml.asp?sec=Movies">
    Movies
</choice>

<choice dtmf="5" next="#exit">Exit</choice>
```

In the <choice> element, you also define DTMF values for each option so that the user can either press the corresponding button on his telephone or speak the option he wants to select. After the user has specified his choice, the program calls the file named xml.asp with the parameter of the option that the user selected. After the <choice> element, you move on to handling different kinds of events that might have occurred during the call.

First, let's tackle a situation in which the user neither speaks anything nor presses the corresponding telephone button to make a selection. In such a case, the system will again ask the user to make a selection and repeat the options for the caller. This is achieved with the VoiceXML <noinput> element.

The next possible case is getting a wrong or invalid input from the user. In such a case, the system informs the user of the unacceptable response and asks him to again enter the correct response from the available options. The code for both the <noinput> and <nomatch> uses is given in the XSL document as follows:

```
<noinput>
   I didn't hear anything. Please choose any section from the
following<enumerate/>
 </noinput>


<nomatch>
   I didn't get that. Please say or type your choice again.<enumerate/>
 </nomatch>
```

Next, if no response is received from the user for a long time, or if the user makes wrong choices repeatedly, the program proceeds to terminate the call to reduce the load on the server. This time you will use the <catch> element to catch the <noinput> and the <nomatch> events to get the number of times these events occurred during the call. If any of these events occurs more than four times during the call, the program terminates the call by throwing an event that contains the commands for hanging up the call for that particular user. The following lines of code do this job:

```
<catch event="nomatch noinput" count="4">

   <prompt> You have exceeded the limits of retry. System will now stop the
application. </prompt>
    <throw event="telephone.disconnect.hangup"/>
     </catch>
```

Now look at the ASP file that you have been using for loading both XML and XSL documents and then performing the transformation for the voice browser that interacts with the caller.

We are using the voice.asp file (Listing 11-2) for loading the home page documents. This file also works as the application's home page for voice browsers.

Listing 11-2: voice.asp

© 2001 Dreamtech Software India Inc.
All Rights Reserved

```
<%@ Language=VBScript %>
<%
  dim xml, xsl

     Set XMLDoc = Server.CreateObject("Microsoft.XMLDOM")
     Set XSLDoc = Server.CreateObject("Microsoft.XMLDOM")
     xml="main.xml"
     xsl="mainvxml.xsl"
```

```
        XSLDoc.async = False
        XMLDoc.async = False


        XMLDoc.load(Server.MapPath(xml))
        XSLDoc.load(Server.MapPath(xsl))

        Response.ContentType = "text/xml"

        Response.Write("<?xml version= ""1.0""?>")
        Response.Write (vbCrlf)
        Response.Write(XMLDoc.transformNode(XSLDoc))

%>
```

One notable line of code in Listing 11-2 is the following:

```
        Response.ContentType = "text/xml"
```

When we started developing this application for voice clients, we faced a major problem in detecting the content type for the voice application, because the browser in this case is a normal telephone, which does not provide any kind of HTTP_USER_AGENT string in the process of calling the server. After a lot of R&D, we decided to keep the content-type setting as "text/xml" when generating the VoiceXML data using XML and XSL. This works fine with the complete application. VoiceXML is an application of XML. That's why we chose this content type while generating the VoiceXML content on-the-fly using Active Server Pages.

Listing 11-3 shows the results of the transformation.

Listing 11-3: Results of the transformation

```
<?xml version= "1.0"?>
<vxml version="1.0">
<link next="http://192.168.1.100/voice.asp">
<grammar type="application/x-gsl">
  main menu
  </grammar>
</link>
<menu>
<prompt>Please say or type one of the following sections to proceed.<enumerate
/> Or say help for further assistance.</prompt>
<choice dtmf="1"
next="http://192.168.1.100/xml.asp?sec=Weather">Weather</choice>
<choice dtmf="2" next="http://192.168.1.100/xml.asp?sec=News">News</choice>
<choice dtmf="3" next="http://192.168.1.100/xml.asp?sec=Email">E-mail</choice>
<choice dtmf="4" next="http://192.168.1.100/xml.asp?sec=Movies">Movies</choice>
<choice dtmf="5" next="#exit">Exit</choice>
<help>
        For weather information of various cities, say 'weather or press 1'.
        For the latest news updates from around the globe, say 'News or press
2'.
        To access your mailbox , say 'Email or press 3'.
        To book movie tickets, say 'Movies or press 4'.
        To exit from this application, say 'Exit or press 5'.
```

```
               To hear this message again, say 'help'.
               To start this application again, say main menu.
        </help>
<noinput>I didn't hear anything. Please choose any section from the
following.<enumerate /></noinput>
<nomatch> I didn't get that. Please say or type your choice again. <enumerate
/></nomatch>
<catch event="nomatch noinput" count="4">
<prompt> You have exceeded the limits of retry. System will now stop the
application. </prompt>
<throw event="telephone.disconnect.hangup" />
</catch>
</menu>
<form id="exit">
<block>Thanks for using this application.</block>
<catch event="cancel">
<throw event="telephone.disconnect.hangup" />
</catch>
</form>
</vxml>
```

Figure 11-1 shows the output of the preceding code on the Motorola Application Development Kit VoiceXML Simulator. If the user does not respond for a long time, the system will again seek a response by sending the message shown in Figure 11-2.

Figure 11-1: The home page prompt (Image of MADK 2.0.Courtesy of Motorola, Inc.)

Figure 11-2: The output of a noinput event (Image of MADK 2.0.Courtesy of Motorola, Inc.)

The home page includes five options to proceed further. These sections are the following:

♦ Weather

♦ News

♦ E-mail

♦ Movies

♦ Exit

When the user selects any of these sections, the application calls the `xml.asp` file with a parameter containing a value that corresponds to the user's selection. In `xml.asp`, you generate the XML data for the selected section based on the parameter passed to you from the application's home page. If the user makes a wrong selection or presses a wrong button on the telephone, the system will give him an error message and ask him to make the choice again (see Figure 11-3).

Figure 11-3: The output of a nomatch event (Image of MADK 2.0.Courtesy of Motorola, Inc.)

One more event is used in the document. This event is for terminating the call by hanging up the phone. Therefore, if the user makes a wrong selection more than four times or does not make any selection from the available categories being reprompted four times, the call will be terminated. Figure 11-4 is the screen shot of the call-termination dialog.

Figure 11-4: The call-termination message (Image of MADK 2.0.Courtesy of Motorola, Inc.)

The following is a possible dialog flow on the application's home page:

1. **System:** Please say or type one of the following sections to proceed: Weather, News, E-mail, Movies, Exit, or say Exit for further assistance.

2. **Caller input:** [Help]

3. **System:** For weather information of various cities, say "weather" or press 1; for the latest news updates from around the globe, say "News" or press 2; to access your mail box, say "E-mail" or press 3; to book movie tickets, say "Movies" or press 4; to exit from this application, say "Exit" or press 5; to hear this message again, say "help"; to start this application again, say "main menu."

4. **Caller input:** [Wrong input]

5. **System:** Fires <nomatch> event.

6. **Caller input:** [No input]

7. **System:** Fires <noinput> event.

8. **Caller input:** [Exit]

9. **System:** Thanks for using this application. [Application terminates.]

This completes the first transformation process of the application's home page. In the next section, you transform the weather section of the application for use by voice clients.

Transform the Weather Section

You first need to generate the XML data for the weather section. You are using the `xml.asp` file to generate the XML data for the weather section of the application, and then you use the `xsl.asp` file to load the XML and XSL documents and perform the transformation task for voice clients.

Listing 11-4 shows the code for `xml.asp`.

Listing 11-4: xml.asp

```
<%@ Language=VBScript %>

<%
dim address,section,fname,ctg,city,emode
ctg =" "
city=" "
emode =" "
section = " "
section = trim(Request.QueryString("sec"))
emode = trim(Request.QueryString("mode"))
ctg = trim(Request.QueryString("category"))
city = trim(Request.QueryString("city"))

set conn = Server.CreateObject("ADODB.Connection")
conn.ConnectionString = "Provider=SQLOLEDB.1;Persist Security Info=False;User
ID=sa;Initial Catalog=portal;Data Source=developers"
conn.Open

Set objrs = Server.CreateObject("ADODB.Recordset")
objrs.CursorType = adOpenStatic

if (section="Weather")  then
  sql = "select city from weather"
  objrs.Open sql,conn

  Set fso =Server.CreateObject("Scripting.FileSystemObject")
  fname = "tweather.xml"

  fullpath = server.MapPath("main")

  Set MyFile=fso.CreateTextFile(fullpath & "/" & fname)

  MyFile.WriteLine("<?xml version='1.0'?>")
  MyFile.WriteLine("<ROOT>")


  while not(objrs.EOF)
    city = objrs.Fields("city")
    MyFile.WriteLine("<city>" & city & "</city>")

  objrs.MoveNext
  wend
```

```
  MyFile.WriteLine("</ROOT>")
    address = "http://192.168.1.100/xsl.asp?name=" & fname
  Response.Redirect(address)
end if
objrs.Close
conn.Close

%>
```

In the preceding code, you have generated a list of all the available cities in the database and then saved this data in XML format using the file system object (FSO) on the server. After generating all the XML data, you call the `xsl.asp` file (Listing 11-5) to load the XML and XSL files and perform the transformation process.

Listing 11-5: xsl.asp

```
<%@ Language=VBScript %>
<%

dim xml, xsl
fname = trim(Request.QueryString("name"))
xml = fname
fname = mid(fname,1,(len(fname) - 4))
xsl = fname & "vxml.xsl"

     Set XMLDoc = Server.CreateObject("Microsoft.XMLDOM")
     Set XSLDoc = Server.CreateObject("Microsoft.XMLDOM")

     XSLDoc.async = False
     XMLDoc.async = False

     XMLDoc.load(Server.MapPath(xml))
     XSLDoc.load(Server.MapPath(xsl))

     Response.ContentType = "text/xml"
     Response.Write("<?xml version= ""1.0""?>")
     Response.Write (vbCrlf)
     Response.Write(XMLDoc.transformNode(XSLDoc))

%>
```

The preceding code is comparatively simple. You use it to load the XML and XSL documents. You are selecting the XSL document based on the XML document that you are going to load for the selected section. Before transformation, you set the content type to "text/xml" for VoiceXML. In the last line, you transform the XML data. Listing 11-6 shows the code for `tweathervxml.xsl`.

Listing 11-6: tweathervxml.xsl

```
<?xml version="1.0" ?>
<xsl:stylesheet xmlns:xsl="http://www.w3.org/TR/WD-xsl">
```

```
<xsl:template match="/">

<vxml version="1.0">

  <link next="http://192.168.1.100/voice.asp">
  <grammar type="application/x-gsl">
  main menu
  </grammar>
  </link>


  <menu>
  <prompt> Welcome to the weather section. </prompt>
  <prompt>Please choose one of the following cities.<enumerate/></prompt>

  <xsl:for-each select="ROOT/city">
  <choice>

    <xsl:attribute name="next">
    http://192.168.1.100/xml.asp?city=<xsl:value-of select="." />
    </xsl:attribute>
    <xsl:value-of select="." />
  </choice>
  </xsl:for-each>

  <noinput>Please say one of the following cities. <enumerate/></noinput>

    <nomatch>
            Sorry, we currently do not provide weather information for the city
you chose.
            Please choose one of the following cities or say main menu to start
the application again.
                <enumerate/>
    </nomatch>


  <catch event="nomatch noinput" count="4">

    <prompt> You have exceeded the limits of retry. System will now stop the
application. Going back to main menu</prompt>
    <goto next="http://192.168.1.100/voice.asp" />

  </catch>



  </menu>
  </vxml>

</xsl:template>

</xsl:stylesheet>
```

Now let's discuss Listing 11-6 step by step. Having generated the <vxml> version info, you provide a link to the user for going back to the application's home page by saying "main menu" anytime during the call.

In the next part of the code, you construct a `<menu>` element and prompt the user to select any one of the cities for which a weather report is available. After prompting the user, you settle down to generate a list of all the cities by fetching the records from the XML document by employing the following lines of code:

```
<xsl:for-each select="ROOT/city">
<choice>

   <xsl:attribute name="next">
   http://192.168.1.100/xml.asp?city=<xsl:value-of select="." />
   </xsl:attribute>
   <xsl:value-of select="." />
</choice>
</xsl:for-each>
```

This code generates a series of `<choice>` elements containing the names of all the cities available in the XML document. You then link these elements with a parameter, which is passed when the user selects a particular choice.

After generating a series of `<choice>` elements, you deal with the two `<noinput>` and `<nomatch>`events for handling the wrong input. You do this if you get no response from the user, by using the following lines of code:

```
<noinput>Please say one of the following cities. <enumerate/></noinput>

  <nomatch>
          Sorry, we currently do not provide weather information for the city
you chose.
          Please choose from one of the following cities, or say main menu to
start the application again.
             <enumerate/>
   </nomatch>
```

After handling both the events, you use the `<catch>` element to catch the number of occurrences of the `<nomatch>` and `<noinput>` events during this document's execution. If the user exceeds the number of permissible retries, you again go back to the application's home page and inform the user.

```
  <catch event="nomatch noinput" count="4">
   <prompt> You have exceeded the limits of retry. System will now stop the
application. Going back to main menu.</prompt>
   <goto next="http://192.168.1.100/voice.asp" />

   </catch>
```

This completes the XSL documents for transforming the `tweather.xml` file. When this code executes on the server, it generates the VXML document in Listing 11-7 on-the-fly.

Listing 11-7: Transformation result of tweather.xml

```
<?xml version= "1.0"?>
<vxml version="1.0">
<link next="http://192.168.1.100/voice.asp">
<grammar type="application/x-gsl">
   main menu
   </grammar>
```

```
</link>
<menu>
<prompt> Welcome to the weather section. </prompt>
<prompt>Please choose one of the following cities.<enumerate /></prompt>

<choice next="http://192.168.1.100/xml.asp?city=Sydney">Sydney</choice>

<choice next="http://192.168.1.100/xml.asp?city=Miami">Miami
</choice>

<choice next="http://192.168.1.100/xml.asp?city=Phoenix">Phoenix
</choice>

<choice next="http://192.168.1.100/xml.asp?city=Tokyo">Tokyo</choice>

<choice next="http://192.168.1.100/xml.asp?city=NewDelhi">NewDelhi</choice>
<choice next="http://192.168.1.100/xml.asp?city=NewYork">NewYork</choice>

<noinput>Please say one of the following cities. <enumerate /></noinput>

<nomatch>
Sorry, we currently do not provide weather information for the city you
chose.Please choose one of the following cities or say main menu to start the
application again.<enumerate />
</nomatch>

<catch event="nomatch noinput" count="4">

<prompt> You have exceed the limits of retry. System will now stop the
application. Going back to main menu.
</prompt>

<goto next="http://192.168.1.100/voice.asp" />
</catch>

</menu>

</vxml>
```

When this file is run on the simulator, it shows the output in Figure 11-5. Figure 11-6 shows the list of cities with which the user is presented. Figures 11-7 and 11-8 show the output in the case of a `noinput` event. Figure 11-9 shows the output of a `nomatch` event.

Figure 11-5: The weather section's welcome message (Image of MADK 2.0.Courtesy of Motorola, Inc.)

Figure 11-6: The prompt that lists the choices of cities (Image of MADK 2.0.Courtesy of Motorola, Inc.)

Figure 11-7: The noinput event message (Image of MADK 2.0.Courtesy of Motorola, Inc.)

Figure 11-8: The prompt for a city name in a noinput event (Image of MADK 2.0.Courtesy of Motorola, Inc.)

Figure 11-9: The error message for a nomatch event (Image of MADK 2.0.Courtesy of Motorola, Inc.)

When the user speaks any of the city names to listen to the weather report for that city, the application again calls the `xml.asp` file with the parameter `city`. This parameter contains the name of the city that the user speaks as the input. Say, for example, the user inputs the city Tokyo. When you give this input to the system, it calls the `xml.asp` file to generate the detailed report based on the input. Listing 11-8 is the code for generating the detailed weather report.

Listing 11-8: Code for a detailed weather report for a specific city

© 2001 Dreamtech Software India Inc.
All Rights Reserved

```
if ((city="Sydney") OR (city="NewDelhi") OR (city="Miami") OR (city="Tokyo") OR
(city="Phoenix") OR (city="NewYork"))   then
   sql = "select * from weather where city = '" & city & "'"
   objrs.Open sql,conn

   Set fso =Server.CreateObject("Scripting.FileSystemObject")
   fname = "tweatherdetail.xml"
   fullpath = server.MapPath("main")

   Set MyFile=fso.CreateTextFile(fullpath & "/" & fname)

   MyFile.WriteLine("<?xml version='1.0'?>")
   MyFile.WriteLine("<ROOT>")


   while not(objrs.EOF)
      lowtemp = objrs.Fields("low_temp")
      hightemp = objrs.Fields("high_temp")
```

```
    srise = objrs.Fields("sunrise")
    sset = objrs.Fields("sunset")
    mrise = objrs.Fields("moonrise")
    mset = objrs.Fields("moonset")
    dhumid = objrs.Fields("dayhumidity")
    nhumid = objrs.Fields("nighthumidity")
    rain = objrs.Fields("rainfall")
    fig = objrs.Fields("figure")

    MyFile.WriteLine("<weather>")
    MyFile.WriteLine("<low_temp>" & lowtemp & "</low_temp>")
    MyFile.WriteLine("<high_temp>" & hightemp & "</high_temp>")
    MyFile.WriteLine("<sunrise>" & srise & "</sunrise>")
    MyFile.WriteLine("<sunset>" & sset & "</sunset>")
    MyFile.WriteLine("<moonrise>" & mrise & "</moonrise>")
    MyFile.WriteLine("<moonset>" & mset & "</moonset>")
    MyFile.WriteLine("<dayhumidity>" & dhumid & "</dayhumidity>")
    MyFile.WriteLine("<nighthumidity>" & nhumid & "</nighthumidity>")
    MyFile.WriteLine("<rainfall>" & rain & "</rainfall>")
    MyFile.WriteLine("<figure>" & fig & "</figure>")
    MyFile.WriteLine("</weather>")
  objrs.MoveNext
  wend

  MyFile.WriteLine("</ROOT>")
    address = "http://192.168.1.100/xsl.asp?name=" & fname
  Response.Redirect(address)
end if
```

The preceding code generates a detailed weather report from the database and saves all the records in XML format in the file `tweatherdetail.xml` (Listing 11-9) on the server. In `xsl.asp`, the system calls the `tweatherdetailvxml.xsl` file to transform this file into a VXML document.

Listing 11-9: tweatherdetailvxml.xsl

```
<?xml version="1.0" ?>
<xsl:stylesheet xmlns:xsl="http://www.w3.org/TR/WD-xsl">
<xsl:template match="/">
<vxml version="1.0">

<form id="weather">
 <block>
  <xsl:apply-templates select="ROOT" />
  <goto next="#call"/>
 </block>
</form>

<form id="call">

 <field name="check">
  <prompt> Say another city for a different city, or say main menu to return to
application home. </prompt>
```

```
    <grammar type="application/x-gsl">
     [  (another city)(main menu)  ]
    </grammar>

    <filled>
     <if cond="check=='another city'">
      <goto next="http://192.168.1.100/xml.asp?sec=Weather"/>
     <elseif cond="check=='main menu'"/>
      <goto next="http://192.168.1.100/voice.asp"/>
     </if>
    </filled>
  </field>
</form>
</vxml>
</xsl:template>

<xsl:template match="ROOT">
<xsl:for-each select="weather">
   low temperature <xsl:value-of select="low_temp" /> degrees celsius
   high temperature <xsl:value-of select="high_temp" />degrees celsius
   Sunrise <xsl:value-of select="sunrise" />A.M.
   Sunset  <xsl:value-of select="sunset" />P.M.
   Moonrise <xsl:value-of select="moonrise" />P.M.
   Rainfall<xsl:value-of select="rainfall" />
</xsl:for-each>
</xsl:template>

</xsl:stylesheet>
```

The preceding code generates a VXML document on-the-fly. You are generating two forms in the document. In the first, you are simply applying a template and then transferring the controls to the next form in the document. In the second form, you prompt the user for his choice. The available choices are the following:

- ♦ Say "main menu" to go back to the home page of the application.
- ♦ Say "another city" to check the weather of a different city.

Here you also define a grammar for accepting the user input to determine which of the two available options was chosen. This is done in the following way:

```
    <grammar type="application/x-gsl">
     [  (another city)(main menu)  ]
    </grammar>
```

When the user gives the input, you use an `if` condition to check the input and call two different files on the basis of the input the user gives.

```
    <filled>
     <if cond="check=='another city'">
      <goto next="http://192.168.1.100/xml.asp?sec=Weather"/>
     <elseif cond="check=='main menu'"/>
      <goto next="http://192.168.1.100/voice.asp"/>
     </if>
    </filled>
```

The next template fetches the values from the XML document and then converts them into VoiceXML. This is achieved by using the following lines of code:

```
<xsl:template match="ROOT">
<xsl:for-each select="weather">
    low temperature <xsl:value-of select="low_temp" /> degrees celsius
    high temperature <xsl:value-of select="high_temp" />degrees celsius
    Sunrise <xsl:value-of select="sunrise" />A.M.
    Sunset  <xsl:value-of select="sunset" />P.M.
    Moonrise <xsl:value-of select="moonrise" />P.M.
    Rainfall<xsl:value-of select="rainfall" />
</xsl:for-each>
</xsl:template>
```

When this transformation takes place on the server side, it generates the VXML document in Listing 11-10 as the output.

Listing 11-10: Transformation result of tweatherdetailvxml.xml

```
<?xml version= "1.0"?>
<vxml version="1.0">
<form id="weather">

<block>
    low temperature 35 Degree Celsius
    high temperature 40Degree Celsius
    Sunrise 06:25:00A.M.
    Sunset  18:25:00P.M.
    Moonrise 07:25:00P.M.
    Rainfall21mm
<goto next="#call" />
</block>
</form>

<form id="call">
<field name="check">
<prompt> Say another city for a different city, or say main menu to return to
application home. </prompt>
<grammar type="application/x-gsl">
    [  (another city)(main menu)  ]
    </grammar>
<filled>
<if cond="check=='another city'">
<goto next="http://192.168.1.100/xml.asp?sec=Weather" />
<elseif cond="check=='main menu'" />
<goto next="http://192.168.1.100/voice.asp" />
</if>
</filled>
</field>
</form>

</vxml>
```

Figures 11-10 and 11-11 show the output on the simulator.

Figure 11-10: The weather report for a city (Image of MADK 2.0.Courtesy of Motorola, Inc.)

Figure 11-11: The transformation of tweatherdetailvxml.xsl (Image of MADK 2.0.Courtesy of Motorola, Inc.)

The following demonstrates a possible dialog flow in the weather section of the application:

1. **System:** Welcome to the weather section. Please choose one of the following: New Jersey, California, Silicon Valley, Tokyo, New Delhi, New York.

2. **Caller input:** [Tokyo]

3. **System:** Low temperature 35 degrees Celsius; high temperature 40 degrees Celsius; sunrise 06:25:00 A.M.; sunset 18:25:00 P.M.; moonrise 07:25:00 P.M.; Rainfall 21mm

4. **System:** Say another city for a different city or say main menu to return to application home.

5. **Caller input:** [main menu]

6. **System:** [Transfers controls to `voice.asp`]

This completes the weather section transformation process for voice clients. In the next section, we will discuss the transformation process of the news section.

Transform the News Section

The news section has four main categories. By the end of this section, the user will be able to listen to the news using the normal telephone network.

Generate XML for the News Section Main Page. First you generate XML data for the news section of the application by using the `xml.asp` file (Listing 11-11). We also use this `xml.asp` file to load the XML and XSL documents and perform the transformation task for voice clients.

Listing 11-11: xml.asp code snippet for generating news section data

© 2001 Dreamtech Software India Inc.
All Rights Reserved

```
if (section="News")   then
  sql = "select DISTINCT category from news"
  objrs.Open sql,conn

  Set fso =Server.CreateObject("Scripting.FileSystemObject")
  fname = "tnews.xml"

  fullpath = server.MapPath("main")

  Set MyFile=fso.CreateTextFile(fullpath & "/" & fname)

  MyFile.WriteLine("<?xml version='1.0'?>")
  MyFile.WriteLine("<ROOT>")


  while not(objrs.EOF)

    category = objrs.Fields("category")

    MyFile.WriteLine("<category>" & category & "</category>")

  objrs.MoveNext
  wend

  MyFile.WriteLine("</ROOT>")
  address = "http://192.168.1.100/xsl.asp?name=" & fname
  Response.Redirect(address)
```

```
end if
```

In the preceding code, you have generated a list of all the categories from the database and then saved this in XML format in the file named `tnews.xml` on the server. Listing 11-12 shows the XSL document that you use to transform this data for voice clients.

Listing 11-12: tnewsvxml.xsl

```
<?xml version="1.0" ?>
<xsl:stylesheet xmlns:xsl="http://www.w3.org/TR/WD-xsl">
<xsl:template match="/">

<vxml version="1.0">

 <link next="http://192.168.1.100/voice.asp">
  <grammar type="application/x-gsl">
  main menu
  </grammar>
</link>


 <menu>
 <prompt> Welcome to News Section. </prompt>
 <prompt>Please choose one of the following sections.<enumerate/></prompt>

 <xsl:for-each select="ROOT/category">
 <choice>

   <xsl:attribute name="next">
   http://192.168.1.100/xml.asp?category=<xsl:value of select="." />
   </xsl:attribute>
   <xsl:value-of select="." />
 </choice>

 </xsl:for-each>

  <noinput>Please say one of the following sections. <enumerate/></noinput>

<nomatch>
    Sorry, we currently do not provide the information for the section you
chose. Please choose one of the following sections or say main menu to start the
application again.<enumerate/>
</nomatch>


  <catch event="nomatch noinput" count="4">

  <prompt> You have exceeded the limits of retry. System will now stop the
application. Going back to main menu.</prompt>
  <goto next="http://192.168.1.100/voice.asp" />

  </catch>


 </menu>
 </vxml>
```

```
</xsl:template>
</xsl:stylesheet>
```

Here it is essential to discuss the style sheet step by step. In the beginning of the document, you are generating a link for the application's home page so that if the user wants to return to the home page during the call, he can do this by simply saying "main menu." The following lines of code easily achieve this particular task:

```
<link next="http://192.168.1.100/voice.asp">
  <grammar type="application/x-gsl">
  main menu
  </grammar>
</link>
```

After defining the link for the application's home page, you now construct a `<menu>` item. Under the menu item, you first pass a welcome message and then speak the series of choices available to the user, which you generate by using the `<xsl:for-each>` and `<xsl:attribute>` elements.

```
<prompt> Welcome to News Section. </prompt>
<prompt>Please choose one of the following sections.<enumerate/></prompt>

<xsl:for-each select="ROOT/category">
<choice>
  <xsl:attribute name="next">
  http://192.168.1.100/xml.asp?category=<xsl:value-of select="." />
  </xsl:attribute>
  <xsl:value-of select="." />
</choice>
</xsl:for-each>
```

After generating the `<choice>` element, you define the `<nomatch>` and `<noinput>` events for this document. Further, you also use the `<catch>` element to count the number of occurrences for both the `<nomatch>` and `<noinput>` elements. If the user exceeds the permitted number of retries, you give him an error message and again load the home page document. All these processes can be seen in the following lines of code:

```
  <noinput>Please say one of the following sections. <enumerate/></noinput>
  <nomatch>
          Sorry, we currently do not provide the information for the section
you chose.
          Please choose one of the following sections or say main menu to start
the application again.
              <enumerate/>
  </nomatch>
  <catch event="nomatch noinput" count="4">
  <prompt> You have exceeded the limits of retry. System will now stop the
application. Going back to main menu.</prompt>
  <goto next="http://192.168.1.100/voice.asp" />
  </catch>
```

When this code executes, it generates a VXML document on-the-fly. Listing 11-13 shows the code for this document.

Listing 11-13: Transformation result of tnews.xml

```
<?xml version= "1.0"?>
<vxml version="1.0">
<link next="http://192.168.1.100/voice.asp">
<grammar type="application/x-gsl">
  main menu
  </grammar>
</link>
<menu>
<prompt> Welcome to the news section. </prompt>
<prompt>Please choose one of the following sections.<enumerate /></prompt>
<choice next="http://192.168.1.100/xml.asp?category=economics">economics
</choice>

<choice next="http://192.168.1.100/xml.asp?category=highlights">highlights
</choice>

<choice next="http://192.168.1.100/xml.asp?category=politics">politics
</choice>

<choice next="http://192.168.1.100/xml.asp?category=sports">sports
</choice>

<noinput>Please say one of the following sections. <enumerate /></noinput>

<nomatch>
Sorry, we currently do not provide the information for the section you chose.
Please choose one of the following sections or say main menu to start the
application again.<enumerate />
</nomatch>

<catch event="nomatch noinput" count="4">
<prompt> You have exceeded the limits of retry. System will now stop the
application. Going back to main menu.</prompt>
<goto next="http://192.168.1.100/voice.asp" />
</catch>
</menu>
</vxml>
```

When you run this code on the simulator, it shows the output in Figures 11-12 and 11-13.

Suppose the caller chooses sports. When the user makes any choice from the available options, we again call the xml.asp file with the category parameter in a query string. This parameter contains the category name the user selected. In this case, it contains the value sports in the parameter category. Next, you'll see how to generate the news details for the given parameter.

Figure 11-12: The transformation of tnews.vxml (Image of MADK 2.0.Courtesy of Motorola, Inc.)

Figure 11-13: Prompting the categories for the user (Image of MADK 2.0.Courtesy of Motorola, Inc.)

Generate XML for Detailed News

In this section, you are generating the news details based on the given parameter. The code in Listing 11-14 is used to generate XML data for the news details section of the application.

Listing 11-14: Code listing for generating news details

```
if ((ctg="highlights") OR (ctg="politics") OR (ctg="economics") OR
(ctg="sports"))   then
  sql = "select * from news where category = '" & ctg & "'"
  objrs.Open sql,conn

  Set fso =Server.CreateObject("Scripting.FileSystemObject")
  fname = "tnewsdetail.xml"

  fullpath = server.MapPath("main")

  Set MyFile=fso.CreateTextFile(fullpath & "/" & fname)

  MyFile.WriteLine("<?xml version='1.0'?>")
  MyFile.WriteLine("<ROOT>")


  while not(objrs.EOF)
    heading = objrs.Fields("heading")
    category = objrs.Fields("category")
    description = objrs.Fields("description")
    newsdate = objrs.Fields("news_date")
    MyFile.WriteLine("<news>")
    MyFile.WriteLine("<category>" & category & "</category>")
    MyFile.WriteLine("<heading>" & heading & "</heading>")
    MyFile.WriteLine("<description>" & description & "</description>")
    MyFile.WriteLine("<news_date>" & newsdate & "</news_date>")
    MyFile.WriteLine("</news>")
  objrs.MoveNext
  wend

  MyFile.WriteLine("</ROOT>")
  address = "http://192.168.1.100/xsl.asp?name=" & fname
  Response.Redirect(address)

end if
```

Now let's analyze the code in Listing 11-14. Here you first check the parameter for the value passed to you and then collect all the records from the database. After getting the records from the database, you now save the data in XML format on the server in the file named tnewsdetail.xml using the file system object (FSO). After saving the data in XML format, you again call the xsl.asp file to transform the data into a VXML document using the tnewsdetailvxml.xsl file (Listing 11-15).

Listing 11-15: tnewsdetailvxml.xsl

```
<?xml version="1.0" ?>
<xsl:stylesheet xmlns:xsl="http://www.w3.org/TR/WD-xsl">
<xsl:template match="/">

<vxml version="1.0">
```

```
 <form id="news">
 <block>
<xsl:apply-templates select="ROOT" />
 <goto next="#call"/>
</block>
</form>

<form id="call">
 <field name="check">
  <prompt> Say news home for a different section, or say main menu to return to
application home. </prompt>


   <grammar type="application/x-gsl">
    [  (news home)(main menu)  ]
   </grammar>

   <filled>
    <if cond="check=='news home'">
     <goto next="http://192.168.1.100/xml.asp?sec=News"/>
    <elseif cond="check=='main menu'"/>
     <goto next="http://192.168.1.100/voice.asp"/>
    </if>
   </filled>

 </field>

</form>


 </vxml>

</xsl:template>

<xsl:template match="ROOT">
<xsl:for-each select="news">
  <xsl:value-of select="heading" />

</xsl:for-each>
</xsl:template>
</xsl:stylesheet>
```

In the preceding code, you are generating a VXML document on-the-fly because of the transformation process of `tweatherdetail.xml`. In this code, you are generating two forms in the VXML document. In the first form, you are simply applying a template and then calling out the second form by using the following lines of code:

```
 <form id="news">
 <block>
<xsl:apply-templates select="ROOT" />
 <goto next="#call"/>
</block>
</form>
```

In the second form, named "call", you first declare a `<field>` element. Then under this `<field>` element, you ask the user to select his choice from the two available options. The options are the following:

- Say "main menu" to go back to the home page of the application.
- Say "new home" to go back to the previous dialog.

You then use the `if` condition to check the user input and, based on that input, transfer control to the desired destination. The following lines of code demonstrate this task:

```
<form id="call">
 <field name="check">
  <prompt> Say news home for a different section, or say main menu to return to
application home. </prompt>
    <grammar type="application/x-gsl">
     [  (news home)(main menu)   ]
    </grammar>
    <filled>
     <if cond="check=='news home'">
      <goto next="http://192.168.1.100/xml.asp?sec=News"/>
     <elseif cond="check=='main menu'"/>
      <goto next="http://192.168.1.100/voice.asp"/>
     </if>
    </filled>
 </field>
</form>
```

In the template named "ROOT", you are simply fetching the headlines of all available news in the category the user selects. Unlike the previous transformation, in which you displayed the full news along with the headings, this time you fetch just the headlines of the news because it saves time in the telephone environment. The following lines of code show the "ROOT" template:

```
<xsl:template match="ROOT">
<xsl:for-each select="news">
  <xsl:value-of select="heading" />


</xsl:for-each>
</xsl:template>
```

When this code executes, it generates the VXML document in Listing 11-16 on-the-fly.

Listing 11-16: Transformation result of tnewsdetail.xml

```
<?xml version= "1.0"?>
<vxml version="1.0">
<form id="news">
<block>
The resurrection of Indian cricket
I was rattled by Harbhajan's spin: Ponting
India defeat Australia by 60 runs
Anand's winning spree halted by Almasi
Mohun Bagan start favourites against Air India
india win
<goto next="#call" />
</block>
</form>
<form id="call">
<field name="check">
```

```
<prompt> Say news home for a different section, or say main menu to return to
application home. </prompt>
<grammar type="application/x-gsl">
    [  (news home)(main menu)  ]
    </grammar>
<filled>
<if cond="check=='news home'">
<goto next="http://192.168.1.100/xml.asp?sec=News"  />
<elseif cond="check=='main menu'" />
<goto next="http://192.168.1.100/voice.asp"  />
</if>
</filled>
</field>
</form>
</vxml>
```

When this code executes on the simulator, it provides the output in Figures 11-14 to 11-16.

Figure 11-14: Speaking news after the user selects a category (Image of MADK 2.0.Courtesy of Motorola, Inc.)

Fig 11-15: Speaking news after the user selects a category

Figure 11-16: Asking for a user selection (Image of MADK 2.0.Courtesy of Motorola, Inc.)

The following is a possible dialog flow in the news section of the application.

1. **System:** Welcome to the news section. Please choose one of the following sections: economics, highlights, politics, sports.

2. **Caller input:** [choice of caller] or

   Caller input: [Noinput]

3. **System:** Please say one of the following sections: economics, highlights, politics, sports

4. **Caller input:** [Wrong Input]

5. **System:** Sorry, we currently do not provide the information for the section you chose. Please choose one of the following sections or say main menu to start the application again: economics, highlights, politics, sports.

6. **Caller input:** [Exceed the limits of retry]

7. **System:** You have exceeded the limits of retry. System will now stop the application. Going back to main menu.

8. **Caller input:** Sports

   [System will prompt all the headlines and call the second form to ask the user's choice.]

9. **System:** Say news home for a different section, or say main menu to return to application home.

10. **Caller input:** main menu

11. **System:** [Transfers control to `voice.asp` by calling it.]

12. **Caller input:** News home

13. **System:** [Goes back to the news section categories page and restarts the news section process again.]

Transform the E-mail Section

By transforming the e-mail section of the application, you enable the user to access his mailbox using the normal telephone network. We are not including the mail-sending option right now. If we provide that option, we have to save a `.wav` or some other kind of audio file on the server and then send this file as an attachment to the user.

Transform the Login Section

Before proceeding further, you have to transform the login section for voice clients. At the time of login, you first check the username and the password for the authentication process. Now let's take a look at `tlogin.xml` (Listing 11-17), the XML file you are using as neutral data for the transformation process.

Listing 11-17: tlogin.xml

© 2001 Dreamtech Software India Inc.
All Rights Reserved

```
<?xml version='1.0'?>
<logininfo>
<username />
<password />
<login />
</logininfo>
```

You have been using this same document throughout this book for all the different transformation processes. This file contains some empty tags. Therefore, you construct the complete VXML document using the XSL document on-the-fly. Listing 11-18 contains the code for `tloginvxml.xsl`.

Listing 11-18: tloginvxml.xsl

```
<?xml version="1.0" ?>
<xsl:stylesheet xmlns:xsl="http://www.w3.org/TR/WD-xsl">
<xsl:template match="/">
<vxml version="1.0">

<link next="http://192.168.1.100/voice.asp">
  <grammar type="application/x-gsl">
  main menu
  </grammar>
</link>

    <var name="uname"/>
    <var name="pwd"/>

<form id="welcome">
 <block>
  Welcome to E-mail Section.
  <goto next="#user"/>
 </block>
</form>

<form id="user">
<field name="uname" type="digits">
   <prompt>Please say your login name</prompt>
   <fillcd>
    <assign name="document.uname" expr="uname"/>
    <goto next="#pass"/>
   </filled>
</field>

</form>

<form id="pass">
   <field name="pwd" type="digits">
        <prompt> please say your password </prompt>
      <filled>
     <assign name="document.pwd" expr="pwd"/>
     <goto next="#call_mail"/>
    </filled>
    </field>
</form>

<form id="call_mail">
<block>
 <var name="loginname"  expr="document.uname"/>
    <var name="password" expr="document.pwd"/>
    <submit next="http://192.168.1.100/receivevxml.asp" method="get"
namelist="loginname password"/>
 </block>
</form>

</vxml>
```

```
</xsl:template>
</xsl:stylesheet>
```

In the preceding code, you generate four forms. At the top of the document, you provide a link to the application's home page and then declare two variables for storing the username and password values. In the first form, you first deliver a welcome message to the user and then transfer control to the next form in the document.

```
<form id="welcome">
 <block>
   Welcome to E-mail Section.
   <goto next="#user"/>
 </block>
</form>
```

In the next form, you ask for the username and then store the value in a document scope variable named `document.uname`. You then call the next form, named `pass`, which you use for getting the value of the password.

```
<form id="user">
<field name="uname" type="digits">
   <prompt>Please say your login name</prompt>
   <filled>
    <assign name="document.uname" expr="uname"/>
    <goto next="#pass"/>
   </filled>
</field>

</form>
```

In the form named `pass`, you prompt the user to provide the password, so that you can authenticate the username and the password with the database. After getting the value from the user, you store this information in a document variable named "pass". You then call the last form in the document for posting to the server all the data you have collected so far for the authentication process.

```
<form id="pass">
    <field name="pwd" type="digits">
        <prompt> please say your password </prompt>
      <filled>
     <assign name="document.pwd" expr="pwd"/>
     <goto next="#call_mail"/>
    </filled>
    </field>
</form>
```

In the last form, named `call_mail`, you declare two variables —`loginname` and `password`— and assign the values from both of the document variables. Then using the `<submit>` element, you post all the data to the server. The file `receivevxml.asp` will handle this data and perform the rest of the task.

```
<form id="call_mail">
<block>
 <var name="loginname"  expr="document.uname"/>
    <var name="password" expr="document.pwd"/>
    <submit next="http://192.168.1.100/receivevxml.asp" method="get"
namelist="loginname password"/>
 </block>
</form>
```

When this code executes, it generates the VXML document in Listing 11-19.

Listing 11-19: Transformation result of tlogin.xml

```
<?xml version= "1.0"?>
<vxml version="1.0">
<link next="http://192.168.1.100/voice.asp">
<grammar type="application/x-gsl">
  main menu
  </grammar>
</link>
<var name="uname" />
<var name="pwd" />
<form id="welcome">
<block>
  Welcome to the e-mail section.
  <goto next="#user" />
</block>
</form>
<form id="user">
<field name="uname" type="digits">
<prompt>Please say your login name.</prompt>
<filled>
<assign name="document.uname" expr="uname" />
<goto next="#pass" />
</filled>
</field>
</form>
<form id="pass">
<field name="pwd" type="digits">
<prompt>Please say your password.</prompt>
<filled>
<assign name="document.pwd" expr="pwd" />
<goto next="#call_mail" />
</filled>
</field>
</form>
<form id="call_mail">
<block>
<var name="loginname" expr="document.uname" />
<var name="password" expr="document.pwd" />
<submit next="http://192.168.1.100/receivevxml.asp" method="get"
namelist="loginname password" />
</block>
</form>
</vxml>
```

When this code runs on the simulator, it gives the output shown in Figures 11-17 to 11-19.

Figure 11-17: The welcome message (Image of MADK 2.0.Courtesy of Motorola, Inc.)

Figure 11-18: Prompting for the username (Image of MADK 2.0.Courtesy of Motorola, Inc.)

Figure 11-19: Prompting for the password (Image of MADK 2.0.Courtesy of Motorola, Inc.)

The following is a possible dialog of the e-mail login section:

1. **System:** Welcome to the e-mail section.

   System: Please say your login name.

2. **Caller input:** [provide login name]

3. **System:** Please say your password.

4. **Caller input:** [provide password]

5. **System:** Post the information to the server.

6. **Caller input:** main menu (optional)

7. **System:** Loads the home page of the application.

Transform the Receive Mail Section for Voice Clients

In the previous section, you transformed the login section of the e-mail system. Now you are going to transform the receive-mail section for voice clients. You are using only one Active Server Pages document (Listing 11-20) for this purpose. In that file, you first authenticate the user information from the mail server. You then transform the message into VoiceXML so the user can listen to his mail using a normal telephone network.

Listing 11-20: receivevxml.asp

© 2001 Dreamtech Software India Inc.
All Rights Reserved

```
<%@LANGUAGE=VBSCRIPT %>

<%
```

```
dim uname,pwd
uname = trim(Request.QueryString("loginname"))
pwd = trim(Request.QueryString("password"))


  Response.ContentType = "text/xml"
  Response.Write("<?xml version= ""1.0""?>")

%>

<vxml version="1.0">
<form>
<block>

<%

Set pop3 = Server.CreateObject( "Jmail.pop3" )

on error resume next
pop3.Connect uname, pwd, "http://192.168.1.100"

c = 1

if (pop3.Count>0) then%>
Total number of messages are<%=pop3.Count%>

 <%

 while (c<=pop3.Count)
  Set msg = pop3.Messages.item(c)

  ReTo = ""
  ReCC = ""

  Set Recipients = msg.Recipients
  delimiter = ", "%>

  Sender is
  <%= msg.FromName %>
  subject of the mail is
  <%=msg.Subject%>
  Message is
  <%= msg.Body %>

  <%c = c + 1
 wend
else
%>
 You do not have mail in your mailbox.
<%
end if
pop3.Disconnect
%>

<submit next="http://192.168.1.100/voice.asp"/>
</block>
```

```
</form>
</vxml>
```

Let's now discuss the preceding code step by step. In the first few lines of code, you declare some variables for future use; in the next line, you declare the content type for the VoiceXML contents that you are going to generate, as seen in the following:

```
<%
dim uname,pwd
uname = trim(Request.QueryString("loginname"))
pwd = trim(Request.QueryString("password"))


  Response.ContentType = "text/xml"
  Response.Write("<?xml version= ""1.0""?>")

%>
```

After this, you start generating the VXML document and use `<form>`and `<block>` elements in some of the subsequent lines of code. After this, you connect the mail server, authenticate the username and password, and get all the messages from the user inbox one by one, by constructing a loop. The following code does this entire task:

```
<%
Set pop3 = Server.CreateObject( "Jmail.pop3" )
on error resume next
pop3.Connect uname, pwd, "http://192.168.1.100"
c = 1
if (pop3.Count>0) then%>
Total number of messages are<%=pop3.Count%>
 <%
 while (c<=pop3.Count)
   Set msg = pop3.Messages.item(c)
   ReTo = ""
   ReCC = ""
   Set Recipients = msg.Recipients
   delimiter = ", "%>
   Sender is
   <%= msg.FromName %>
   subject of the mail is
   <%=msg.Subject%>
   Message is
   <%= msg.Body %>
   <%c = c + 1
 wend
else
%>
```

When this code executes on the server, it prompts all the messages in the user's inbox, one by one, by using the installed TTS engine on the server. If the user doesn't have any mail in his inbox, you give the user a message saying so.

When this code executes on the simulator, it shows the output on the screen in Figure 11-20.

Figure 11-20: Speaking an e-mail message to the user (Image of MADK 2.0.Courtesy of Motorola, Inc.)

Transform the Movie Section

After you transform the movie section, this application will be completely compliant for voice clients. By the end of this transformation, the user will be able to get information about the movies that are being screened in various theaters in the city. The user will also be able to book the tickets for the movies using the normal telephone.

Generate XML Data for Theater Listings

When the user enters the movie section, the system first gives a listing of all the available theaters from the database. For performing this task, you will first generate the XML data for the theater listing and then transform all the data into VoiceXML documents. Listing 11-21 shows the code you use to do this.

Listing 11-21: Code to generate theater listings

© 2001 Dreamtech Software India Inc.
All Rights Reserved

```
if (section="Movies") then

  sql = "select * from hall"
  objrs.Open sql,conn

  Set fso =Server.CreateObject("Scripting.FileSystemObject")
  fname = "tmovies.xml"

  fullpath = server.MapPath("main")

  Set MyFile=fso.CreateTextFile(fullpath & "/" & fname)
```

```
   MyFile.WriteLine("<?xml version='1.0'?>")
   MyFile.WriteLine("<ROOT>")


   while not(objrs.EOF)
      hallid = objrs.Fields("hall_id")
      hallname = objrs.Fields("hall_name")
      add = objrs.Fields("address")
      MyFile.WriteLine("<hall>")
      MyFile.WriteLine("<hall_id>" & hallid & "</hall_id>")
      MyFile.WriteLine("<hall_name>" & hallname & "</hall_name>")
      MyFile.WriteLine("<address>" & add & "</address>")
      MyFile.WriteLine("</hall>")
   objrs.MoveNext
   wend


   MyFile.WriteLine("</ROOT>")
   address = "http://192.168.1.100/xsl.asp?name=" & fname
   Response.Redirect(address)

end if
```

In the preceding code, you check for the section the user selects and then collect all the data from the database. In the next step, you save all the data in an XML format in a file named `tmovies.xml` on the server using the file system object (FSO). After saving all records in XML format, you call the `xsl.asp` file to perform the transformation process.

Write XSL for Theater Listings

Now you prepare an XSL document to transform the data into VoiceXML format, so that it becomes compatible for the voice clients during the transformation process. Listing 11-22 contains the code for the file `tmoviesvxml.xsl` that you will use in the transformation process.

Listing 11-22: tmoviesvxml.xsl

© 2001 Dreamtech Software India Inc.
All Rights Reserved

```
<?xml version="1.0" ?>
<xsl:stylesheet xmlns:xsl="http://www.w3.org/TR/WD-xsl">
<xsl:template match="/">


<vxml version="1.0">

   <link next="http://192.168.1.100/voice.asp">
   <grammar type="application/x-gsl">
    main menu
   </grammar>
  </link>

  <menu>
   <prompt> Welcome to movie section. </prompt>
   <prompt>Please select any one from the following:
theatres<enumerate/></prompt>
   <xsl:for-each select="ROOT/hall">
    <choice>
     <xsl:attribute name="next">
```

```
        http://192.168.1.100/movielist.asp?hall=<xsl:value-of select="hall_id" />
     </xsl:attribute>
     <xsl:value-of select="hall_name" />
   </choice>
  </xsl:for-each>


  <noinput>Please say one of <enumerate/></noinput>
   <nomatch>
           Sorry, we currently do not provide the information for the theater
you chose.
           Please choose one of the following theatres or say main menu to start
the application again.
              <enumerate/>
   </nomatch>


 </menu>
 </vxml>


</xsl:template>


</xsl:stylesheet>
```

In the beginning of Listing 11-21, you provide a link for the application's home page using the `<link>` element. In the next step, you start constructing a `<menu>` list that presents a welcome message to the user. In the next message, you ask the user to choose any of the theaters from the list. Next, you build a loop using the `<xsl:for-each>` element for fetching all the theater names from the XML document. Finally, you close the loop, as follows:

```
  <prompt> Welcome to the movie section. </prompt>
  <prompt>Please select any one from the following theaters<enumerate/></prompt>
  <xsl:for-each select="ROOT/hall">
   <choice>
    <xsl:attribute name="next">
      http://192.168.1.100/movielist.asp?hall=<xsl:value-of select="hall_id" />
    </xsl:attribute>
    <xsl:value-of select="hall_name" />
   </choice>


  </xsl:for-each>
```

After this, you write the code for handling an incorrect response or no response from the user by using the `<noinput>` and `<nomatch>` elements.

```
  <noinput>Please say one of <enumerate/></noinput>
   <nomatch>
           Sorry, we currently do not provide the information for the theater
you chose.
           Please choose from one of the following theaters or say main menu to
start the application again.
              <enumerate/>
   </nomatch>
```

When this code performs a transformation on the server, it generates the following VXML document in Listing 11-23 on-the-fly.

Listing 11-23: Transformation result of tmovies.xml

```
<?xml version= "1.0"?>
<vxml version="1.0">
<link next="http://192.168.1.100/voice.asp">
<grammar type="application/x-gsl">
   main menu
  </grammar>
</link>
<menu>
<prompt> Welcome to the movie section. </prompt>
<prompt>Please select any one from the following theaters<enumerate /></prompt>
<choice next="
    http://192.168.1.100/movielist.asp?hall=1
">
pvr1
</choice>
<choice next="
    http://192.168.1.100/movielist.asp?hall=2
">
pvr2
</choice>
<choice next="
    http://192.168.1.100/movielist.asp?hall=3
">
pvr3
</choice>
<choice next="
    http://192.168.1.100/movielist.asp?hall=4
">
pvr4
</choice>
<choice next="
    http://192.168.1.100/movielist.asp?hall=5
">
Satyam
</choice>
<choice next="
    http://192.168.1.100/movielist.asp?hall=6
">
Priya
</choice>
<choice next="
    http://192.168.1.100/movielist.asp?hall=7
">
Liberty
</choice>
<choice next="
    http://192.168.1.100/movielist.asp?hall=8
">
Sheela
</choice>
<choice next="
```

```
        http://192.168.1.100/movielist.asp?hall=9
">
Rachna
</choice>
<choice next="
        http://192.168.1.100/movielist.asp?hall=10
">
Plaza
</choice>
<noinput>Please say one of <enumerate /></noinput>
<nomatch>
        Sorry, we currently do not provide the information for the theater
you chose.
        Please choose one of the following theaters or say main menu to start
the application again.
            <enumerate />
</nomatch>
</menu>
</vxml>
```

When this code executes on the server using the simulator, it shows the output shown in Figures 11-21 and 11-22.

Figure 11-21: The movie section's welcome message (Image of MADK 2.0.Courtesy of Motorola, Inc.)

Figure 11-22: The movie section's welcome message (continued) (Image of MADK 2.0.Courtesy of Motorola, Inc.)

After the user selects one of the theaters from the available list, we call the `movielist.asp` file (Listing 11-24) with the hall ID as the parameter. In this file, you are generating a list of the currently screened movies at the theatre selected by the user.

Listing 11-24: movielist.asp

© 2001 Dreamtech Software India Inc.
All Rights Reserved

```
<%@ Language=VBScript %>

<%
dim address,section,fname

id = trim(Request.QueryString("hall"))
session("theatreid")= id

set conn = Server.CreateObject("ADODB.Connection")
conn.ConnectionString = "Provider=SQLOLEDB.1;Persist Security Info=False;User
ID=sa;Initial Catalog=portal;Data Source=developers"
conn.Open

Set objrs = Server.CreateObject("ADODB.Recordset")
objrs.CursorType = adOpenStatic

Set rsmovie = Server.CreateObject("ADODB.Recordset")
rsmovie.CursorType = adOpenStatic

  sqlmovies = "select DISTINCT(status.movie_id),movie.movie_name from
movie,status where status.movie_id=movie.movie_id and status.hall_id=" & id
  rsmovie.Open sqlmovies,conn
```

```
Set fso =Server.CreateObject("Scripting.FileSystemObject")
fname = "movielist.xml"


fullpath = server.MapPath("main")


Set MyFile=fso.CreateTextFile(fullpath & "/" & fname)

MyFile.WriteLine("<?xml version='1.0'?>")
MyFile.WriteLine("<ROOT>")
MyFile.WriteLine("<hall>")
 MyFile.WriteLine("<hall_id>" & id & "</hall_id>")

while not (rsmovie.EOF)
 MyFile.WriteLine("<movie>")
 moviename = rsmovie.Fields("movie_name")
 MyFile.WriteLine("<movie_name>" & moviename & "</movie_name>")
 movieid = rsmovie.Fields("movie_id")
 MyFile.WriteLine("<movie_id>" & movieid & "</movie_id>")
 MyFile.WriteLine("</movie>")
rsmovie.MoveNext
wend
MyFile.WriteLine("</hall>")
MyFile.WriteLine("</ROOT>")
conn.Close

address = "http://192.168.1.100/xsl.asp?name=" & fname
Response.Redirect(address)
%>
```

In this code, you select the names of all the movies based on the hall ID passed to you by the previous document. `Hall ID` represents the hall selected by the user as the choice. After collecting all the data from the database, you save it in XML format on the server in the file `movielist.xml` using the following lines of code:

```
sqlmovies = "select DISTINCT(status.movie_id),movie.movie_name from
movie,status where status.movie_id=movie.movie_id and status.hall_id=" & id
 rsmovie.Open sqlmovies,conn

Set fso =Server.CreateObject("Scripting.FileSystemObject")
fname = "movielist.xml"
```

After saving the list of the movies on the server, you transform this file by again calling the `xsl.asp` file with the XML document named as a parameter in the query string.

Write XSL for the Movie Listings

In this section, you are preparing an XSL document for transforming the movie-listing data for voice clients. Listing 11-25 shows the code of the file `movielistvxml.xsl` that you use for this transformation.

Listing 11-25: movielistvxml.xsl

```
<?xml version="1.0" ?>
```

```
<xsl:stylesheet xmlns:xsl="http://www.w3.org/TR/WD-xsl">
<xsl:template match="/">

<vxml version="1.0">

<link next="http://192.168.1.100/voice.asp">
  <grammar type="application/x-gsl">
  main menu
  </grammar>
</link>


 <menu>
 <prompt>Now say one of following movie names: <enumerate/></prompt>


 <xsl:for-each select="ROOT/hall/movie">
 <choice>
   <xsl:attribute name="next">
   http://192.168.1.100/status.asp?movie=<xsl:value-of select="movie_id" />
   </xsl:attribute>
   <xsl:value-of select="movie_name" />
 </choice>


 </xsl:for-each>
 <noinput>Please say one of <enumerate/></noinput>

</menu>
 </vxml>

</xsl:template>
</xsl:stylesheet>
```

In the preceding code, you generate a VXML document. In the first few lines of the document, you provide a link to the application's home page by typing the following lines of code:

```
<link next="http://192.168.1.100/voice.asp">
  <grammar type="application/x-gsl">
  main menu
  </grammar>
</link>
```

In the next step, you construct a <menu> list that contains the names of all the movies. You fetch this list from the XML document. For fetching all the records, you use a looping structure so that you can fetch the names of all the available movies from the document.

When this code executes on the server, it generates the VXML document in Listing 11-26.

Listing 11-26: Transformation result of movielist.xml

```
<?xml version= "1.0"?>
<vxml version="1.0">
```

```
<link next="http://192.168.1.100/voice.asp">
<grammar type="application/x-gsl">
  main menu
  </grammar>
</link>

<menu>

<prompt>Now say one of the following movie names: <enumerate /></prompt>
<choice next="
    http://192.168.1.100/status.asp?movie=1
">
Life Is Beautiful (Drama)
</choice>

<choice next="
    http://192.168.1.100/status.asp?movie=3
">
Five-Man Army (Action)
</choice>

<noinput>Please say one of <enumerate /></noinput>

</menu>

</vxml>
```

When you run this code on the simulator, it shows the output in Figure 11-23.

Figure 11-23: Prompting the movie names (Image of MADK 2.0.Courtesy of Motorola, Inc.)

Generate XML Data for Showtimes

When the user selects any one of the available movies spoken by the system, you call the `status.asp` file (Listing 11-27) with the movie ID as a parameter in the query string to check the showtimes for the selected movie.

Listing 11-27: status.asp

```
<%@ Language=VBScript %>
<%
dim address,section,fname
movieid = trim(Request.QueryString("movie"))
id = session.Contents(1)
session("mid")= movieid

set conn = Server.CreateObject("ADODB.Connection")
conn.ConnectionString = "Provider=SQLOLEDB.1;Persist Security Info=False;User
ID=sa;Initial Catalog=portal;Data Source=developers"
conn.Open

Set rsstatus = Server.CreateObject("ADODB.Recordset")
rsstatus.CursorType = adOpenStatic

  Set fso =Server.CreateObject("Scripting.FileSystemObject")
  fname = "status.xml"

  fullpath = server.MapPath("main")

  Set MyFile=fso.CreateTextFile(fullpath & "/" & fname)

  MyFile.WriteLine("<?xml version='1.0'?>")
  MyFile.WriteLine("<ROOT>")
  MyFile.WriteLine("<hall>")
 MyFile.WriteLine("<hall_id>" & id & "</hall_id>")
  MyFile.WriteLine("<movie>")
  MyFile.WriteLine("<movie_id>" & movieid & "</movie_id>")
   sqlstatus = "select
movie_date,showtime,balcony,rear_stall,middle_stall,upper_stall  from status
where movie_id=" & movieid & "and hall_id=" & id
  rsstatus.Open sqlstatus,conn
  while not (rsstatus.EOF)
     MyFile.WriteLine("<status>")
     moviedate = rsstatus.Fields("movie_date")
     showtime = rsstatus.Fields("showtime")
     balcony = rsstatus.Fields("balcony")
     rearstall = rsstatus.Fields("rear_stall")
     middlestall = rsstatus.Fields("middle_stall")
     upperstall = rsstatus.Fields("upper_stall")
     MyFile.WriteLine("<movie_date>" & moviedate & "</movie_date>")
     MyFile.WriteLine("<showtime>" & showtime & "</showtime>")
     MyFile.WriteLine("<balcony>" & balcony & "</balcony>")
     MyFile.WriteLine("<rear_stall>" & rearstall & "</rear_stall>")
     MyFile.WriteLine("<middle_stall>" & middlestall & "</middle_stall>")
     MyFile.WriteLine("<upper_stall>" & upperstall & "</upper_stall>")
```

```
      MyFile.WriteLine("</status>")
      rsstatus.MoveNext
  wend
  MyFile.WriteLine("</movie>")
  sstatus.Close
  MyFile.WriteLine("</hall>")
  MyFile.WriteLine("</ROOT>")
  conn.Close
  address = "http://192.168.1.100/xsl.asp?name=" & fname
  Response.Redirect(address)
%>
```

In the preceding code, you generate data for the showtimes and show the dates of the selected movie at the selected theater so that the user can make his choice for a particular showtime.

You collect all the data from the database on the basis of the theater and the movie name the user selects and then save all the data in XML format in a file named `status.xml` using the file system object (FSO) on the server.

Prepare XSL Documents for Show Status

To transform the `status.xml` file for prompting the show date on the user's voice client, you use the `statusvxml.xsl` file (Listing 11-28).

Listing 11-28: statusvxml.xsl

© 2001 Dreamtech Software India Inc.
All Rights Reserved

```
<?xml version="1.0" ?>
<xsl:stylesheet xmlns:xsl="http://www.w3.org/TR/WD-xsl">

<xsl:template match="/">
<vxml version="1.0">
 <link next="http://192.168.1.100/voice.asp">
  <grammar type="application/x-gsl">
   main menu
  </grammar>
 </link>
 <menu>
 <prompt>Say one of the show dates: <enumerate/></prompt>
 <xsl:for-each select="ROOT/hall/movie/status">
 <choice>
  <xsl:attribute name="next">
  http://192.168.1.100/showlist.asp?mdate=<xsl:value-of select="movie_date" />
   </xsl:attribute>
   <xsl:value-of select="movie_date" />

 </choice>
 </xsl:for-each>
  <noinput>Please say one of <enumerate/></noinput>
 </menu>
 </vxml>
</xsl:template>
</xsl:stylesheet>
```

In the preceding code, you simply generate a VXML document. At the top of the document, you provide a link to the application's home page by using the `<link>` element, as follows:

```
<link next="http://192.168.1.100/voice.asp">
  <grammar type="application/x-gsl">
    main menu
  </grammar>
</link>
```

In the next step, you generate a `<menu>` list containing the `<choice>` element to prompt the show dates to the user. When the user selects one of the available dates from the list, you call the `showlist.asp` file with the show date in a parameter named `mdate` in the query string.

```
<prompt>Say one of the show date: <enumerate/></prompt>
<xsl:for-each select="ROOT/hall/movie/status">
<choice>
  <xsl:attribute name="next">
  http://192.168.1.100/showlist.asp?mdate=<xsl:value-of select="movie_date" />
  </xsl:attribute>
  <xsl:value-of select="movie_date" />

</choice>
```

Listing 11-29 shows the transformation result of `statusvxml.xsl`, and Figure 11-24 shows the screen output.

Listing 11-29: Transformation result of statusvxml.xsl

```
<?xml version= "1.0"?>
<vxml version="1.0">
<link next="http://192.168.1.100/voice.asp">
<grammar type="application/x-gsl">
    main menu
  </grammar>
</link>
<menu>
<prompt>Say one of the show dates: <enumerate /></prompt>
<choice next="
    http://192.168.1.100/showlist.asp?mdate=17 May
">
17 May
</choice>
<choice next="
    http://192.168.1.100/showlist.asp?mdate=18 May
">
18 May
</choice>
<noinput>Please say one of <enumerate /></noinput>
</menu>
</vxml>
```

After the user speaks the show date, control is passed to the `showlist.asp` file (Listing 11-30). You are using the `showlist.asp` file to generate the XML data on the showtimes of the selected movie on the particular day the user selects.

Figure 11-24: Prompting the show dates (Image of MADK 2.0.Courtesy of Motorola, Inc.)

Listing 11-30: showlist.asp

© 2001 Dreamtech Software India Inc.
All Rights Reserved

```
<%@ Language=VBScript %>

<%
dim address,section,fname
moviedate = trim(Request.QueryString("mdate"))
id = session.Contents(1)
movieid = session.Contents(2)
session("date_movie")=moviedate

set conn = Server.CreateObject("ADODB.Connection")
conn.ConnectionString = "Provider=SQLOLEDB.1;Persist Security Info=False;User
ID=sa;Initial Catalog=portal;Data Source=developers"
conn.Open

Set objrs = Server.CreateObject("ADODB.Recordset")
objrs.CursorType = adOpenStatic

Set rsmovie = Server.CreateObject("ADODB.Recordset")
rsmovie.CursorType = adOpenStatic

Set rsstatus = Server.CreateObject("ADODB.Recordset")
rsstatus.CursorType = adOpenStatic
```

```
Set fso =Server.CreateObject("Scripting.FileSystemObject")
fname = "showlist.xml"

fullpath = server.MapPath("main")

Set MyFile=fso.CreateTextFile(fullpath & "/" & fname)

MyFile.WriteLine("<?xml version='1.0'?>")
MyFile.WriteLine("<ROOT>")
MyFile.WriteLine("<hall>")
MyFile.WriteLine("<hall_id>" & id & "</hall_id>")
MyFile.WriteLine("<movie>")
MyFile.WriteLine("<movie_id>" & movieid & "</movie_id>")

sqlstatus = "select movie_date,showtime from status where movie_id=" & movieid
& "and hall_id=" & id
rsstatus.Open sqlstatus,conn
while not (rsstatus.EOF)
    MyFile.WriteLine("<status>")
    moviedate = rsstatus.Fields("movie_date")
    showtime = rsstatus.Fields("showtime")
    MyFile.WriteLine("<movie_date>" & moviedate & "</movie_date>")
    MyFile.WriteLine("<showtime>" & showtime & "</showtime>")
    MyFile.WriteLine("</status>")
    rsstatus.MoveNext
wend
MyFile.WriteLine("</movie>")
rsstatus.Close
MyFile.WriteLine("</hall>")
MyFile.WriteLine("</ROOT>")
conn.Close
address = "http://192.168.1.100/xsl.asp?name=" & fname
Response.Redirect(address)
%>
```

This code is similar to the status.asp file. However, there are some changes that we have incorporated to generate the showlist.xml file containing the show date and showtime data in XML format. In this code, you select the show dates and the showtimes from the status table in the database based on the movie ID. Movie ID, as mentioned earlier, represents the movie name the user selects.

```
sqlstatus = "select movie_date,showtime from status where movie_id=" & movieid
& "and hall_id=" & id
```

After collecting all the records, you save the data in XML format in the file showlist.xml on the server.

Write XSL to Prompt the Movie Showtimes

You already have all the data in the showlist.xml file about the show dates and showtimes for the movie the user selected. Now you write an XSL document to prompt the showtimes to the user. Listing 11-31 contains the code for showlistvxml.xsl, which you use to transform the showlist.xml file into a VoiceXML document.

Listing 11-31: showlistvxml.xsl

```
<?xml version="1.0" ?>
<xsl:stylesheet xmlns:xsl="http://www.w3.org/TR/WD-xsl">
<xsl:template match="/">
<vxml version="1.0">

 <link next="http://192.168.1.100/voice.asp">
  <grammar type="application/x-gsl">
   main menu
  </grammar>
 </link>

 <menu>
  <prompt>Say one of the show times: <enumerate/></prompt>
  <xsl:for-each select="ROOT/hall/movie/status">
  <choice>
   <xsl:attribute name="next">
   http://192.168.1.100/tickets.asp?stime=<xsl:value-of select="showtime" />
   </xsl:attribute>
   <xsl:value-of select="showtime" />
  </choice>
  </xsl:for-each>

  <noinput>Please say one of <enumerate/></noinput>
  <nomatch>
          Sorry, we currently do not provide the information for the movie you
chose.
          Please choose from one of the following movies or say main menu to
start the application again.
             <enumerate/>
  </nomatch>

 </menu>
 </vxml>

</xsl:template>
</xsl:stylesheet>
```

In the preceding code, you first provide a link for the application's home page using the <link> element. In the next part, you construct a loop using the <xsl:for-each> element and fetch the showtimes from the XML document using the <xsl:value-of select> element, as follows:

```
<xsl:for-each select="ROOT/hall/movie/status">
<choice>
 <xsl:attribute name="next">
 http://192.168.1.100/tickets.asp?stime=<xsl:value-of select="showtime" />
 </xsl:attribute>
 <xsl:value-of select="showtime" />
</choice>
</xsl:for-each>
```

After fetching the values, you write the code for both the <nomatch> and <noinput> events as follows:

```
<noinput>Please say one of <enumerate/></noinput>
<nomatch>
```

```
            Sorry, we currently do not provide the information for the movie you
chose.
            Please choose from one of the following movies or say main menu to
start the application again.
            <enumerate/>
  </nomatch>
```

Listing 11-32 shows the transformation result of showlist.xml, and Figure 11-25 shows the screen output.

Listing 11-32: Transformation result of showlist.xml

```xml
<?xml version= "1.0"?>
<vxml version="1.0">
<link next="http://192.168.1.100/voice.asp">
<grammar type="application/x-gsl">
   main menu
  </grammar>
</link>
<menu>
<prompt>Say one of the show times: <enumerate /></prompt>
<choice next="
   http://192.168.1.100/tickets.asp?stime=11.30  a.m.
">
11.30 a.m.
</choice>
<choice next="
   http://192.168.1.100/tickets.asp?stime=2.00  p.m.
">
2.00 p.m.
</choice>
<noinput>Please say one of <enumerate /></noinput>
<nomatch>
            Sorry, we currently do not provide the information for the movie
Choosen by you.
            Please choose one of the following movies or say main menu to start
the application again.
            <enumerate />
</nomatch>
</menu>
</vxml>
```

Figure 11-25: Prompting showtimes to the user (Image of MADK 2.0.Courtesy of Motorola, Inc.)

When the user enters the desired showtimes, we call the `tickets.asp` file to generate XML data for the available tickets from the database so that user knows the status of available tickets for a particular showing on a particular day for a selected movie.

Generate XML Data for the Booking Status

In this section, you generate XML data for the available number of tickets in different categories for the movie the user selects. You perform this task in the file `tickets.asp`, shown in Listing 11-33.

Listing 11-33: tickets.asp

© 2001 Dreamtech Software India Inc.
All Rights Reserved

```
<%@ Language=VBScript %>
<%
dim address,section,fname
showtime = trim(Request.QueryString("stime"))
id = session.Contents(1)
movieid = session.Contents(2)
moviedate = session.Contents(3)
session("showtime")=showtime

set conn = Server.CreateObject("ADODB.Connection")
conn.ConnectionString = "Provider=SQLOLEDB.1;Persist Security Info=False;User
ID=sa;Initial Catalog=portal;Data Source=developers"
conn.Open

Set rsstatus = Server.CreateObject("ADODB.Recordset")
```

```
rsstatus.CursorType = adOpenStatic

  Set fso =Server.CreateObject("Scripting.FileSystemObject")
  fname = "tickets.xml"

  fullpath = server.MapPath("main")

  Set MyFile=fso.CreateTextFile(fullpath & "/" & fname)

  MyFile.WriteLine("<?xml version='1.0'?>")
  MyFile.WriteLine("<ROOT>")
  MyFile.WriteLine("<hall>")
  MyFile.WriteLine("<hall_id>" & id & "</hall_id>")
  MyFile.WriteLine("<movie>")
  MyFile.WriteLine("<movie_id>" & movieid & "</movie_id>")

  sqlstatus = "select balcony,rear_stall,middle_stall,upper_stall  from status
where movie_id=" & movieid & "and hall_id=" & id & "and movie_date='" &
moviedate & "' and showtime='" & showtime & "'"
  rsstatus.Open sqlstatus,conn
  while not (rsstatus.EOF)
      MyFile.WriteLine("<status>")
      balcony = rsstatus.Fields("balcony")
      rearstall = rsstatus.Fields("rear_stall")
      middlestall = rsstatus.Fields("middle_stall")
      upperstall = rsstatus.Fields("upper_stall")
      MyFile.WriteLine("<balcony>" & balcony & "</balcony>")
      MyFile.WriteLine("<rear_stall>" & rearstall & "</rear_stall>")
      MyFile.WriteLine("<middle_stall>" & middlestall & "</middle_stall>")
      MyFile.WriteLine("<upper_stall>" & upperstall & "</upper_stall>")
      MyFile.WriteLine("</status>")
      rsstatus.MoveNext
  wend
  MyFile.WriteLine("</movie>")
  rsstatus.Close
  MyFile.WriteLine("</hall>")
  MyFile.WriteLine("</ROOT>")

  conn.Close
  address = "http://192.168.1.100/xsl.asp?name=" & fname
  Response.Redirect(address)
%>
```

In the preceding code, you select the records from the database to know the number of available tickets for all the categories for the movie the user selects.

You write a SQL query, in which you fetch the values for the balcony, the rear_stall, the middle_stall, and the upper_stall from the status table based on the movieid, hallid, movie date, and showtime the user selects at different prompts during the procedure.

```
sqlstatus = "select balcony,rear_stall,middle_stall,upper_stall  from status
where movie_id=" & movieid & "and hall_id=" & id & "and movie_date='" &
moviedate & "' and showtime='" & showtime & "'"
```

After collecting all the records, you save all the data in XML format in the file tickets.xml using the file system object (FSO), so that in the next step you can transform this data into VoiceXML.

Write XSL for the Booking Status

Now you prepare an XSL document for transforming the `tickets.xml` file into VoiceXML data so that the user can listen to all the data using the voice client. This file is shown in Listing 11-34.

Listing 11-34: ticketsvxml.xsl

```xml
<?xml version="1.0" ?>
<xsl:stylesheet xmlns:xsl="http://www.w3.org/TR/WD-xsl">
<xsl:template match="/">

<vxml version="1.0">
   <var name="stall"/>
    <var name="ticket"/>
    <var name="tel"/>


<form id="info">
  <block>
   Number of tickets available
    <xsl:for-each select="ROOT/hall/movie/status">
       Balcony      <xsl:value-of select="balcony" />
       Upper Stall <xsl:value-of select="upper_stall" />
       Middle Stall <xsl:value-of select="middle_stall" />
       Rear stall <xsl:value-of select="rear_stall" />
     </xsl:for-each>
  <goto next="#tstall"/>
  </block>

</form>

<form id="tstall">
 <field name="stall">
   <prompt>Please say a stallname for ticket booking , balcony or middlestall or
upperstall or rearstall</prompt>
    <grammar type="application/x-jsgf">
            balcony | upperstall | middlestall | rearstall
        </grammar>
  <filled>
   <assign name="document.stall" expr="stall"/>
   <goto next="#tic"/>
  </filled>
</field>

</form>

<form id="tic">

   <field name="ticket" type="digits">
       <prompt> please say number of tickets you want to book. </prompt>
   <filled>
    <assign name="document.ticket" expr="ticket"/>
```

```
          <goto next="#phone"/>
      </filled>
      </field>

</form>


<form id="phone">

    <field name="tel" type="digits">
        <prompt> please say your telephone number. </prompt>
    <filled>
      <assign name="document.tel" expr="tel"/>
      <goto next="#call_asp"/>
    </filled>
      </field>

</form>



<form id="call_asp">

  <block>
     <var name="bookstall"  expr="document.stall" />
     <var name="ticketno" expr="document.ticket"/>
     <var name="telno" expr="document.tel"/>
     <submit next="http://192.168.1.100/booking.asp"  method="get"
namelist="bookstall ticketno telno"/>
  </block>

</form>
 </vxml>

</xsl:template>
</xsl:stylesheet>
```

This is quite a big XSL document, so we will undertake a step-by-step discussion of it. At the top of the file, you declare some variables for the user in the document using the `<var>` element.

```
    <var name="stall"/>
     <var name="ticket"/>
     <var name="tel"/>
```

Next, you generate a `<form>` for prompting the available number of tickets in each category in the database by fetching their values from the XML document. After prompting all the data to the user, you call the next form by using the `<goto>` statement.

```
<form id="info">
  <block>
   Number of tickets available
    <xsl:for-each select="ROOT/hall/movie/status">
      Balcony      <xsl:value-of select="balcony" />
      Upper Stall <xsl:value-of select="upper_stall" />
      Middle Stall <xsl:value-of select="middle_stall" />
      Rear stall <xsl:value-of select="rear_stall" />
    </xsl:for-each>
```

```
    <goto next="#tstall"/>
   </block>
</form>
```

In the second form, you ask the user his choice of the name of the stall in which he wants to book tickets for a particular movie. You also use some inline grammar to determine the correct input for the stall name using the Java Speech Grammar Format (JSGF).

```
<form id="tstall">
 <field name="stall">
   <prompt>Please say a stallname for ticket booking , balcony or middlestall or
upperstall or rearstall</prompt>
    <grammar type="application/x-jsgf">
            balcony | upperstall | middlestall | rearstall
         </grammar>
  <filled>
   <assign name="document.stall" expr="stall"/>
   <goto next="#tic"/>
  </filled>
 </field>
</form>
```

In the third form, you ask the user for the number of tickets he wants to book for the selected movie, and the stall name selected in the previous form. After collecting the values from the user, you call the next form in the document, named `phone`.

```
<form id="tic">

   <field name="ticket" type="digits">
       <prompt> please say number of tickets you want to book. </prompt>
   <filled>
    <assign name="document.ticket" expr="ticket"/>
    <goto next="#phone"/>
   </filled>
   </field>

</form>
```

In the fourth form, named `phone`, you ask the user for his phone number for tracking the user's identity in the database. After collecting the phone number from the user, you call the last form in the document named `CALL_ASP`, which you use to post all such data on the server.

```
<form id="phone">
    <field name="tel" type="digits">
        <prompt> please say your telephone number. </prompt>
    <filled>
     <assign name="document.tel" expr="tel"/>
     <goto next="#call_asp"/>
    </filled>
    </field>
</form>
```

In the fifth and last form in the document `call_asp`, you declare some variables for storing the values and then post all the data to the server using the `<submit>` element.

```
<form id="call_asp">
<block>
    <var name="bookstall"  expr="document.stall" />
```

```
    <var name="ticketno" expr="document.ticket"/>
    <var name="telno" expr="document.tel"/>
    <submit next="http://192.168.1.100/booking.asp" method="get"
namelist="bookstall ticketno telno"/>
 </block>
</form>
```

While calling the action file `booking.asp`, you send all the data as different parameters in the query string to book the tickets. The transformation results of `tickets.xml` is shown in Listing 11-35, and the output of all the processes is shown in Figures 11-26 to 11-28.

Listing 11-35: Transformation result of tickets.xml

```
<?xml version= "1.0"?>
<vxml version="1.0">
<var name="stall" />
<var name="ticket" />
<var name="tel" />
<form id="info">
<block>
    Number of tickets available

        Balcony     4
        Upper Stall 18
        Middle Stall 25
        Rear stall 4
<goto next="#tstall" />
</block>
</form>
<form id="tstall">
<field name="stall">
<prompt>Please say a stallname for ticket booking , balcony or middlestall or
upperstall or rearstall</prompt>
<grammar type="application/x-jsgf">
            balcony | upperstall | middlestall | rearstall
        </grammar>
<filled>
<assign name="document.stall" expr="stall" />
<goto next="#tic" />
</filled>
</field>
</form>
<form id="tic">
<field name="ticket" type="digits">
<prompt> please say number of tickets you want to book. </prompt>
<filled>
<assign name="document.ticket" expr="ticket" />
<goto next="#phone" />
</filled>
</field>
</form>
<form id="phone">
<field name="tel" type="digits">
```

```
<prompt> please say your telephone number. </prompt>
<filled>
<assign name="document.tel" expr="tel" />
<goto next="#call_asp" />
</filled>
</field>
</form>
<form id="call_asp">
<block>
<var name="bookstall" expr="document.stall" />
<var name="ticketno" expr="document.ticket" />
<var name="telno" expr="document.tel" />
<submit next="http://192.168.1.100/booking.asp" method="get" namelist="bookstall
ticketno telno" />
</block>
</form>
</vxml>
```

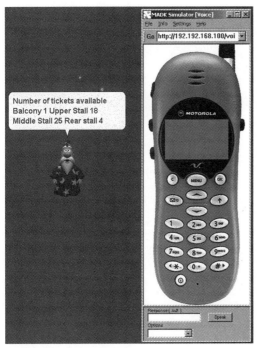

Figure 11-26: Prompting the number of available tickets to the user (Image of MADK 2.0.Courtesy of Motorola, Inc.)

Figure 11-27: Asking for the stall name from the user (Image of MADK 2.0.Courtesy of Motorola, Inc.)

Figure 11-28: Asking for the number of tickets from the user (Image of MADK 2.0.Courtesy of Motorola, Inc.)

Book the Tickets

In this section, we discuss the `booking.asp` file (Listing 11-36) that performs the task of ticket booking with the checking process and redirects the control to the final dialogs of the movie section.

Listing 11-36: booking.asp

```
<%@ Language=VBScript %>
<%
dim stall,ticket,fname,email
stall = trim(Request.QueryString("bookstall"))
ticket = trim(Request.QueryString("ticketno"))
phone = trim(Request.QueryString("telno"))
hallid = session.Contents(1)
movieid = session.Contents(2)
showdate = trim(session.Contents(3))
showtime = trim(session.Contents(4))

if stall="upperstall" then
 stall="upper_stall"
elseif stall="middlestall" then
 stall="middle_stall"
elseif stall="rearstall" then
 stall="rear_stall"
end if

stall=UCase(stall)

dim conn,objrs,tsql,balcony,rear,middle,upper,total
set conn = Server.CreateObject("ADODB.Connection")
conn.ConnectionString = "Provider=SQLOLEDB.1;Persist Security Info=False;User
ID=sa;Initial Catalog=portal;Data Source=developers"
conn.Open

Set objrs = Server.CreateObject("ADODB.Recordset")
tsql="select balcony,rear_stall,middle_stall,upper_stall from status where
hall_id=" & hallid & "and movie_id=" & movieid & "and showtime='" & showtime &
"'and movie_date='" & showdate &"'"
objrs.CursorType = adOpenStatic
objrs.Open tsql,conn

while not (objrs.EOF )
 if (stall = "BALCONY") then
  total = UCase(trim(objrs.Fields("Balcony")))
 end if

 if (stall = "REAR_STALL") then
  total = UCase(trim(objrs.Fields("rear_stall")))
 end if

 if (stall = "MIDDLE_STALL") then
  total = UCase(trim(objrs.Fields("middle_stall")))
 end if
```

```
 if (stall = "UPPER_STALL") then
   total = UCase(trim(objrs.Fields("upper_stall")))
 end if

 balcony = objrs.Fields("Balcony")
 rear = objrs.Fields("rear_stall")
 middle = objrs.Fields("middle_stall")
 upper = objrs.Fields("upper_stall")

objrs.MoveNext
wend

if (CInt(total) < CInt(ticket) ) then
 Response.Write "wrong"
 Response.Redirect "http://192.168.1.100/xsl.asp?name=bookfail.xml"
else
 if (stall = "BALCONY") then
   balcony = CInt(balcony) - CInt(ticket)
 end if

 if (stall = "MIDDLE_STALL") then
   middle = CInt(middle) - CInt(ticket)
 end if

 if (stall = "REAR_STALL") then
   rear = CInt(rear) - CInt(ticket)
 end if

 if (stall = "UPPER_STALL") then
   upper = CInt(upper) - CInt(ticket)
 end if
 conn.Execute("update status set balcony=" & balcony & ",rear_stall=" & rear &
",middle_stall=" & middle & ",upper_stall=" & upper & "where hall_id=" & hallid
& "and movie_id=" & movieid & "and showtime='" & showtime & "'")
 Response.Redirect "http://192.168.1.100/xsl.asp?name=bookpass.xml"

end if

%>
```

Let's discuss the code step by step. In the very beginning of the document, you declare some variables for later use in the document and store values in all the variables passed to you by the `tickets.xml` transformation process.

```
dim stall,ticket,fname,email
stall = trim(Request.QueryString("bookstall"))
ticket = trim(Request.QueryString("ticketno"))
phone = trim(Request.QueryString("telno"))
hallid = session.Contents(1)
movieid = session.Contents(2)
showdate = trim(session.Contents(3))
showtime = trim(session.Contents(4))
```

Next, you check the number of tickets available in the database for the category the user selected, with the number of tickets the user enters as the input to the system in two `if` conditions. If the number of tickets the user enters is more than is available in the database, you redirect control to the xsl.asp file with the parameter NAME holding the value bookfail.xml.

If the second condition is true, you update the database to reflect the changes and redirect control to xsl.asp with the same parameter holding the value bookpass.xml.

```
if (stall = "BALCONY") then
 total = UCase(trim(objrs.Fields("Balcony")))
end if

if (stall = "REAR_STALL") then
 total = UCase(trim(objrs.Fields("rear_stall")))
end if

if (stall = "MIDDLE_STALL") then
 total = UCase(trim(objrs.Fields("middle_stall")))
end if

if (stall = "UPPER_STALL") then
 total = UCase(trim(objrs.Fields("upper_stall")))
end if

balcony = objrs.Fields("Balcony")
rear = objrs.Fields("rear_stall")
middle = objrs.Fields("middle_stall")
upper = objrs.Fields("upper_stall")


objrs.MoveNext
wend


if (CInt(total) < CInt(ticket) ) then
 Response.Write "wrong"
 Response.Redirect "http://192.168.1.100/xsl.asp?name=bookfail.xml"
else
 if (stall = "BALCONY") then
  balcony = CInt(balcony) - CInt(ticket)
 end if

 if (stall = "MIDDLE_STALL") then
  middle = CInt(middle) - CInt(ticket)
 end if

 if (stall = "REAR_STALL") then
  rear = CInt(rear) - CInt(ticket)
 end if

 if (stall = "UPPER_STALL") then
  upper = CInt(upper) - CInt(ticket)
 end if
 conn.Execute("update status set balcony=" & balcony & ",rear_stall=" & rear &
",middle_stall=" & middle & ",upper_stall=" & upper & "where hall_id=" & hallid
& "and movie_id=" & movieid & "and showtime='" & showtime & "'")
 Response.Redirect "http://192.168.1.100/xsl.asp?name=bookpass.xml"

end if
```

Prompt the Confirmation Dialog

The final dialog has two possible conditions. If the `booking.asp` file calls the `bookfail.asp` file to give the error message to the user, you choose the file `bookfailvxml.xsl` (Listing 11-37) to transform the `bookfail.xml` document.

Listing 11-37: bookfailvxml.xsl

```
<?xml version="1.0" ?>
<xsl:stylesheet xmlns:xsl="http://www.w3.org/TR/WD-xsl">
<xsl:template match="/">
<vxml version="1.0">
<form>
<block>
 <xsl:value-of select="ROOT"/>
 <submit next="http://192.168.1.100/voice.asp"/>
</block>
</form>
</vxml>
</xsl:template>
</xsl:stylesheet>
```

When this code is executed on a server and transformed to the `bookfail.xml` file, it generates the document in Listing 11-38 and shows the output in Figure 11-29.

Listing 11-38: Transformation result of bookfail.xml

```
<?xml version= "1.0"?>
<vxml version="1.0">
<form>
<block>
Please Try Again!
<submit next="http://192.168.1.100/voice.asp" />
</block>
</form>
</vxml>
```

If the system successfully reserves the tickets, we transform `bookpass.xml` file (Listing 11-39) with the `bookpass.xsl` file to prompt the confirmation dialog.

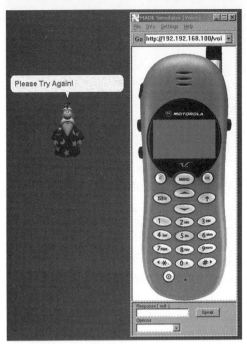

Figure 11-29: Prompting the error message (Image of MADK 2.0.Courtesy of Motorola, Inc.)

Listing 11-39: bookpass.xsl

© 2001 Dreamtech Software India Inc.
All Rights Reserved

```
<?xml version="1.0" ?>
<xsl:stylesheet xmlns:xsl="http://www.w3.org/TR/WD-xsl">

<xsl:template match="/">

<vxml version="1.0">

<form>
<block>
 <xsl:value-of select="ROOT"/>
 <submit next="http://192.168.1.100/voice.asp"/>
</block>
</form>

</vxml>

</xsl:template>

</xsl:stylesheet>
```

Listing 11-40 is the transformation result of the `bookpass.xml` file. Figure 11-30 shows the output on the simulator when this code executes.

Listing 11-40: Transformation result of bookpass.xml

```xml
<?xml version= "1.0"?>

<vxml version="1.0">

<form>

<block>
Congratulations! Your tickets have been booked.
<submit next="http://192.168.1.100/voice.asp" />
</block>

</form>

</vxml>
```

Figure 11-30: Showing the booking confirmation dialog (Image of MADK 2.0.Courtesy of Motorola, Inc.)

Summary

In this chapter, you have learned how to build dynamic interactive voice-response systems based on VoiceXML using the XML- and XSLT-based approach. This chapter also covered all the transformation processes.

Chapter 12

Developing Cross-Platform GUIs for Desktop Applications

This chapter focuses on one of the emerging technologies that has just hit the information technology market. This is a comparatively new markup language called XUL (XML User Interface Language). This chapter focuses on the following topics:

♦ An introduction to XUL

♦ A description of XUL elements

♦ An introduction to XPCOM architecture

Introduction to Cross-Platform GUI Desktop Applications

So far in this book, we have discussed how to build multiplatform Web applications. However, desktop application's development also has a big share in the software development process, and it also faces problems of non-compliance across different platforms. There are some desktop-application development options, such as Java-based development, for cross-platform compliance; but it still requires hours of coding and effort to make your desktop application run across the platforms. In all desktop applications, there are some common user interfaces like menus, windows, dialog boxes, and so forth; and every time you are going to write code for a different platform your also need to rewrite the code for the user interface. XUL provides a way to overcome this problem by writing the user interface in the form of XML data, so that it can be used on almost any platform.

A platform can be defined in terms of hardware, such as mainframe computers, the Macintosh platform, the Sunsparc platform, the IBM PC platform, and so on. Platforms can also be defined in terms of the operating system, such as DOS, Windows, Sun Solaris, UNIX, Linux, OS/2, and so on. A GUI-based application allows the user to operate the computer using the mouse instead of the command prompts that older versions of UNIX used. A GUI application is based on WIMP properties (windows, icon, mouse, and pointer). The user uses the windows and graphical widgets for performing the desired tasks instead of typing commands at the command prompt. So a cross-platform application can be defined as an application that can be used on different kinds of platforms (hardware or software). For example, HTML is a cross-platform application.

A cross-platform GUI application has its own advantages, such as the following:

♦ You are not required to develop a multiple set of source codes for the GUI to make it run on different operating systems. Consequently, development is faster.

♦ A cross-platform GUI itself takes care of the integration of the user interface for each target platform.

♦ Development of a GUI can be started earlier because it targets all the platforms on which the application is going to run in the future.

♦ A GUI provides a consistent look and feel of the application on different kinds of operating systems.

♦ By making the GUI cross platform, the number of the application's target users increases.

One disadvantage of cross-platform GUIs is that the complexity of developing them makes it difficult to gain expertise in XUL development.

The Mozilla XPFE Project

XPFE stands for Cross-Platform Front End, and it is Mozilla's User Interface Component developed by Mozilla, Inc. This is one of the best frameworks available for GUIs. The purpose of XPFE is to develop applications such as browsers, instant messengers, chat clients, IDEs, and other desktop applications that can work on cross platforms. The intention of the project is not only to build a language for providing cross-platform GUIs, but also to provide functionalities so that it becomes easier for the developer to build full-fledged applications such as browsers.

Netscape is building a version of Navigator that can run on various platforms such as Windows, UNIX, and Macintosh. Developing such applications calls for enormous efforts from programmers. To overcome this problem, the project began by developing a user-interface architecture that can be implemented cross platform using minimal efforts and time. All of these efforts have resulted in XUL (XML User Interface Language). XUL defines the standard layout for UI components like trees, menus, toolbars, input controls, and so on. Further, to define the layout, XUL uses RDF, which is an abbreviation for Resource Description Format.

Project Overview

This project has commenced so that nonprogrammers can design and develop a user interface for the product by using the various W3C standards such as HTML and RDF with the combination of JavaScript and some other components. In-depth knowledge of C and C++ is not required to develop an interface using this approach, although it might be helpful in various stages of development.

The major project goals are the following:

♦ To develop a cross-platform user interface so that desktop applications' interfaces becomes as easy to build as Web pages.

♦ To support the common user interface needs of applications such as Navigator, Messenger, and other desktop applications.

♦ To reduce the programming efforts by developing a cross-platform user interface separate from the development of the logical part of the application.

♦ To provide support so that the user interface can be customized using downloadable chrome.

♦ To engage Net users in the development of a new user-interface language for XML programmers and facilitate the building of cross-platform applications.

Main Features

Some of the main features of XPFE are as follows:

♦ The available options are loosely related facilities so that you can pick and choose them easily.

♦ The Common User Interface object is platform-independent API.

♦ Uses an XML-based stream to load windows, dialogs, menus, and other widgets quickly.

♦ Is based on existing standards of Web development such as XML, RDF, HTML, CSS, DOM, and so on.

♦ Provides the Model-View-Controller design pattern for XP implementation.

♦ Implements a user interface and core services in C++, C, and JavaScript.

♦ Easily intermixed with Web languages such as HTML and JavaScript.

The XUL Interface for Desktop Applications

XUL provides a language to describe an HTML-based user interface. It provides many more widgets such as tree controls, scroll bars, splitters, buttons, and windows for generating a GUI. This application provides several benefits in spite of being a cross-platform application. The benefits are the following:

- ♦ Allows the developer to define cross-platform and cross-device user interfaces based on HTML.
- ♦ Can be used to build efficient and small applications that can be downloaded quickly because of their small size.

XML-based User Interface Language, popularly known as XUL (pronounced "zool"), is used to develop cross-platform GUI applications. It is a more specific and rich version of HTML. HTML elements such as `<html>`, `<head>`, `<body>`, `<p>`, `<form>`, and `<table>` are used to develop an HTML document. On the other hand, XUL uses elements like `<window>`, `<menu>`, `<button>`, and `<scrollbar>` that represent the interface, or *widgets*. XUL is part of Mozilla's Open Source browser project and is itself not framed as a standard; rather, it is a technology invented by Netscape and Mozilla.org. All XUL applications can be implemented and tested on the Mozilla browser. To install a Mozilla browser, you can download installer.exe (for Windows-based systems) or other files based on the OS from `http://www.mozilla.org/releases`. You can run the file and follow the steps for the installation.

The basic purpose of creating XUL is to make Mozilla's browser faster and easier. XUL has functions and features that are similar to XML.

XUL Features and Functions

XUL, as mentioned earlier, is a markup language that is based on XML. It can include any HTML elements. The standard event model facilitates XUL to use any scripting language (like JavaScript) to write the user interface code. Some of the primary features are as follows:

- ♦ XUL is based on XML and it can also easily incorporate HTML elements.
- ♦ Any scripting language, such as JavaScript, can be used to write user interface code with the help of the standard event model.
- ♦ Dynamic modification of user interfaces is possible by using the standard DOM interfaces.
- ♦ The user interface communicates with back-end components using language-independent XPCOM.
- ♦ It can be used in combination with various toolkits (potentially KDE/GNOME agnostic).
- ♦ The appearance of the application can be changed using CSS (cascading style sheets).

Most of the time, it becomes tiring for the developer to develop an application for a particular platform and then convert it for a different platform. Thus, a number of solutions have been created for cross-platform application development. Java is a good example; Java files can run on various platforms using JVM (Java Virtual Machine) and so on. The objective for developing XUL is to build a portable user interface that can run on various platforms.

To develop, compile, and debug an application for even one platform takes a long time. So developing the same application for cross-platform use takes much longer. XUL makes it easy and fast to implement and modify a user interface.

Different Types of User Interfaces Can Be Made with XUL

XUL has the ability to develop most elements that are found in modern graphical user interfaces. It is generic enough to develop various applications for the special needs of certain devices, and strong enough to develop sophisticated cross-platform applications.

Some of the user interfaces that you can create by using XUL are the following:

- ♦ Forms containing input-control elements such as input boxes, radio buttons, checkboxes, popup menus, and so on.

- ♦ Toolbars containing buttons, content, textboxes, and so on, like the toolbar that appears at the top of Mozilla's browser.

- ♦ Menu bars and popup menus.

- ♦ Dialog boxes.

- ♦ Hierarchical tree structures to store information.

- ♦ Shortcuts that can be applied through keyboards.

The Architecture of XUL

Let's first understand the architecture of XUL.

- ♦ The extension of XUL applications is `.xul`; these consist of XML files.
- ♦ The XUL files define the content of the application.
- ♦ RDF (Resource Description Framework) files contain additional application data.
- ♦ CSS files are used to provide formatting, style, and some behavior for the XUL application.
- ♦ XUL supports scripting languages such as JavaScript.
- ♦ XUL carries full support to multimedia files, such as PNG images and other audio-visual files.
- ♦ The XUL application referred to as "chrome" carries all the information and the recommendation of W3C for all file types. It defines behavior, look, and content of the user interface.
- ♦ The best example of a XUL application is the Mozilla browser itself.
- ♦ The `chrome` subdirectory is in the main directory of both the Mozilla and Navigator browsers.
- ♦ The `chrome` directory is further divided in the subdirectory `content` containing the XUL, JS, and RDF files, `skin` containing CSS files, and `locale` containing DTD files.

To develop a Web application using XUL, you can save `.xul` and other files such as CSS in the same folder, or you can create a folder containing the directory structure, like `chrome`. These `.xul` files can be viewed using the Mozilla browser. However, it is advisable to store the files in the `chrome` directory or a directory containing the `chrome` structure; otherwise, it may lose some functionality, such as localization.

All XUL files are saved with the `.xul` extension. The structure of a XUL file is described as follows:

```
<?xml version="1.0"?>
<?xml-stylesheet href="chrome://global/skin/" type="text/css"?>

<window
    id="findfile-window"
    title="Find Files"
    xmlns="http://www.mozilla.org/keymaster/gatekeeper/there.is.only.xul">
 ...
</window>
```

Let's go through the syntax line by line.

```
<?xml version="1.0"?>
```

All XUL files start with this tag, similar to how HTML files start with `<HTML>` tags. It is used to identify the file as an XML file. The value `1.0` indicates the version of XML. You can change the version of XUL when new versions are released in the future.

```
<?xml-stylesheet href="chrome://global/skin/" type="text/css"?>
```

The appearance of XUL files can be changed using CSS. This tag is used to define the `.css` file, which changes the appearance of XUL file. It is advisable to import the style-sheet file in the global/skin directory. The XUL file will have the appearance defined in `global.css` if no style file is specified in this line.

```
<window
```

As the body of an HTML file starts with a `<body>` tag, a XUL file uses the `<window>` tag for the same purpose. The `<window>` tag has various attributes, such as `id`, `title`, and so on. The attributes can be placed either on a different line or on the same line.

```
id="findfile-window"
```

The `id` attribute gives a unique name to window elements and uses all the elements in the window. It helps the script identify the window referred to from the XUL file. A XUL file can have one or more windows.

```
title="Find Files"
```

This is used to give a title to the window that appears on the title bar. In the preceding line, the `Find Files` is the title of the window.

```
xmlns="http://www.mozilla.org/keymaster/gatekeeper/there.is.only.xul">
```

This stands for XML namespace — a XUL feature that is used to indicate how the Mozilla browser will interpret the XUL file's tags. Here, it doesn't indicate anything, but it will be used in the future to provide guidelines to Mozilla to interpret the XUL tags. This name may be changed in the future, because we still await the release of the final version. The other Windows elements such as buttons, input controls, toolbars, title bars, menu bars, and so on can be placed in this section.

```
</window>
```

The preceding is the closing tag for the window element.

A Working Model of XUL

There are several methods of opening a window of a XUL file. One of the easiest methods is the `window.open` function of JavaScript. This function has almost the same arguments as are required to open a window for HTML, except one additional argument called "`chrome`". This argument tells the `window.open` function that it is a XUL window. The syntax for opening a window using JavaScript is as follows:

```
window.open(url,windowname,flags);
window.open(url,windowname,flags,secondary_url);
```

where the `flags` argument contains the flag named `"chrome"` to indicate that it is a XUL file. If you want to open a new browser window and specify the URL of the page to load in the window, you need to give the `secondary_url`. For example:

```
window.open("chrome://navigator/content/windowtoopen.xul", "btitle",
"chrome,width=400,height=400");
```

Let's start creating a XUL file like the Find File dialog and save it is as `find.xul`. You can either save it in the `chrome` directory of the Mozilla browser or save it in the directory with the same structure as chrome. It will look like the following:

```
<?xml version="1.0"?>
<?xml-stylesheet href="chrome://global/skin/" type="text/css"?>
```

```
<window
    id="find-window"
    title="Find Files"
    xmlns="http://www.mozilla.org/keymaster/gatekeeper/there.is.only.xul">

</window>-
```

Add buttons to a window

Let's now add some buttons in the window. Here we'll add the Find and Cancel buttons to the window. The `button` tag has different attributes such as `id`, `class`, `src`, and so on. This works similar to HTML, also. The two main properties associated with a `button` tag are an image and a label. You need to specify either of them or both. The button can be used in toolbars, dialog boxes, and so on, such as a dialog box containing the two buttons Find and Cancel.

The button tag has the following syntax:

```
<button
    id="b1"
    class="dialog"
    label="Yes"
    src="image.gif"
    orient="horizontal"
    default="true"
    disabled="true"
    accesskey="y"/>
```

The different attributes are optional and given as follows:

- ♦ `id`—gives the unique identification to the button so that it can be easily identified in the group of buttons available in Windows. These buttons can be referred to in the style sheet or scripts by their IDs.

- ♦ `class`—specifies the style of the style sheet. If you are using the style sheet to give a different appearance to your button, you can define the class that the button appears in. This works in the same fashion in HTML.

- ♦ `label`—specifies the value that appears on the button, such as OK, Cancel, Find, New, Open, and so on. If no value is defined in this attribute, the button remains blank.

- ♦ `src`—used to specify the image on the button. If no value is defined, it represents that no image will appear on the button.

- ♦ `orient`—specifies whether the child elements of the window are oriented horizontally or vertically. The default value is `horizontal`.

- ♦ `default`—on HTML forms, the Submit button is a default button that sends the data when the user presses Enter, regardless of whether it has the focus. To define the default button in XUL, we have set the value to `true`. A window can have one button with a `true` value, in other words, the default button for the window. The other value can be set to `false`. For example, if you are making a dialog window with OK and Cancel, usually the OK button has the `true` value.

- ♦ `disabled`—specifies whether the button will perform the function. If the value of this attribute is set to `true`, the button is disabled. The default value is `false`.

- ♦ `accesskey`—used to specify a letter to be set as a shortcut key. You should show this letter in the label of the button, and it should be underlined. When the user presses the Alt key (or a similar key, depending on the platform) and the shortcut key, the button is activated from anywhere in the window.

Listing 12-1 shows one example of adding buttons.

Listing 12-1: button.xul

```
<?xml version="1.0"?>
<?xml-stylesheet href="chrome://global/skin/" type="text/css"?>

<window
    id="find-window"
    title="Find Files"
    xmlns="http://www.mozilla.org/keymaster/gatekeeper/there.is.only.xul">

<button id="Normal" label="Normal"/>
<button id="Disabled" label="Disabled" disabled="true"/>
<button id="Default" label="Default" default="true"/>

</window>
```

Save this file as `button.xul` in the res/samples. When you view it in the Mozilla browser, you see the output shown in Figure 12-1.

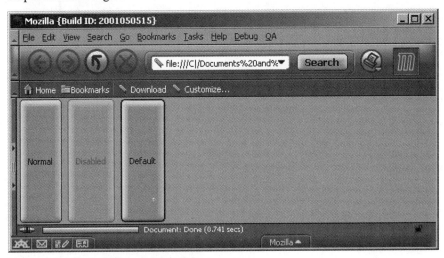

Figure 12-1: A simple button example in XUL

Let's consider another example (Listing 12-2).

Listing 12-2: findfile.xul

```
<?xml version="1.0"?>
<?xml-stylesheet href="chrome://global/skin/" type="text/css"?>
<window
    id="findfile"
    title="Find Files"
    xmlns="http://www.mozilla.org/keymaster/gatekeeper/there.is.only.xul">

  <button id="find " label="Find" default="true"/>
  <button id="cancel " label="Cancel"/>
```

```
</window>
```

If you save this file as `findfile.xul` and view it in the Mozilla browser, you get the output in Figure 12-2.

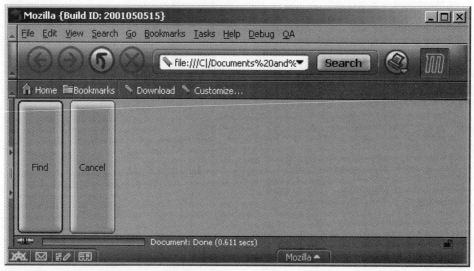

Figure 12-2: The Find File example, containing the Find and Cancel buttons

In this example, you have used the Find button as a default button so that the value of the default is set to `true`. For the Cancel button, the default attribute is left off, which shows that the default value is set to `false`. Notice that there is a thicker border on the Find button, which represents that it is the default button.

Add labels and images to a window

You can use the element text in XUL to put labels into a window. Consider the example in Listing 12-3.

Listing 12-3: label.xul

© 2001 Dreamtech Software India Inc.
All Rights Reserved

```
<?xml version="1.0"?>
<?xml-stylesheet href="chrome://global/skin/" type="text/css"?>
<window
    id="Label"
    title="Using Label"
    xmlns="http://www.mozilla.org/keymaster/gatekeeper/there.is.only.xul">
<text value="hi this is an example of label"/>
</window>
```

This file can be saved as `label.xul` and viewed in a Mozilla browser to get the output in Figure 12-3.

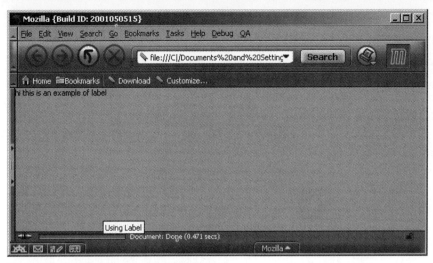

Figure 12-3: Creating a label in the window element

You can also use HTML tags within a XUL file, as seen in Listing 12-4.

Listing 12-4: example2.xul

```
<?xml version="1.0"?>
<?xml-stylesheet href="chrome://global/skin/" type="text/css"?>
<window
    id="HTMLtag"
    title="Using HTML"
    xmlns="http://www.mozilla.org/keymaster/gatekeeper/there.is.only.xul">
<html>
Using HTML with in XUL file
</html>
</window>
```

The output of the preceding code is shown in Figure 12-4.

Figure 12-4: Using HTML and XUL together

Images

You can display images in a window, as in HTML. Consider Listing 12-5.

Listing 12-5: example3.xul

```
<?xml version="1.0"?>
<?xml-stylesheet href="chrome://global/skin/" type="text/css"?>
<window
    id="img"
    title="Using Images"
    xmlns="http://www.mozilla.org/keymaster/gatekeeper/there.is.only.xul">
<img src="images/go.gif"/>
</window>
```

Here images is the name of the directory, which carries a go.gif image. If the image is in the same folder, you don't have to give the path. The output of the preceding example is shown in Figure 12-5.

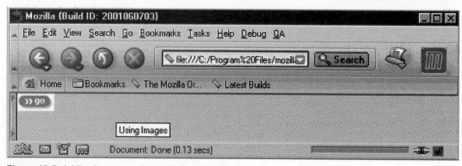

Figure 12-5: Adding images such as .gifs in a XUL file

Input Controls in XUL

XUL contains elements similar to HTML form elements. Let's understand the input controls in XUL one by one.

Textbox

This element is the same as the Text field in HTML. The syntax structure of a text-entry field is as follows:

```
<textbox
    id="text1"
    class="tclass"
    value="Enter your name"
    disabled="true"
    type="password"
    multiline="true"
    maxlength="30">
```

The different attributes of a textbox are the following:

◆ id—gives the unique identification to the textbox so that it can be easily identified in the group of buttons available in Windows. These textboxes can be referred to in style sheets or scripts by their IDs.

- `class`—specifies the style of the style sheet. If you are using the style sheet to give a different appearance to your textbox, you can define the class that the button appears in. This works the same with HTML, also.

- `value`—specifies the default text that appears at the time the page loads.

- `disabled`—possible values are either `true` or `false`. `True` represents the disabled button and `false` represents the enabled button.

- `type`—you can specify the type of textbox; for example, `password` creates a textbox similar to a password entry box in HTML.

- `maxlength`—specifies the maximum number of characters allowed in the text field.

- `multiline`—when set to `true`, allows the user to enter the value on multiple lines instead of one. The default value is `false`. This is the same as `textarea` in HTML.

Radio buttons and checkboxes

These are special types of input control boxes, which allow the user to select either one option (radio buttons) or more than one option (checkboxes).

The syntax structure of checkboxes is as follows:

```
<checkbox
    id="c1"
    imgalign="top"
    accesskey="c"
    checked="true"
    label=" Reading"
    src="image.gif"
    disabled="true"/>
```

The syntax structure of radio buttons is as follows:

```
<radiogroup
    id="r1"
    imgalign="top"
    accesskey="r"
    checked="true"
    label=" Male"
    src="true"
    disabled="true"/>
```

The different attributes of check boxes and radio buttons are the following:

- `id`—gives a unique identification to checkboxes or radio buttons so that they can be easily identified in the group of checkboxes or radio buttons available in Windows. These can be referred to in the style sheet or scripts by their IDs.

- `imgalign`—specifies the alignment of checkboxes or radio buttons, which appear on the top in window.

- `accesskey`—used to specify a letter to be set as a shortcut key. You should show this letter in the label of the button, and it should be underlined. When the user presses the Alt key (or a similar key depending on the platform) and the shortcut key, the button is activated from anywhere in the window.

- `checked`—specifies whether the radio button or checkboxes are checked by default. The default value is `false`.

- `label`—specifies the value that appears in the window, such as male, female, reading, music, and so on.

- `src`—used to specify the image explicitly.
- `disabled`—default value, either `true` or `false`, representing that the text entry is disabled.

Drop-down lists

You create drop-down lists in XUL by using the `menulist` element. This is same as the `select` option in HTML.

The syntax structure of popup menu is as follows:

```
<menulist id="mn1" src="image.gif" accesskey="m">
  <menupopup>
    <menuitem label="Car" value="car" />
    <menuitem label="Taxi" value="taxi" />
    <menuitem selected="true" label="Bus" value="Bus"/>
    <menuitem label="Train" value="train" />
  </menupopup>
```

The different subelements and attributes of `menupopup` are the following:

- `menuitem`—specifies the different options of the drop-down list, such as `car`, `taxi`, `bus`, `train`, and so on.
- `id`—gives the unique identification to `menupopup` so that it can be easily identified in the group of popup menus available in Windows. These can be referred to in the style sheet or scripts by their IDs.
- `accesskey`—used to specify a letter to be set as a shortcut key. You should show this letter in the label of the tag button, and it should be underlined. When the user presses the Alt key (or a similar key, depending on the platform) and the shortcut key, the button is activated from anywhere in the window.
- `selected`—specifies which option will be selected by default. The default value is `false`.
- `label`—specifies the value that appears in the window, such as male, female, reading, music, and so on.
- `src`—used to specify the image explicitly.
- `value`—represents the value that will transfer after submitting the form.

Listing 12-6 is a simple example to help you understand all the preceding elements in XUL:

Listing 12-6: example4.xul

```
<?xml version="1.0"?>
<?xml-stylesheet href="chrome://global/skin/" type="text/css"?>
<window
  id="input-control" title="Using input-control"
  xmlns:html="http://www.w3.org/1999/xhtml"
  xmlns="http://www.mozilla.org/keymaster/gatekeeper/there.is.only.xul"
orient="vertical">



  <text value="Enter Your Name"/>
  <textbox id="t1" multiline="false"/>

  <text value="Enter a password"/>
```

```
<textbox type="password" maxlength="8" id="p1"/>

<text value="Enter Your Hobbies"/>
<checkbox id="cr" checked="true" label="Reading"/>
<checkbox id="cm" label="Music"/>

<text value="Enter Your Sex"/>
<radio id="orange" checked="false" label="Male"/>
<radio id="violet" checked="true" label="Female"/>

<text value="Enter Your country"/>
<menulist value="Bus" id="mn1" accesskey="m" maxsize="10">
<menupopup>
   <menuitem label="USA" value="USA"/>
   <menuitem label="UK" value="UK"/>
   <menuitem selected="true" label="India" value="India"/>
   <menuitem label="Russia" value="Russia"/>
</menupopup>
</menulist>


</window>
```

The result of the preceding code is shown in Figure 12-6.

Figure 12-6: Creating a form that contains input controls such as entry fields, radio buttons, checkboxes, drop-down lists, and so on.

Work with HTML and HTML Elements

In addition to the available XUL elements, HTML tags can be added directly into the XUL file. However, remember that you should always use lowercase while adding HTML tags. This is because XUL is a case-sensitive language. It is not advisable to use HTML elements with XUL because you can do everything within the XUL file that you can do using HTML tags. In this section, we will describe how to use them anyway. You are required to declare that you are using HTML elements in a XUL file, which helps Mozilla distinguish between HTML tags and XUL elements. For this, you can use the following attributes of the window:

```
xmlns:html="http://www.w3.org/1999/xhtml"
```

The preceding is a declaration tag for HTML. It is the same as the one that you use for XUL. This declaration says that this document uses HTML tags. Let's add some HTML tags in the old window for XUL:

```
<?xml version="1.0"?>
<?xml-stylesheet href="chrome://global/skin/" type="text/css"?>
<window
    id="findfile"
    title="Find Files"
    xmlns:html="http://www.w3.org/1999/xhtml"

xmlns="http://www.mozilla.org/keymaster/gatekeeper/there.is.only.xul">

  <button id="find-button" class="dialog" label="find" default="true"/>
  <button id="cancel-button" class="dialog" label="cancel"/>

</window>
```

You will get the same output as you did from using the XUL elements, as shown in Figure 12-7.

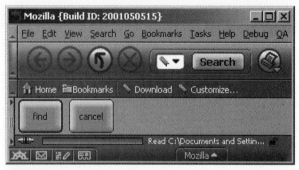

Figure 12-7: Creating a XUL file with the help of HTML elements

Following are the important points that you should adhere to while using HTML with a XUL file:

♦ HTML: is always used before every HTML tag.

♦ You should use lowercase for writing HTML tags.

♦ The values for all the attributes must come under double codes.

♦ A trailing slash should always be included at the end of every tag.

All HTML tags can be used in XUL files, even tags such as <head> and <body>. Consider the following examples:

Listing 12-7: example5.xul

© 2001 Dreamtech Software India Inc.
All Rights Reserved

```
<?xml version="1.0"?>
<?xml-stylesheet href="chrome://global/skin/" type="text/css"?>
```

```
<window
    id="findfile"
    title="Find Files"
    xmlns:html="http://www.w3.org/1999/xhtml"
    xmlns="http://www.mozilla.org/keymaster/gatekeeper/there.is.only.xul"
    orient="vertical">
<html>
<html:br/>
<html:input type="checkbox" value="true"/>Male
<html:table border="1">
  <html:tr>
    <html:td>
      A simple table
    </html:td>
  </html:tr>
</html:table>
</html>
</window>
```

The output of the preceding example is shown in Figure 12-8.

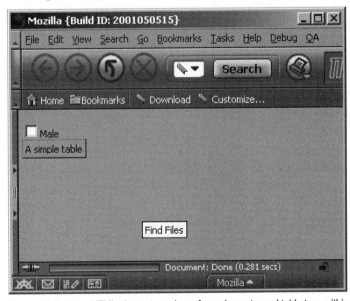

Figure 12-8: Using HTML elements such as, form elements, and table tags within a XUL file

XUL elements can also be placed in HTML. Consider the following example:

Listing 12-8: example6.xul

```
<?xml version="1.0"?>
<?xml-stylesheet href="chrome://global/skin/" type="text/css"?>
<window
    id="findfile"
    title="Find Files"
    xmlns:html="http://www.w3.org/1999/xhtml"
```

```
xmlns="http://www.mozilla.org/keymaster/gatekeeper/there.is.only.xul"
    orient="vertical">
<html>
<html:br/>

<html:p>
  Search for:
<html:br/>
  <html:input id="t1"/>
  <button id="n1" label="OK"/>
</html:p>

</html>
</window>
```

This provides the output shown in Figure 12-9.

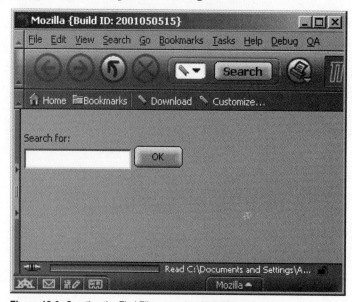

Figure 12-9: Creating the Find File example with HTML elements

Add elements to the Find File dialog

Consider the previous example. Now you'll add more text using HTML and more buttons using XUL with the Cancel label. You can use the HTML `div` tag to split the elements onto the next line, as in Listing 12-9.

Listing 12-9: example7.xul

```
<?xml version="1.0"?>
<?xml-stylesheet href="chrome://global/skin/" type="text/css"?>
<window
    id="findfile"
    title="Find Files"
```

```
    xmlns:html="http://www.w3.org/1999/xhtml"
    xmlns="http://www.mozilla.org/keymaster/gatekeeper/there.is.only.xul"
    orient="horizontal">

<html:div>
  <html:div>
    Enter your string for search criterion below and click the Find button to
begin the search.
</html:div>
  <text value="Search the following:"/>
  <textbox id="t1"/></html:div>
  <button id="bf" class="dialog" label="Find" default="true"/>
  <button id="bc" class="dialog" label="Cancel"/>
</window>
```

This provides the output in Figure 12-10.

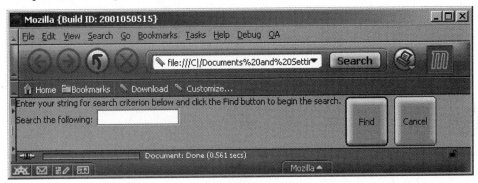

Figure 12-10: Creating the Find File application using HTML elements

Use Spring for Spacing

Browser resolution is always a problem for someone developing a cross-platform interface because resolutions vary from system to system. Hence, it is extremely important for the developer to understand the basic layout of the design. Normally, different languages (used for the development of cross-platform applications) do have the facility for resizing and repositioning the window element depending on the platform. XUL has the built-in capability of automatic resizing and repositioning of window elements on the basis of screen resolution. In previous examples, the window elements were resized and repositioned automatically depending on the size or resolution of the browser. XUL has a feature called "Box Model," which enables you to divide the window into different boxes that hold the elements. Another important window element is spring, which is used to put space in a window. It helps to increase and decrease the size of the window elements as the user resizes the window. This helps the developer place window elements anywhere in the window, like the bottom, the top right, and so on. The simplest syntax for spring is the following:

```
<spring flex="1"/>
```

The spring attribute flex is used to define the flexibility in the spring element and make it stretchy. Let's add a spring element to the previous Find File example:

Listing 12-10: example8.xul

© 2001 Dreamtech Software India Inc.
All Rights Reserved

```
<?xml version="1.0"?>
```

```
<?xml-stylesheet href="chrome://global/skin/" type="text/css"?>
<window
    id="findfile"
    title="Find Files"
    xmlns:html="http://www.w3.org/1999/xhtml"
    xmlns="http://www.mozilla.org/keymaster/gatekeeper/there.is.only.xul"
    orient="horizontal">
<html:div>
    <html:div>
    Enter your string for search criteria below and click the Find button to
begin
        the search.
</html:div>
  <text value="Search the following:"/>
  <textbox id="t1"/></html:div>
  <spring flex="1"/>
  <button id="bf" class="dialog" label="Find" default="true"/>
  <button id="bc" class="dialog" label="Cancel"/>
</window>
```

Now the Find and Cancel buttons will be placed at the right side of the window, as shown in Figure 12-11.

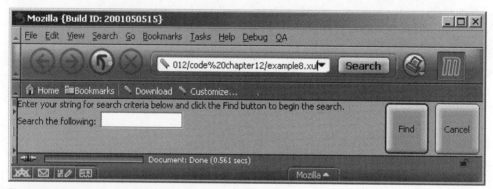

Figure 12-11: Using the spring element to provide spacing in the XUL window

You can assign any value to the `flex` attribute, like 2, 3, and so on. For example, say that a window has two buttons, one with a flex value of 1 and other with a flex value of 2. The second button will be twice the size of the first button. If we assign 0 to the `flex` attribute, that element cannot be flexible. The `flex` attribute can also be placed inside the window's elements — even in HTML elements, as seen in Listing 12-11.

Listing 12-11: example9.xul

```
<?xml version="1.0"?>
<?xml-stylesheet href="chrome://global/skin/" type="text/css"?>
<window
    id="flex"
    title="Using Flex"
    xmlns:html="http://www.w3.org/1999/xhtml"
    xmlns="http://www.mozilla.org/keymaster/gatekeeper/there.is.only.xul"
    orient="horizontal">
```

```
    <button label="Find" flex="10"/>
    <button label="Cancel" flex="1"/>

</window>
```

In this example, the Find button is ten times bigger than the Cancel button, as shown in Figure 12-12.

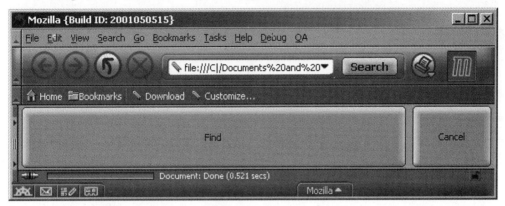

Figure 12-12: Using different flex values to provide flexibility within a window element

Use the Box Model in XUL

Boxes are the main layout form for XUL files. They allow you to divide the window into a number of boxes. The box model can align either horizontally or vertically by using the `orient` attribute. The basic syntax for the box model is as follows:

```
<box orient="horizontal">
  ...
</box>
```

Here the `orient` attribute specifies the layout of the box—either horizontal or vertical. Consider Listing 12-12.

Listing 12-12: example10.xul

© 2001 Dreamtech Software India Inc.
All Rights Reserved

```
<?xml version="1.0"?>
<?xml-stylesheet href="chrome://global/skin/" type="text/css"?>
<window
    id="boxmodel"
    title="Using Box Model"
    xmlns:html="http://www.w3.org/1999/xhtml"
    xmlns="http://www.mozilla.org/keymaster/gatekeeper/there.is.only.xul"
    orient="horizontal">

<box orient="vertical">
  <button id="b1" label="Ignore"/>
  <button id="b2" label="Retry"/>
  <button id="b3" label="Cancel"/>
</box>

<box orient="horizontal">
  <button id="b4" label="Ignore"/>
```

```
  <button id="b5" label="Retry"/>
  <button id="b6" label="Cancel"/>
</box>

</window>
```

Figure 12-13 shows the output of the preceding code.

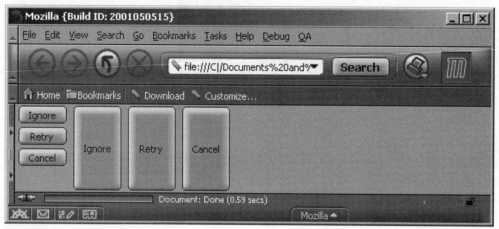

Figure 12-13: Creating different button examples using the box model

You can add a box within a box to give the better alignment of your interface (Listing 12-13).

Listing 12-13: example11.xul

```
<?xml version="1.0"?>
<?xml-stylesheet href="chrome://global/skin/" type="text/css"?>
<window
    id="boxmodel"
    title="Using Box Model"
    xmlns:html="http://www.w3.org/1999/xhtml"
    xmlns="http://www.mozilla.org/keymaster/gatekeeper/there.is.only.xul"
    orient="horizontal">


<box orient="vertical">
  <box orient="horizontal">
    <text for="username" value="User Name:"/>
    <textbox id="t1"/>
  </box>
  <box orient="horizontal">
    <text for="pass" value="Password:"/>
    <textbox id="p1"/>
  </box>
  <button id="b1" label="GO"/>
  <button id="b2" label="Refresh"/>
</box>
</window>
```

You will get two boxes containing four elements, which are aligned horizontally, and the parent box containing the remaining Go and Refresh buttons aligned vertically, as shown in Figure 12-14.

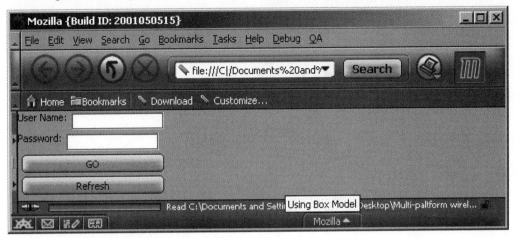

Figure 12-14: Creating an example with a box within a box

In Listing 12-14, you will use the box model in the previous Find File example. A box with vertical orientation will be used around all the elements, and a box with vertical orientation will be used around the textbox and the buttons.

Listing 12-14: example12.xul

```
<?xml version="1.0"?>
<?xml-stylesheet href="chrome://global/skin/" type="text/css"?>
<window
    id="findfile"
    title="Find Files"
    xmlns:html="http://www.w3.org/1999/xhtml"
    xmlns="http://www.mozilla.org/keymaster/gatekeeper/there.is.only.xul"
    orient="horizental">
<box orient="vertical" flex="1">
 <html>
    Enter your string for search criteria below and click the Find button to
begin
      the search.
 </html>
 <box orient="horizontal">
  <text value="Search the following:"/>
  <textbox id="t1"/>
 </box>
 <box orient="horizontal">
  <spring flex="1"/>
  <button id="bf" class="dialog" label="Find" default="true"/>
  <button id="bc" class="dialog" label="Cancel"/>
 </box>
</box>
</window>
```

Figure 12-15 shows the result of the preceding code.

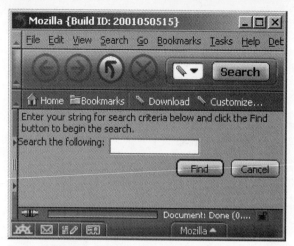

Figure 12-15: Using different orientation attributes of the box element

You can use another element, `<vbox>`, which is a box element with default vertical orientation. This is equivalent to `<box orient="vertical">`.

Now you have seen the orientation of the box. The other attributes of the box element are `autostretch` and `valign`. If the value `never` is provided in the `autostretch` attribute, it means that the elements of that box never stretch with respect to the size of the window. The `valign` attribute has four values. They are the following:

- ♦ `top`
- ♦ `middle` (default position)
- ♦ `bottom`
- ♦ `baseline`

Consider the Listing 12-15.

Listing 12-15: example13.xul

© 2001 Dreamtech Software India Inc.
All Rights Reserved

```xml
<?xml version="1.0"?>
<?xml-stylesheet href="chrome://global/skin/" type="text/css"?>
<window
    id="boxpos"
    title="Adding box position"
    xmlns:html="http://www.w3.org/1999/xhtml"
    xmlns="http://www.mozilla.org/keymaster/gatekeeper/there.is.only.xul"
    orient="horizontal">

<box>
  <button label="OK"/>
  <button label="Cancel"/>
</box>
<box autostretch="never" valign="top">
  <button label="Ignore"/>
  <button label="Retry"/>
</box>
```

```
</window>
```

In the preceding code, the `autostretch` attribute contains the `never` value. So the Ignore and Retry buttons are never resized on the basis of screen resolution. The size of these buttons will depend on the label and will increase or decrease depending on the label size. The `valign` attribute with a value of `top` will position the Ignore and Retry buttons on the top of the window. This is shown in Figure 12-16.

Figure 12-16: Providing positioning to different box elements in the window

Different Attributes of the Window Element

You can also set the size of the window elements with the help of the following attributes:

♦ `width`—not required with flexible elements because they change their width depending on the size of the window. This attribute specifies the width of the window element and can be used as the initial width of the window's elements.

♦ `min-width`—used with flexible elements and specifies the minimum width of the window's element.

♦ `max-width`—used with flexible elements and specifies the maximum width of the window's elements.

♦ `height`—not required with flexible elements because they change their height depending on the size of the window. This attribute specifies the height of the window element and can be used as an initial height for window's elements.

♦ `min-height`—used with flexible elements and specifies the minimum height of the window's elements.

♦ `max-height`—used with flexible elements and specifies the maximum height of the window's elements.

Listing 12-16 uses these attributes in the HTML `img` tag.

Listing 12-16: example14.xul

```
<?xml version="1.0"?>
<?xml-stylesheet href="chrome://global/skin/" type="text/css"?>

<window
```

```
          id="button size"
          title="specifying button size"
          xmlns:html="http://www.w3.org/1999/xhtml"
          xmlns="http://www.mozilla.org/keymaster/gatekeeper/there.is.only.xul"
          orient="horizontal">


<button label="Click me" width="100" height="100"/>
<button label="Click me" max-width="200" max-height="200"/>
<button label="C" min-width="100" miin-height="100"/>


</window>
```

In this example, all the buttons are flexible elements, the first using the `width` and `height` properties, the second using the `max-width` and `max-height` properties, and the last one using the `min-width` and `min-height` properties. You get the output in Figure 12-17.

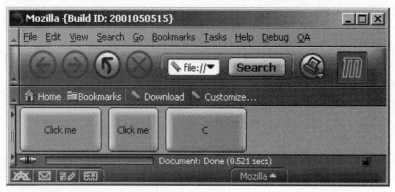

Figure 12-17: Giving sizes to the boxes

It is always advisable to use these attributes as a value with the `style` attribute, as seen in the following code:

```
<button label="First" style="width: 100px;"/>
<button label="Second" style="width: 100em; height: 10px;"/>
<button label="Third" flex="1" style="min-width: 50px;"/>
<button label="Fourth" flex="1" style="min-height: 2ex;
   max-height: 100px"/>
<textbox flex="1" style="max-width: 10em;"/>
<html style="max-width: 50px">This is some boring but simple wrapping
text.</html>
```

Crop Attribute and Text and Buttons

If you are using a button or text that displays text that is bigger than the specific size of the button or text and window, the `crop` property allows you to specify how the text should be cropped. The possible values for `crop` are the following:

♦ `left`—If the label of the button is bigger than the maximum width of the button, some of the label is not visible from the left side.

♦ `right`—Some of the button label is not visible from the right side.

♦ `center`—The middle section of the button label is visible only in the browser.

♦ `none`—The default value, which specifies that the full button label is visible.

Listing 12-17 shows an example of using the `crop` attribute.

Listing 12-17: example15.xul

© 2001 Dreamtech Software India Inc.
All Rights Reserved

```
<?xml version="1.0"?>
<?xml-stylesheet href="chrome://global/skin/" type="text/css"?>

<window
    id="crop"
    title="Using crop"
    xmlns:html="http://www.w3.org/1999/xhtml"
    xmlns="http://www.mozilla.org/keymaster/gatekeeper/there.is.only.xul"
    orient="vertical">

<button label="hi now i m using crop with button" crop="right" flex="1"
style="max-width: 50px;"/>


</window>
```

The preceding code provides the output in Figure 12-18.

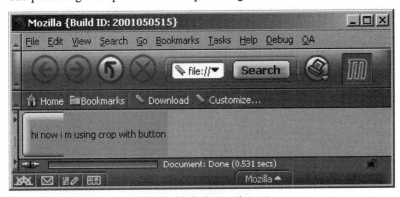

Figure 12-18: Using the crop attribute with the button element

The Progress Bar with XUL

A progress bar (or progress meter) shows the ongoing status of any task, in other words, how much of a task has been completed and how much is left for a particular operation. The progress bar is visible on the bottom-left corner of the Mozilla browser when you open any document. It shows the progress in reading that particular document. In XUL, the following types of progress meters are available:

- ♦ **Determinate progress meters:** Used in case the completion time of the task is known to the developer. The progress meter will start filling up to show the progress of the operation. At the end of the operation, it will show the full details. You can use this type of progress bar for the Download File dialog because you can estimate the completion time by looking at the file size and the speed of the network.

- ♦ **Indeterminate progress meters:** They are used if the developer doesn't know the completion time of the task.

Following are the attributes of a progress bar:

- ♦ id—gives a unique identification to the progress bar.
- ♦ mode—used to specify whether a progress meter is of determinate or indeterminate type.

♦ value—shows the current value and is used only for a determinate progress meter. You can use any value between 0% and 100%. You can use any scripting language, such as JavaScript, to show the change in the value of the progress bar as the task progresses.

♦ align—used to specify the direction of the progress meter: either horizontal (the default value) or vertical. If it is horizontal, the progress bar progresses from left to right; if it is vertical, the progress bar progresses from bottom to top.

♦ label—allows the insertion of some text such as "loading…" and so on in the progress meter.

Listing 12-18 will help you understand the progress meter:

Listing 12-18: example16.xul

```
<?xml version="1.0"?>
<?xml-stylesheet href="chrome://global/skin/" type="text/css"?>

<window
    id="progressbar"
    title="Using Progress Bar"
    xmlns:html="http://www.w3.org/1999/xhtml"
    xmlns="http://www.mozilla.org/keymaster/gatekeeper/there.is.only.xul"
    orient="vertical">
<titledbox orient="horizontal">
  <title><text value="Criteria for Search"/></title>

<spring style="height: 10px"/>
<box orient="horizontal">
  <menulist id="searchtype">
    <menupopup>
      <menuitem label="Name"/>
      <menuitem label="Size"/>
      <menuitem label="Date Modified"/>
    </menupopup>
  </menulist>

  <spring style="width: 10px;"/>
  <menulist id="searchmode">
    <menupopup>
      <menuitem label="Incuded"/>
      <menuitem label="Is Not Incuded"/>
    </menupopup>
  </menulist>
  <spring style="width: 10px;"/>
  <textbox id="find-text" flex="1" style="min-width: 15em;"/>
</box>

</titledbox>

<box orient="horizontal">
  <button label="Find"/>
  <button label="Cancel"/>

</box>
```

```
<box orient="horizontal">
<progressmeter
    id="pm1"
    mode="determined"
    value="0%"
    align="horizontal"
    label="Wait"/>
</box>
</window>
```

Here we used another window element called `titledbox`, which is used to combine the elements and borders that will be drawn, to show them in the same group. The output of the preceding code is shown in Figure 12-19.

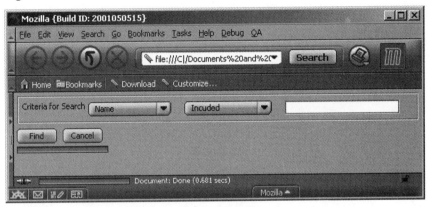

Figure 12-19: Providing a progress bar to the XUL application

Scroll Bars in XUL

If you are displaying a large document, image, or any other object that doesn't fit in the window, you need a scroll bar element that allows the user to move from left to right and from top to bottom in the document, to get a complete vision of the object. Likewise, we also have horizontal and vertical scroll bars in the browser.

The `<scrollbar>` tag has the following syntax structure:

```
<scrollbar
    id="identifier"
    align="horizontal"/>
    curpos="10"
    maxpos="100"
    increment="2"
    pageincrement="5"/>
```

The attributes are optional and are as follows:

- ◆ `id`—gives a unique identification to the scroll bar.
- ◆ `align`—used to specify the progress of the scroll bar either in the horizontal direction (the default value) or the vertical direction. In the case of horizontal direction, the scroll bar progresses from left to right; in the vertical case, it progresses from bottom to top.
- ◆ `curpos`—specifies the current position of the slider (the thumb of the scroll bar). You can give it any value between 0 and the value of `maxpos`. 0 is the default value for `curpos`.
- ◆ `maxpos`—specifies the maximum position of the slider. `100` is the default value.

♦ increment—specifies how much the scroll bar moves when the user clicks on the scroll bar's arrows. 1 is the default value here.

♦ pageincrement—specifies how much movement is allowed (the changes in the current position of the slider) to the scroll bar while the user clicks on the scroll bar's arrows or anywhere on the body of the scroll bar. 10 is the default value here.

The other elements of the scroll bar are as follows:

♦ scrollbarbutton—used to represent the arrow buttons at the end of the scroll bar. The possible values of scrollbarbutton are increment and decrement for the direction of the scroll bar.

♦ slider—used to represent the area of the slider (that is, thumb), which moves across the scroll bar.

A scroll bar consists of three elements: a slider bar and two arrow buttons, as shown in Listing 12-19.

Listing 12-19: example17.xul

```
<?xml version="1.0"?>
<?xml-stylesheet href="chrome://global/skin/" type="text/css"?>

<window
    id="scrollbar"
    title="Using Scroll Bar"
    xmlns:html="http://www.w3.org/1999/xhtml"
    xmlns="http://www.mozilla.org/keymaster/gatekeeper/there.is.only.xul"
    orient="vertical">
<scrollbar id="scid" curpos="0" maxpos="100"
    increment="1" pageincrement="10">
  <scrollbarbutton type="decrement"/>
  <slider flex="1"/>
  <scrollbarbutton type="increment"/>
</scrollbar>
</window>
```

The preceding code gives the output in Figure 12-20.

Tab Boxes in XUL

Tab boxes are a collection of tabs that can appear at the top of a window. The user gets different options by clicking the different tabs in the tab boxes. It requires five new elements. They are the following:

♦ tabbox—the outer box, which collects all the tabs.

♦ tabs—the inner box, which collects the individual tabs (the tabs row).

♦ tab—a specific tab member of the inner or outer box that brings the tab page to the front when the user clicks on it.

♦ tabpanels—holds the child element tab panel. You can place multiple tab panels within this element.

♦ tabpanel—represents the body of a single page in the tab box. Used to hold the content of different pages in different tab panels. For example, the first tab panel holds the content of the first tab, the second tab panel holds the contents of the second tab, and so on.

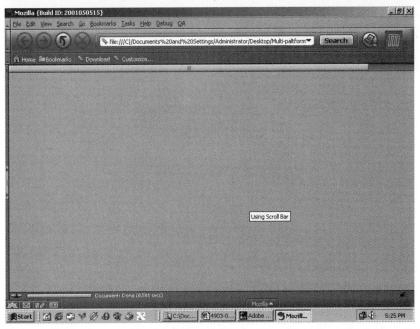

Figure 12-20: Creating a scroll bar for a XUL application that has a bigger image than the box will hold

The basic syntax structure for tab boxes is as follows:

```
<tabbox orient="vertical" id="tablist">
  <tabs orient="horizontal">
    -- tabs go here --
  </tabs>
  <tabpanels>
    -- tabpanels go here --
  </tabpanels>
</tabbox>
```

The attributes of tab boxes are as follows:

- ♦ id—gives a unique identification to the tab box so that it can be easily identified in the group of tab boxes available in Windows. These tab boxes can be referred to in the style sheet or scripts by their IDs.

- ♦ orient—specifies whether the tab boxes in Windows are horizontal or vertical direction. By default, the alignment is horizontal.

Let's understand this by looking at Listing 12-20.

Listing 12-20: example18.xul

```
<?xml version="1.0"?>
<?xml-stylesheet href="chrome://global/skin/" type="text/css"?>
<window
id="windw"
title="Windows 2"
xmlns:html="http://www.w3.org/1999/xhtml"
xmlns="http://www.mozilla.org/keymaster/gatekeeper/there.is.only.xul">
```

```
<tabbox orient="vertical" flex="10">
<tabs orient="horizontal" >
<tab label="hello"/>
<tab label="Authoring"/>
<tab label="Discussion Group"/>
<tab label="Mails"/>
<tab label="Collaborative"/>
<tab label="Services"/>
<tab label="Online Stuff"/>
<tab label="Status"/>
<tab label="Sharing"/>
<tab label="Conversion"/>
<tab label="Conversion"/>
</tabs>
<tabpanels>
<tabpanel id="mailtab" flex="1">
<checkbox label="This is a Simple Checkbox label"/>
</tabpanel>
<tabpanel id="newstab">
<html:table  >
<html:tr >
<html:td>
<box orient="horizontal" flex="1">
<progressmeter value="25%" style="margin: 10px; bgcolor: red; hieght: 10px;
width: 50px;" />
<spring flex="1"/>
</box>
</html:td>
</html:tr>
</html:table>
</tabpanel>
<tabpanel id="mailtab">
<checkbox label="Automatically check for mail"/>
</tabpanel>
<tabpanel id="newstab">
<box autostretch="never" valign="top">
<button label="Yes"/>
<button label="No"/>
</box>
<box autostretch="never" valign="top">
</box>
</tabpanel>
</tabpanels>
</tabbox>
</window>
```

The preceding code gives the output in Figure 12-21.

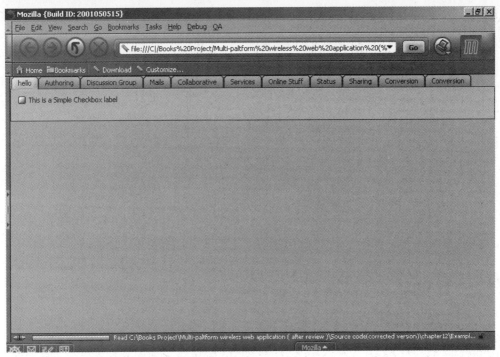

Figure 12-21: Creating a XUL application containing tab boxes

Using the Toolbar in XUL

A toolbar is a type of box that contains rows of buttons as well as other window elements. The best example is the toolbar at the top of Mozilla's browser, which holds the different buttons and input boxes (containing the URL). You can place a toolbar either horizontally or vertically in a window. Because it is a type of box, you can place the toolbar anywhere in the window: the top, the bottom, the middle, and so on.

A set of toolbars can be placed at the top of a window. Consider the toolbar example in Listing 12-21.

Listing 12-21: example19.xul

```
<?xml version="1.0"?>
<?xml-stylesheet href="chrome://global/skin/" type="text/css"?>

<window
    id="toolbox"
    title="Using Toolbox"
    xmlns:html="http://www.w3.org/1999/xhtml"
    xmlns="http://www.mozilla.org/keymaster/gatekeeper/there.is.only.xul"
    orient="vertical">

<toolbox>
  <toolbar id="toolbar">
    <button label="Back"/>
    <button label="Forward"/>
  </toolbar>
```

```
</toolbox>

</window>
```

The output of this example is shown in Figure 12-22.

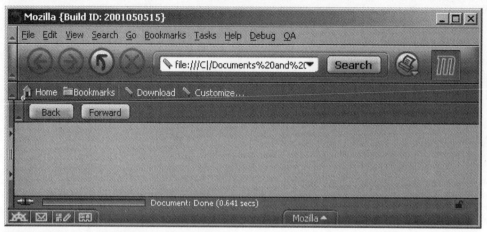

Figure 12-22: Creating a toolbar in a XUL application

By default, a toolbar is aligned horizontally. However, it can also be aligned vertically by giving the value of `orient` as `vertical`. The following example gives vertical orientation to the box. The toolbox orientation is horizontal, as in the case of multiple toolbars; the buttons will appear next to each other instead of on top of one another.

```
<toolbox orient="horizontal">
  <toolbar id="toolbar" orient="vertical">
    <box orient="vertical"/>
      <button label="Back"/>
      <button label="Forward"/>
    </box>
  </toolbar>
</toolbox>
```

The menu bar

A menu bar is just like a toolbar; however, a XUL document can contain only one menu bar. It is specified with the `<menubar>` tag in the XUL page. A menu bar consists of five elements. They are the following:

- ♦ `menubar`—contains the row of menus.
- ♦ `menu`—specifies the names of the menus on the menu bar and menu popup as child elements.
- ♦ `menupopup`—carries the list of menu options that appears when the user clicks on any menu title.
- ♦ `menuitem`—the list of items that are available in `menupopup`.
- ♦ `menuseparator`—required to set up two menu items with each other.

These elements can be placed in the menu bar or separately. Let's go through Listing 12-22 so you can understand the menu bar:

Listing 12-22: example20.xul

```
<?xml version="1.0"?>
<?xml-stylesheet href="chrome://global/skin/" type="text/css"?>

<window
    id="menubar"
    title="Using Menubar"
    xmlns:html="http://www.w3.org/1999/xhtml"
    xmlns="http://www.mozilla.org/keymaster/gatekeeper/there.is.only.xul"
    orient="vertical">

<menubar id="mb1">
  <menu id="fm" label="File">
    <menupopup id="fp">
      <menuitem label="New"/>
      <menuitem label="Open"/>
      <menuitem label="Save"/>
      <menuitem label="Print"/>
      <menuseparator/>
      <menuitem label="Exit"/>
    </menupopup>
  </menu>
  <menu id="en" label="Edit">
    <menupopup id="ep">
      <menuitem label="Cut"/>
      <menuitem label="Copy"/>
      <menuitem label="Paste"/>
    </menupopup>
  </menu>
</menubar>

</window>
```

This example shows you the `menuitem` of the file menu when you click on the file and `menuitem` of edit menu when you click on edit. The output of this code is shown in Figure 12-23.

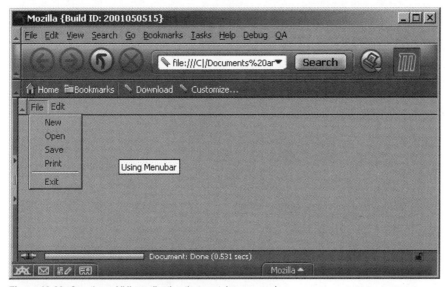

Figure 12-23: Creating a XUL application that contains a menu bar

The attributes of the menu element work much like those of the button element.

◆ id—gives a unique identification to the menu title button.

◆ label—specifies the value that appears on the button, such as File, Edit, and so on.

◆ disabled—specifies whether the menu item will perform the function. If the value of this attribute is set to true, the button is disabled. The default value is false.

◆ accesskey—used to specify a letter to be set as a shortcut key. You should show this letter in the label of the menu item and also underline it. When the user presses the Alt key (or a similar key, depending on the platform) and the shortcut key, the button is activated from anywhere in the window.

◆ open—specifies whether the menu is open. The script will automatically check the value to determine whether it is open, so you normally don't need to set the value yourself.

◆ menuactive—used to set the focus on the menu item. The script automatically checks whether the menu item is focused, so you do not need to set the value yourself.

Submenus in a Menu Bar

The submenu has its own list of options within a menu. Consider the example in Listing 12-23.

Listing 12-23: example21.xul

```
<?xml version="1.0"?>
<?xml-stylesheet href="chrome://global/skin/" type="text/css"?>

<window
    id="menubar"
    title="Using Menubar"
    xmlns:html="http://www.w3.org/1999/xhtml"
    xmlns="http://www.mozilla.org/keymaster/gatekeeper/there.is.only.xul"
    orient="vertical">

<menubar id="mb1">
  <menu id="fm" label="File">
    <menupopup id="fp">

      <menu id="fn" label="New">
        <menupopup id="new-popup">
         <menuitem label="Page"/>
         <menuitem label="Web"/>
       </menupopup>
      </menu>
      <menuitem label="Open"/>
      <menuitem label="Save"/>
      <menuitem label="Print"/>
      <menuseparator/>
      <menuitem label="Exit"/>
    </menupopup>
  </menu>
  <menu id="en" label="Edit">
    <menupopup id="ep">
     <menuitem label="Cut"/>
     <menuitem label="Copy"/>
```

```
        <menuitem label="Paste"/>
      </menupopup>
    </menu>
  </menubar>
</window>
```

The preceding code gives the output in Figure 12-24.

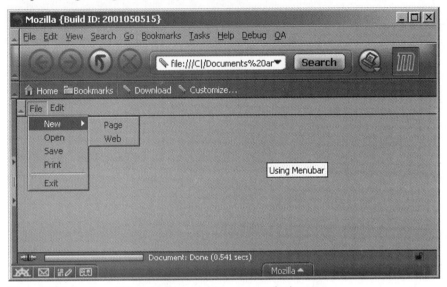

Figure 12-24: Creating a XUL application containing menus and submenus

Popup Menus in XUL

The three types of popup menus are described in the following section. The difference lies in their appearance.

- ♦ **Plain popups**— These appear when the user clicks the left mouse button on any element. They are similar to the menus in the menu bar, but they can be placed anywhere and can hold any type of content.

- ♦ **Context popups**— These appear on clicking the right mouse button. This click button varies from platform to platform. In the case of a Macintosh computer, it appears when the user presses the Ctrl key and the mouse button together.

- ♦ **Tooltips**—appear on the `onmouseover` effect on any element. It normally provides a description of the button, the toolbar image, and so on.

An example of code for a popup menu is shown in Listing 12-24.

Listing 12-24: example22.xul

```
<?xml version="1.0"?>
<?xml-stylesheet href="chrome://global/skin/" type="text/css"?>

<window
    id="popup"
    title="Using Popup Menu"
```

```
    xmlns:html="http://www.w3.org/1999/xhtml"
    xmlns="http://www.mozilla.org/keymaster/gatekeeper/there.is.only.xul"
    orient="vertical">

<popupset>
  <popup id="pc">
    <menuitem label="Cut"/>
    <menuitem label="Copy"/>
    <menuitem label="Paste"/>
  </popup>
</popupset>

<box context="pc">
  <text value="Click for checking the pop up menu"/>
</box>

</window>
```

The output of the preceding code is shown in Figure 12-25.

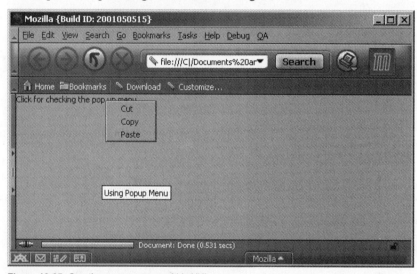

Figure 12-25: Creating popup menus within XUL

In the preceding example, you used the `context` attribute to associate the box with a popup by providing the same ID. Similarly, you can associate more popups with their respective elements, as in the popup example for a Tooltip shown in Listing 12-25.

Listing 12-25: example23.xul

```
<?xml version="1.0"?>
<?xml-stylesheet href="chrome://global/skin/" type="text/css"?>

<window
    id="tooltip"
    title="Using Tool tip"
    xmlns:html="http://www.w3.org/1999/xhtml"
    xmlns="http://www.mozilla.org/keymaster/gatekeeper/there.is.only.xul"
```

```
      orient="vertical">

<popupset>
  <popup id="st" style="background-color: #FFFFC0;">
    <text value="Click the button to save your stuff"/>
  </popup>
</popupset>

<button label="Save" tooltip="st"/>

</window>
```

The output of the preceding code is shown in Figure 12-26. As you can see, the text "Click the button to save your stuff" appears when you move the mouse over the button.

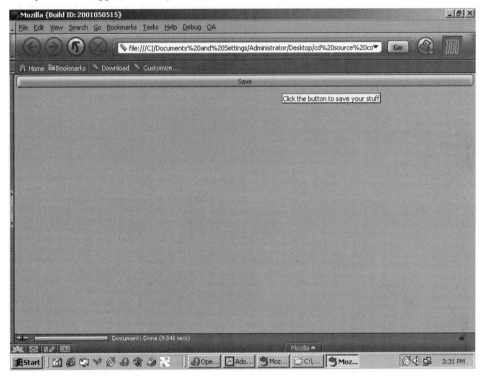

Figure 12-26: Creating a Tooltip in the XUL file

Grids in XUL

To display data in a tabular format in HTML, you use the `<table>` tag. In XUL you use the `grid` element for the same purpose. Elements in the grid are aligned in a row to display data, just like tables in HTML. You can insert buttons, labels, and input controls inside the grid.

The `<row>` tag is used to declare a set of rows in a grid. You can place the content in the row that you want to display. Similarly, the `<column>` tag is used to declare the content of the columns. Both the rows and columns are the child elements of the `grid` element.

Look at the example of the `grid` element in Listing 12-26.

Listing 12-26: example24.xul

```
<?xml version="1.0"?>
<?xml-stylesheet href="chrome://global/skin/" type="text/css"?>

<window
    id="Grid"
    title="Using Grid"
    xmlns:html="http://www.w3.org/1999/xhtml"
    xmlns="http://www.mozilla.org/keymaster/gatekeeper/there.is.only.xul"
    orient="vertical">

<grid flex="1">
  <columns>
    <column flex="2"/>
    <column flex="1"/>
  </columns>

  <rows>
    <row>
      <button label="OK"/>
      <button label="Cancel"/>
    </row>
    <row>
      <button label="Ignore"/>
      <button label="Retry"/>
    </row>
  </rows>
</grid>

</window>
```

This example has two rows and two columns. The column is declared with a `<column>` tag and is given a `flex` attribute. Each row holds two window's elements (buttons). The output for the preceding example is shown in Figure 12-27.

Figure 12-27: Creating an application using a grid for aligning buttons

You can use the box element within the grid and the box, which contains many different elements. (See Listing 12-27.)

Listing 12-27: example25.xul

```
<?xml version="1.0"?>
<?xml-stylesheet href="chrome://global/skin/" type="text/css"?>

<window
    id="Grid"
    title="Using Grid"
    xmlns:html="http://www.w3.org/1999/xhtml"
    xmlns="http://www.mozilla.org/keymaster/gatekeeper/there.is.only.xul"
    orient="vertical">

<grid flex="1">
  <columns>
    <column/>
    <column flex="1"/>
  </columns>

  <rows>
    <row>
      <text value="Name of the document:"/>
      <textbox flex="1"/>
    </row>
    <row>
      <text value="Enter path or click browse button:"/>
      <box flex="1">
        <textbox flex="1"/>
        <button label="Browse..."/>
      </box>
    </row>
  </rows>
</grid>

</window>
```

The preceding code yields the output in Figure 12-28.

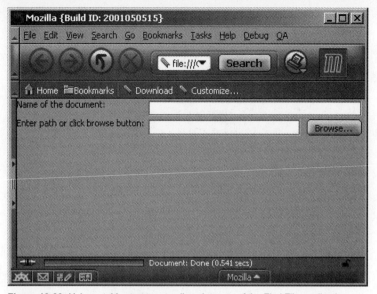

Figure 12-28: Using a table structure to align elements of the Find File application

The Tree Element in XUL

The `tree` element is used to create a tabular or hierarchical list in XUL. It is one of the most complicated elements of XUL. It creates a list of items like hierarchical lists or tables. The tree can be created with the help of the `tree` element or the `outliner` element. If `max rows` contains the text, it is advisable to use the `outliner` element or the `tree` element because it is more flexible to represent the data.

The different tree elements are described as follows:

- ◆ `tree`—just like the HTML `<table>` tag. Used to start a tree. All the contents are held by the `<tree>` tag.
- ◆ `treehead`—used to give a title header to the tree, like the `<thead>` tag used in HTML.
- ◆ `treeitem`—required to contain the top-level row and all its descendants.
- ◆ `treechildren`—holds the contents of the `tree` element, just like the `tbdoy` tag in HTML.
- ◆ `treerow`—holds the content of the single row, just like the `tr` tag in HTML.
- ◆ `treecell`—holds the content of the single column, just like the `td` tag in HTML.
- ◆ `treecol`—used to declare a column of the `tree` element.
- ◆ `treecolgroup`—holds the `treecol` elements.

The attributes of the `tree cell` element are described as follows:

- ◆ `label`—the content that appears in the cell.
- ◆ `align`—aligns the content in the cell. Values are placed left, right, or center.
- ◆ `crop`—allows the cropping of the text if the text is too small for the cell. Possible values are `none`, `left`, `right`, or `center`.

Consider the example in Listing 12-28.

Listing 12-28: example26.xul

```
<?xml version="1.0"?>
<?xml-stylesheet href="chrome://global/skin/" type="text/css"?>

<window
    id="Grid"
    title="Using Grid"
    xmlns:html="http://www.w3.org/1999/xhtml"
    xmlns="http://www.mozilla.org/keymaster/gatekeeper/there.is.only.xul"
    orient="vertical">

<tree>

  <treecolgroup>
    <treecol flex="1"/>
    <treecol flex="1"/>
  </treecolgroup>

  <treehead>
    <treerow>
      <treecell label="From"/>
      <treecell label="Subject"/>
    </treerow>
  </treehead>

  <treechildren flex="1">
    <treerow>
      <treecell label="sun@sunindia.com"/>
      <treecell label="Hello to all"/>
    </treerow>
    <treerow>
      <treecell label="raj@rajindia.com"/>
      <treecell label="Welcome to all"/>
    </treerow>
  </treechildren>
</tree>

</window>
```

The preceding example gives the output in Figure 12-29.

Figure 12-29: Creating a tree structure within a XUL file

Event Handlers in XUL

In XUL, you can use a scripting language such as JavaScript to provide functionality to the window, toolbar, and so on. You can embed a `<script>` tag in XUL, as in the following:

```
<script src="samplefile.js"/>
```

Event handlers in HTML and XUL work in the same manner. XUL has all the event handlers of HTML, as well as some additional event handlers. The main event handlers used in XUL are shown in Table 12-1.

Table 12-1: XUL Event Handlers

Events	Description
onclick	Fired in response to the click of the mouse button (on release) on any element of the window. Generally it is used with the `button` element.
Events	Description
onmousedown	Fired in response to the press of the mouse button (even before release) on any element of the window.
onmouseup	Fired in response to the release of a mouse button on any element of with window.
onmouseover	Fired as the pointer to the mouse when it is moved onto any element of the window.
onmousemove	Fired as the pointer of the mouse when it is moved over any element of the window.
onmouseout	Fired as the pointer of the mouse when it is moved off of any element of the window.
oncommand	Fired on the selection of the menu item. It is preferred over mouse events because the user can select the option by using the mouse or by using the shortcut key or the keyboard shortcut.
onkeypress	Fired when a key is released (first, the key is pressed and then released) when an element has the focus. Normally, it is used to handle the events for accessing the key or the keyboard shortcut.
onkeydown	Fires when a key is pressed (even before it is released) while an element has the focus.
onkeyup	Fired on the release of the key while an element has the focus.
onfocus	Fired as the focus is changed by using the Tab key or mouse over a window element. Fired as the element is getting the focus.
onblur	Fired when an element loses the focus, either when the user presses the Tab key or clicks the mouse.
onload	Fired as the user loads (opens) the window in the browser.
onunload	Fired when the user unloads (closes) the window.

Consider the example in Listing 12-29 for an `onclick` event on `menuitem` and on the `button` element:

Listing 12-29: example27.xul

```
<?xml version="1.0"?>
<?xml-stylesheet href="chrome://global/skin/" type="text/css"?>
```

```
<window
    id="menubar"
    title="Using Menubar"
    xmlns:html="http://www.w3.org/1999/xhtml"
    xmlns="http://www.mozilla.org/keymaster/gatekeeper/there.is.only.xul"
    orient="vertical">

<menubar id="mb1">
  <menu id="fm" label="File">
    <menupopup id="fp">
      <menuitem label="New"/>
      <menuitem label="Close" accesskey="c" oncommand="window.close();"/>
    </menupopup>
    </menu>
  </menubar>

<button id="cancel-button" class="dialog" label="Cancel"
    style="width: 8ex" onclick="window.close();"/>

</window>
```

Figure 12-29 shows the output for the above example. When the user clicks the Close menu item or the button, the window closes.

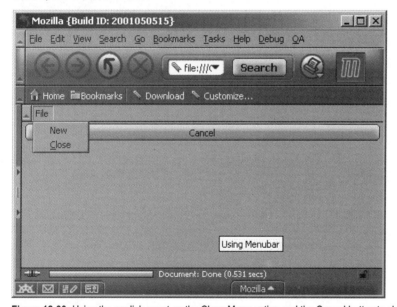

Figure 12-30: Using the onclick event on the Close Menu option and the Cancel button to close the window

XPtoolkit

XPtoolkit provides developers with a way to work with XUL. A number of loosely related facilities are the members of XPtoolkit. A developer can choose the various facilities available for the development of cross-platform applications in XPtoolkit. This does not, however, mean that all the facilities that are platform independent fall under XPtoolkit. For example, JavaScript is a distinct service for writing a browser-independent client-side script that can work on all the browsers in the world. It is also important to know that you can't eliminate platform-specific code from the application, even by using XPtoolkit.

A kernel of code will help the client avoid overlapping the services while using the toolkit. The kernel of code does all nonuser interface things for the application, such as speaking with HTTP or implementing a database. XPtoolkit also has a scriptable interface that is defined with XPCOM and XPIDL. These scripts will help the interface to query and set any parameter, or furthermore, to issue any command.

Application-specific forms require hunks of user-interface description, which means hunks of user-interface machinery. These descriptions are stored as individual files, as resources, as database entries, or even remotely to be accessed by applications with the help of URLs.

Hunks of user-interface machinery, such as how it behaves, how it looks, structure, and localizable strings, are called a *package*. Navigator, Messenger, Preferences, Bookmarks, and Composer in Netscape Communicator are some examples of packages. Packages can communicate with each other and can share functionality.

The seven primary components of a package are as follows:

♦ **Content**—An XML-based description of the user interface structure. This description typically uses two types of namespaces—HTML (defined as XTML in the working draft) and XUL (XML-based User Interface Language)—to define the UI. XUL provides additional widgets, features, and capabilities over HTML elements, such as toolbars, menu bars, and so on.

♦ **Appearance**—The appearance of the user interface can be changed with the help of a CSS. You can specify the style to XUL elements as you specify it in HTML. XUL supports CSS1 and some properties of CSS2 also.

♦ **Behavior**—The behavior of the package can be defined with the help of scripting languages such as JavaScript, services, and the AOM (Application Object Model).

♦ **Services**—Services are the workhorses of an application. They perform the actual application-specific tasks like printing, fetching the data, or sequencing some DNA. The widgets are used to send the services; they are controlled and directed with messages. For example, when the user clicks the Print button, it opens the print dialog box.

♦ **The application object model**—A collection of facilities that provides a platform and language-independent interface so that programs and scripts can access and carry out updates on the content, style, or structure of the document. Like DOM, it helps define a shared environment on document elements in case of script or services. An AOM works in the same way to define the environment for XPtoolkit components. AOM is a major part in the architecture of XPtoolkit.

♦ **Locale**—The fourth component of the package; includes all localizable strings and other necessary information required for the appearance and behavior of the package.

♦ **Platform**—Contains the platform-specific information. The last two components—locale and platform—overlap with the first three components because both of these components are necessary to provide the required and appropriate appearance and behavior to the user interface.

XPCOM Interfaces

The Mozilla browser uses XPCOM as the Object model. XUL is used to develop user interfaces with the combination of XBL and JavaScript. However, sometimes it is difficult to perform quite a number of things with JavaScript. For example, if you want to create a mail application, JavaScript doesn't provide the functionality to connect with the mail server to send and retrieve the mail messages. You can use the native code for this problem but you need a way for the script to call these native codes. Here you require XPCOM.

XPCOM stands for Cross-Platform Component Object Model. It is a framework for cross-platform applications or modular software wherein the developer can write XPCOM components using scripting languages such as JavaScript, or other languages such as C, C++, and Perl. In the future, the developer can also use Python to write the XUL components. An application requires a set of core XPCOM libraries to selectively load and manipulate XPCOM components.

XPCOM supports all the platforms that can run a C++ compiler, such as the following:

♦ Win 32 (NT/9x)

♦ UNIX and Linux

♦ HP-UX

♦ Solaris

♦ AIX

♦ OpenVMS

♦ MacOS

Some applications that use XPCOM are already available on the market, such as the following:

♦ Komodo (IDE)

♦ Chatzilla (IRC client)

♦ Jabberzilla (instant messaging)

♦ NS6 (Netscape's branded browser)

♦ Mozilla (the browser)

XPConnect

XPConnect is a service that coordinates between native code and JavaScript. It uses the dynamically built `Wrapper` object to get access to virtually transparent use of foreign objects; and it uses no generated code. Usually, the object interfaces must be declared in XPIDL (Cross Platform Interface Definition Language) without any restrictions for participating objects. However, all these objects must follow the rules and conventions proposed by XPCOM.

XPConnect has certain features worth considering:

♦ The objects are accessible universally, irrespective of linguistic borders; and it permits mapping between reference counting and automatic garbage collection.

♦ XPConnect generates cross-platform, typelib-based code instead of generated code.

♦ It allows mapping between XPCOM `out` params and JavaScript for returning the values and type conversions of params. Results are automatically updated with automatic wrapper creation if so demanded.

♦ XPConnect provides support for specialized JavaScript behavior by using nsIXPCScriptable. It also facilitates a one-time declaration of interfaces in XPIDL for headers and typelibs.

♦ XPConnect has multiple interfaces per object support.

♦ Finally, here JavaScript gets a reflection of the XPCOM object repository, and arbitrary JavaScript properties are supported by XPConnect on wrapped native XPCOM objects.

XBL

XBL stands for eXtensible Bindings Language. It is a markup language that defines some special elements, or bindings, for XUL widgets. XBL is an XML-based language, so certain features show up in its elements and attributes. For example, XBL defines new content for a XUL widget, adds event handlers to a XUL widget, and defines newer interface properties and methods. The paucity of space is a limiting feature here, so we definitely will not undertake a comprehensive discussion on XBL.

How XBL Works

Mozilla uses the XBL file to define the bindings for it in XML. It is located in `xulbindings.xml`. When you try to refer to individual files, you refer to them from the CSS file, which is also called the skin. The skin is loaded from the actual XUL file. "Skinning" a XUL file facilitates defining new content and element additions. Combining the extensibility of XBL and the dynamic quality of cascading style sheets is definite proof for the extent of XUL's stability.

Element classes and individual elements are skinned in the XUL file whose style is defined in the CSS file. The emergence of XBL has facilitated pointing for these XUL elements that contain new content, interfaces, or handlers.

XBL Binding

Similar to XUL, XBL also contains elements that are like parent-child relationships. `<bindings>` is the root element present in an XBL file, which facilitates the definition of any other `<binding>` element. Table 12-2 shows the principal child elements of an XBL binding.

Table 12-2: XBL Binding Child Elements

Element	Description
`<content>`	Defines the child elements that are anonymous for a particular binding.
`<interface>`	Describes some additional properties and methods with which you can associate the bound element.
`<event>`	Used to define events that are new and can be called from XUL.
`<handler>`	Defines the event handlers that are called when suitable events are thrown in the interface.

Consider the following example to help you understand the process in which XBL provides behavior to the window's elements. Here you require three files: a XUL file, a CSS file, and an XBL file (Listings 12-30 to 12-32).

XUL (`box.xul`):

Listing 12-30: box.xul

© 2001 Dreamtech Software India Inc.
All Rights Reserved

```
<?xml version="1.0"?>
<?xml-stylesheet href="chrome://global/skin/" type="text/css"?>
<?xml-stylesheet href="chrome://example/skin/box.css" type="text/css"?>

<window
    xmlns="http://www.mozilla.org/keymaster/gatekeeper/there.is.only.xul">
  <box class="xulbuttons"/>
</window>
```

CSS (`box.css`):

Listing 12-31: box.css

```
box.xulbuttons {
    -moz-binding: url('chrome://example/skin/box.xbl#bind1');
}
```

XBL (`box.xbl`):

Listing 12-32: box.xbl

```
<?xml version="1.0"?>
<bindings xmlns="http://www.mozilla.org/xbl"

xmlns:xul="http://www.mozilla.org/keymaster/gatekeeper/there.is.only.xul">
  <binding id="bind1">
    <content>
      <xul:button label="find"/>
      <xul:button label="Cancel"/>
    </content>
  </binding>
</bindings>
```

Figure 12-31 shows the output of the preceding code listings.

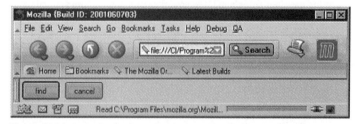

Figure 12-31: A simple example using XML, CSS, and XBL files together

Netscape Gecko Engine

Gecko Engine is an Internet browser engine developed by Netscape. It supports all the Web standards such as HTML 4.0, XML, CSS, and DOM. It is small, fast, and modular. Netscape Gecko has been designed considering future requirements of Internet browsers.

Some features of Netscape Gecko Engine are the following:

♦ It fully supports Web standards such as HTML 4.0, XML, XML Namespaces, RDF, style sheets, DOM, and scripting, such as JavaScript.

♦ It supports cross-platform, cross-device interfaces such as XUL and XP Widgets Library; its architecture supports XPCOM and XPConnect.

♦ It's extensible and supports embedded technologies such as Web Shell APIs, ActiveX Control, and a JavaScript interpreter engine.

♦ It supports other standards and features such as Data Transport Protocols, Multilingual Character Data, Image Data, Java Support, and plug-in APIs so that the browser can be fully functional.

Industry-Leading Web Standards Support

Gecko supports all the Web standards such as markup languages (HTML, XML, and so on) and scripting languages (such as JavaScript). It enables the developer to create the user interface and Web applications for high performance speed. The developer can create entire user interfaces, including the desktop, menus, toolbar, icons, scroll bars, and so on. Gecko supports the following Web standards:

- ♦ **Markup Language and Framework:** Used to display and describe data.

- ♦ **HTML:** Used to support the latest Hypertext Markup Standard (HTML 4.0) and includes support for style sheets.

- ♦ **XML:** Supports Extensible Markup Language (XML 1.0) to describe the data structure. It helps the developer build rich and powerful Web applications with better performance and a superior user experience.

- ♦ **XML Namespaces :** Helps developers use XML data across organizations and helps them develop simpler and more powerful applications.

- ♦ **RDF:** Supports RDF (resource description framework) for describing resources of all kinds, their properties, and the relationships between them. RDF is an XML application from W3C.

- ♦ **W3C DOM:** Supports DOM Level 1 and some features of DOM Level 2 such as event handling and setting style-sheet properties. DOM helps the user access all the elements of the Web using scripting languages such as JavaScript.

- ♦ **Cross-platform, cross-device user interfaces:** Netscape provides advanced support for developing cross-platform, cross-device user interfaces and applications.

- ♦ **XUL (XML-based User Interface Language):** Netscape invented XUL so that cross-platform and cross-device applications can be developed faster and more easily. XUL is used for the development of the design and layout of the user interface. It also helps in the development of objects and features.

- ♦ **Gecko's Extensible Architecture (XPCOM [Cross-Platform Component Object Model]):** Netscape adopted some features of COM, called XPCOM, for implementing and developing applications across the platform.

- ♦ **XPConnect (Controlling binary components from scripts):** Coordinates between XPCOM and JavaScript. It is the same as applets and plugins and is added using JavaScript.

- ♦ **Multilingual Character Data:** Gecko responds to the content in all languages, which helps with global distribution of content and applications.

- ♦ **Image Data:** Gecko supports all widely used image formats like GIF, JPEG, PNG, and XBM.

The Benefits of the Netscape Gecko Engine

The benefits of using the Gecko browser vary depending on who the user is—a vendor, a developer, or an end user. Let's now undertake a brief analysis of what the Gecko browser has to offer its target users.

Application and device vendors can use the engine to provide fast and full-featured Web browsing to the end user. Further, the support of industry-standard, small-size, modular architecture and cross-platform support can help application and device vendors develop Web applications, desktop applications, and so on.

On the other hand, Web content and application developers without additional skills can use the Gecko engine to develop applications , because it is based on XML. Using DOM will help them build an application to exchange data with XML. CSS helps developers separate cleanly the description and display of data. The cross-platform and cross-device support helps them build better applications for the user instead of developing different applications for different platforms. The free availability enables developers to embed the features without additional costs.

Finally, for end users, the engine endeavors to provide fast, full-featured, powerful, cross-platform, cross-device applications such as browsers. It's also easy for end users to install and maintain these applications.

Summary

XUL is used to develop user interfaces for cross-platform and cross-device Web applications. The Mozilla Web browser was developed using XUL. The user interface component of Mozilla is called XPFE (Cross Platform Front End). It helps the XML programmer develop a user interface that runs on different platforms. It supports all HTML elements, scripting languages like JavaScript, and style sheets, which can also be used to change the appearance. The user interface may have form elements like input boxes, radio buttons, checkboxes, drop-down menus, and so on. Further, you can also create a menu bar, a toolbar, and so on within a XUL file. You can also implement a progress bar for showing the progress of loading an application, like the one in the lower-right corner of the Mozilla browser as you open any Web page. XPToolkit is a collection of loosely related facilities that is used for writing the client-side script, like JavaScript. You can also use the grid and the table structure to align your application's window elements. XUL supports all the events available in JavaScript and also provides some extra events. XPCOM (Cross Platform Component Object Model) is a framework for application or modular software running on cross-platforms. XPConnect is a service that coordinates between native code and JavaScript. XBL stands for eXtensible Binding Language and is used to provide behavior to XUL widgets. Gecko Engine is the next-generation Internet browser engine developed by Netscape to help the developer to develop cross-platform and cross-device applications that have smaller size and faster performance.

Appendix

What's on the CD-ROM

This appendix provides information about this book's companion CD-ROM, found on the inside back cover of the book. For the latest information, please refer to the ReadMe file located in the root directory of the CD.

System Requirements

Make sure that your computer meets the minimum system requirements listed in this section. If your computer doesn't match up to these requirements, you might have a problem using the contents of the CD.

For Microsoft Windows 9*x* or Windows 2000, your computer must have a CD-ROM drive that is double-speed (2x) or faster.

CD Contents

The CD-ROM contains source code examples, applications, and an electronic version of the book. Following is a summary of the contents of the CD-ROM:

Source Code

The folder named "Source Code" is categorized into different folders named according to the chapter numbers. The source code of the case studies and the programs is contained in the folders for the corresponding chapters. Following is a list of the folders in the Source Code folder:

- ◆ **Chapter-1:** This folder contains a folder named "examples", which contains source code for all the examples from Chapter 1.
- ◆ **Chapter-3:** This folder contains the folder named "examples", which contains source code for all the examples from Chapter 3.
- ◆ **Chapter-4:** This folder contains the source code of the complete application that we built for the PC browser. This includes all the XML and XSL files with the ASP files we are using.
- ◆ **Chapter-5:** This folder contains the source code of the complete transformation for WAP clients. This includes all the XML and XSL files, with the ASP files we are using.
- ◆ **Chapter-6:** This folder contains two folders, named "case study" and "examples." The "case study" folder contains the complete source code of the case study. In the folder named "examples," you can browse all the examples from the chapter.
- ◆ **Chapter-7:** This folder contains the source code of the complete transformation for HDML devices. This includes all the XML and XSL files, with the ASP files we are using.
- ◆ **Chapter-8:** This folder contains two folders, named "case study" and "examples." The "case study" folder contains the complete source code of the case study. In the folder named "examples," you can browse all the examples from the chapter.

♦ **Chapter-9:** This folder contains the source code of the complete transformation for i-mode clients. This includes all the XML and XSL files, with the ASP files we are using.

♦ **Chapter-10:** This folder contains two folders, named "case study" and "examples." The "case study" folder contains the complete source code of the case study. In the folder named "examples," you can browse all the examples from the chapter. All the examples and the case study in this chapter were tested using the Motorola Application Development Kit.

♦ **Chapter-11:** This folder contains the source code of the complete transformation of the application for voice clients. This includes all the XML and XSL files, with the ASP files we are using.

♦ **Chapter-12:** This folder contains the folder named "examples", which contained all the examples from the chapter. You can browse all the examples in XUL format. All the examples in this folder were tested on the Mozilla browser.

Applications

The following applications are on the CD-ROM:

♦ The Mozilla folder contains the Mozilla browser.

♦ The Adobe folder contains Adobe Acrobat Reader.

Troubleshooting

If you face any difficulty in installing or using the CD-ROM programs, try the following solutions:

♦ **Turn off any anti-virus software that you may have running.** Installers sometimes mimic virus activity and can make your computer incorrectly believe that it is being infected by a virus. (Be sure to turn the anti-virus software back on later.)

♦ **Close all running programs.** The more programs you're running, the less memory is available to the other programs. Installers also typically update files and programs; if you keep other programs running, installation may not work properly.

If you are still having trouble with the CD, please call Hungry Minds Customer Service. The phone number is (800) 762-2974. If you are not in the United States, please call (317) 572-3994 or e-mail at techsupdum@hungryminds.com. Please note that Hungry Minds will provide technical support only for installation and other general quality-control items. For technical support on the applications themselves, please consult the program's vendor or author.

Index

Numbers & Symbols

Hungry Minds, Inc.
End-User License Agreement

READ THIS. You should carefully read these terms and conditions before opening the software packet(s) included with this book ("Book"). This is a license agreement ("Agreement") between you and Hungry Minds, Inc. ("HMI"). By opening the accompanying software packet(s), you acknowledge that you have read and accept the following terms and conditions. If you do not agree and do not want to be bound by such terms and conditions, promptly return the Book and the unopened software packet(s) to the place you obtained them for a full refund.

1. **License Grant.** HMI grants to you (either an individual or entity) a nonexclusive license to use one copy of the enclosed software program(s) (collectively, the "Software") solely for your own personal and non-commercial purposes on a single computer (whether a standard computer or a workstation component of a multi-user network). The Software is in use on a computer when it is loaded into temporary memory (RAM) or installed into permanent memory (hard disk, CD-ROM, or other storage device). HMI reserves all rights not expressly granted herein.

2. **Ownership.** HMI is the owner of all right, title, and interest, including copyright, in and to the compilation of the Software recorded on the disk(s) or CD-ROM ("Software Media"). Copyright to the individual programs recorded on the Software Media is owned by the author or other authorized copyright owner of each program. Ownership of the Software and all proprietary rights relating thereto remain with HMI and its licensers.

3. **Restrictions on Use and Transfer.**

    (a) You may only (i) make one copy of the Software for backup or archival purposes, or (ii) transfer the Software to a single hard disk, provided that you keep the original for backup or archival purposes. You may not (i) rent or lease the Software, (ii) copy or reproduce the Software through a LAN or other network system or through any computer subscriber system or bulletin-board system, or (iii) modify, adapt, or create derivative works based on the Software.

    (b) You may not reverse engineer, decompile, or disassemble the Software. You may transfer the Software and user documentation on a permanent basis, provided that the transferee agrees to accept the terms and conditions of this Agreement and you retain no copies. If the Software is an update or has been updated, any transfer must include the most recent update and all prior versions.

4. **Restrictions on Use of Individual Programs.** You must follow the individual requirements and restrictions detailed for each individual program in Appendix A of this Book. These limitations are also contained in the individual license agreements recorded on the Software Media. These limitations may include a requirement that after using the program for a specified period of time, the user must pay a registration fee or discontinue use. By opening the Software packet(s), you will be agreeing to abide by the licenses and restrictions for these individual programs that are detailed in Appendix A and on the Software Media. None of the material on this Software Media or listed in this Book may ever be redistributed, in original or modified form, for commercial purposes.

5. **Limited Warranty.**

    (a) HMI warrants that the Software and Software Media are free from defects in materials and workmanship under normal use for a period of sixty (60) days from the date of purchase of this Book. If HMI receives notification within the warranty period of defects in materials or workmanship, HMI will replace the defective Software Media.

    (b) **HMI AND THE AUTHOR OF THE BOOK DISCLAIM ALL OTHER WARRANTIES, EXPRESS OR IMPLIED, INCLUDING WITHOUT LIMITATION IMPLIED WARRANTIES OF MERCHANTABILITY AND FITNESS FOR A PARTICULAR PURPOSE, WITH RESPECT TO THE SOFTWARE, THE PROGRAMS, THE SOURCE CODE CONTAINED THEREIN, AND/OR THE TECHNIQUES DESCRIBED IN THIS BOOK. HMI DOES NOT WARRANT THAT THE FUNCTIONS CONTAINED IN THE SOFTWARE WILL MEET YOUR REQUIREMENTS OR THAT THE OPERATION OF THE SOFTWARE WILL BE ERROR FREE.**

(c) This limited warranty gives you specific legal rights, and you may have other rights that vary from jurisdiction to jurisdiction.

6. Remedies.

    (a) HMI's entire liability and your exclusive remedy for defects in materials and workmanship shall be limited to replacement of the Software Media, which may be returned to HMI with a copy of your receipt at the following address: Software Media Fulfillment Department, Attn.: *Multi-Platform Wireless Web Applications: Cracking the Code*, Hungry Minds, Inc., 10475 Crosspoint Blvd., Indianapolis, IN 46256, or call 1-800-762-2974. Please allow four to six weeks for delivery. This Limited Warranty is void if failure of the Software Media has resulted from accident, abuse, or misapplication. Any replacement Software Media will be warranted for the remainder of the original warranty period or thirty (30) days, whichever is longer.

    (b) In no event shall HMI or the author be liable for any damages whatsoever (including without limitation damages for loss of business profits, business interruption, loss of business information, or any other pecuniary loss) arising from the use of or inability to use the Book or the Software, even if HMI has been advised of the possibility of such damages.

    (c) Because some jurisdictions do not allow the exclusion or limitation of liability for consequential or incidental damages, the above limitation or exclusion may not apply to you.

7. U.S. Government Restricted Rights. Use, duplication, or disclosure of the Software for or on behalf of the United States of America, its agencies and/or instrumentalities (the "U.S. Government") is subject to restrictions as stated in paragraph (c)(1)(ii) of the Rights in Technical Data and Computer Software clause of DFARS 252.227-7013, or subparagraphs (c) (1) and (2) of the Commercial Computer Software - Restricted Rights clause at FAR 52.227-19, and in similar clauses in the NASA FAR supplement, as applicable.

8. General. This Agreement constitutes the entire understanding of the parties and revokes and supersedes all prior agreements, oral or written, between them and may not be modified or amended except in a writing signed by both parties hereto that specifically refers to this Agreement. This Agreement shall take precedence over any other documents that may be in conflict herewith. If any one or more provisions contained in this Agreement are held by any court or tribunal to be invalid, illegal, or otherwise unenforceable, each and every other provision shall remain in full force and effect.